Historical & Cultural Astronomy

Historical & Cultural Astronomy

More information about this series at http://www.springer.com/series/15156

Dirk L. Couprie

When the Earth Was Flat

Studies in Ancient Greek and Chinese Cosmology

Springer

Dirk L. Couprie
Department of Philosophy
Faculty of Philosophy and Arts,
University of West Bohemia
Pilsen, Czech Republic

Amsterdam, The Netherlands

ISSN 2509-310X ISSN 2509-3118 (electronic)
Historical & Cultural Astronomy
ISBN 978-3-030-07292-6 ISBN 978-3-319-97052-3 (eBook)
https://doi.org/10.1007/978-3-319-97052-3

This Springer imprint is published by the registered company Springer Nature Switzerland AG
The registered company address is: Gewerbestrasse 11, 6330 Cham, Switzerland

To the memory of my brother Jan (1924–2012), who could calculate in the dozenal system

This book was supported by the Czech Grant Agency Project, GACR 15-08890S.

Foreword

Whether or not there was an actual relationship between early Chinese thought and early Greek thought is difficult to determine. But we can really claim that there is analogical relationship between early Chinese thought and early Greek thought. Even with regard to the concepts of Yin and Yang which have been regarded by most scholars for a long time and up to now as very peculiar Chinese ideas, we can find that there is analogical relationship with the concepts of the hot and the cold and the dry and the wet in early Greek thought.[1] Therefore, I strongly believe that, if we systematically and profoundly try to do comparative research into early Chinese thought and early Greek thought, we should meet many surprising discoveries. And this will help us not only to find many comparable thoughts between the two, but also to solve the puzzles which have remained for understanding because of the lost and the lack of documents on either side.

Dirk L. Couprie's new book, *When the Earth Was Flat: Studies in Ancient Greek and Chinese Cosmology*, is a great attempt in this aspect. In this volume, he continues the research of his earlier work, *Heaven and Earth in Ancient Greek Cosmology*, and makes a detailed and deep study of flat earth cosmology. While in the previous book he paid more attention to how spherical earth cosmology replaced flat earth cosmology and was accepted in early Greek thought, in the new book he devotes much more discussion to the question of how flat earth cosmology has explained all kinds of puzzles of astronomy. It is precisely in this context that he devotes an independent and special part to the study of Chinese flat earth cosmology, called the *gai tian* system, which was contained in an ancient Chinese work, *Zhou Bi Suan Jing*. He makes use of his extensive knowledge of ancient astronomy when studying its thoughts related to flat earth cosmology deeply and professionally, so that although I am a Chinese person, I become a layman with respect to him regarding this Chinese ancient work. But, I dare say, it is by his new book that he illustrates how a comparative research between early Chinese thought and early

[1]See Nie Minli (2016).

Greek thought may provide useful material to remedy the deficiency of literature on both sides.

I got acquainted with Dirk Couprie in Thessaloniki, Greece, in July 2014, when I attended the Fourth Conference of International Association for Presocratic Studies. In that conference, Dirk Couprie read his paper, "The Paths of The Celestial Bodies According to Anaximenes," and my paper was "Yin and Yang, and the Hot and the Cold." In my paper, I compared the cosmological concepts of the hot and the cold in Pre-Socratics with the concepts of the Yin and Yang in Pre-Qin and paid special attention to the analogies between Lao Tzu and Anaximander. I did not anticipate that this paper would arouse the interest of Dirk Couprie. We had an intense discussion about Anaximenes's cosmology and ancient Chinese cosmology such as the *gai tian* system and the *hun tian* system, the former of which claims the idea of flat earth and the latter the idea of spherical heavens. I felt shy that as a Chinese person, I could not provide him with more knowledge than he already possessed about ancient Chinese cosmology. After that meeting, we met again in February 2016 at the 10th London Ancient Science Conference. At that conference, we talked about Anaximander's cosmology, especially about the puzzle in this sentence, "a kind of sphere of flame from this was formed round the air surrounding the earth" (Ps.-Plutarch *Strom.* 2; DK 12 A 10). He told me that he intended to publish a book about flat earth cosmology based on the comparative study of early Greek cosmology and early Chinese cosmology.

Now, I receive the last draft of his new book, *When the Earth Was Flat: Studies in Ancient Greek and Chinese Cosmology*, which is just what he promised in 2016. I believe this work will be a new beginning in the comparative studies of ancient Western thought and ancient Chinese thought, which tells us about the similarities and differences between their cosmologies. Dirk Couprie tends to stress the differences. In my opinion, however, to inquire for the similarities will give us even more inspired ideas than to inquire for the differences.

Renmin University of China in Beijing,
Beijing, China

Nie Minli

Reference

Nie Minli, "Yin and Yang, and the Hot and the Cold," in *Frontiers of Philosophy in China*, Volume 11, Number 1, March 2016, pp. 73–87.

Spherical Versus Flat

Corey McCorkle, *Yayoi* (2005), Middelheim Park, Antwerp. What do we see here? A sphere, or a disk as flat as the paper on which it is printed?

Eratosthenes said: The earth is a sphere. The sun is very far away, so that its rays reach the earth parallel to each other. At noon at the summer solstice in Alexandria, the angle of the shadow line from the top of the gnomon is seven degrees, but a gnomon in Syene casts no shadow. Therefore, I can calculate that the circumference of the earth is about fifty times the distance between Alexandria and Syene.

Anaxagoras could have argued: The earth is flat. At noon at the summer solstice in Alexandria, the shadow is about one-eighth of the length of a gnomon. When I extend the line from the end of the shadow to the top of the gnomon upward, the point where it crosses the perpendicular in Syene is the place where the sun stands. Therefore, I can calculate that the distance of the sun is no more than about eight times the distance between Alexandria and Syene.

Aristotle said: during a lunar eclipse, the shape of the shadow of the earth on the moon is always curved. This proves that the earth is a sphere.

Anaxagoras could have argued: The earth is flat. The line between the light and the dark parts of the moon during an eclipse is always curved. This proves that an eclipse of the moon cannot be caused by the shadow of the earth. There must be invisible heavenly bodies between us and the moon that cause lunar eclipses.

Ptolemy said: The earth must be spherical, because on a flat earth, it would be the same time everywhere, and the heavenly bodies would rise and set for all people at the same time.

Chinese astronomers argued: The earth is flat. The celestial bodies orbit in a flat plane parallel to the earth's surface. The sun shines successively on different parts of the earth, causing the differences in time.

Acknowledgements

In the field of ancient cosmology, I learned most from Dmitri Panchenko, whose thoughtful and inspiring articles always opened new horizons of interpretation. I thank my colleague and friend Radim Kočandrle for intensive participation in two projects at the University of West Bohemia. I owe him the idea of a new interpretation of one of the most intriguing subjects of Anaximander's thinking, the ἄπειρον, on which we wrote an article and a book.[2] Since the early 1990s I have had numerous conversations with my old friend Robert Hahn on various topics of pre-Socratic cosmology. Alexander Mourelatos, Daniel Graham, Andrew Gregory, Livio Rossetti, Christopher Cullen, Carlo Rovelli, Jaap Mansfeld, and David Runia responded accurately to my questions and reacted kindly to my sometimes unusual opinions. Georg Wöhrle was kind enough to adequately answer my questions and remarks on his splendid volumes with the doxography of the Ionian philosophers. István Baksa kindly helped me with a difficulty in a text of Olympiodorus. Dmitri Panchenko, already mentioned, and Nie Minli inspired me to intensively study quite another version of flat earth cosmology in the intriguing ancient Chinese so-called *gai tian* system. I owe much to Christopher Cullen's magnificent book *Astronomy and Mathematics in Ancient China: Zhou bi suan jing*, with which he made the *gai tian* system accessible to those who, like me, cannot read Chinese. Without his book on the *Zhou bi* the second part of my book could not have been written. Nie Minli promoted my work on the *gai tian* cosmology in China, and Lyu Xingyu was kind enough to publish my first study on this intriguing subject in a journal of the Tsinghua University. I thank Bart Zevenhek for his information concerning the applicability of the shadow rule. My brother Leendert D. Couprie helped me with several drawings and calculations. My wife Heleen Pott, although she is working in a quite different field of philosophy, always listened to my wild ideas and made not only encouraging but also most relevant comments. I hope these eminent scholars will forgive me when I felt I had good reasons to occasionally or sometimes

[2]See Couprie and Kočandrle (2013) and Kočandrle and Couprie (2017). See also Chap. 8, 155–158.

fundamentally disagree with their opinions. I thank Steve Hoerner for his information about bronze foundry. I am also grateful to Daniel R. Strebe, in particular for the azimuthal equidistant projection of the earth with the South Pole in the center, which he kindly offered to make and which now closes this book.

The cover of this book shows a detail of the triptych The Seven Wonders of the World, painted by my friend Hans Exterkate.

This is also the place to make a correction which Otta Wenskus brought to my notice. In *Heaven and Earth*, when discussing the ancient dates of the autumnal setting of the Pleiades, I wrote on pp. 18–19: "Wenskus has tried to explain the data (…) by suggesting that in the case of Anaximander we must suppose that he had very sharp eyes and was able to see the Pleiades set less than half an hour before sunrise." My use of the words "we must suppose that he had sharp eyes" was due to my misunderstanding of the German expression "es ist Ansichtssache," which means "it is a matter of opinion." What Wenskus points out is that the Pleiades already become invisible (φαῦσις) before they actually set at the horizon (δύσις).[3]

References

Couprie, D.L. and R. Kočandrle, Anaximander's Boundless Nature. *PEITHO, Examina Antiqua* 4 (2013), 63–91.

Kočandrle, R. and D.L. Couprie, *Apeiron: Anaximander on Generation and Destruction*. New York: Springer 2017.

Wenskus, O., *Astronomische Zeitangaben von Homer bis Theophrast*. Stuttgart: Franz Steiner Verlag 1990.

[3]Wenskus (1990, p. 60).

Contents

List of Abbreviations

DK H. Diels, and W. Kranz, *Die Fragmente der Vorsokratiker*. Zürich/ Hildesheim: Weidmann 1951^{6} (references to DK numbers).

Dox H. Diels, *Doxographi Graeci*. Berlin: Walter de Gruyter 1879 (references to page numbers).

GG D.E. Gershenson, and D.A. Greenberg, *Anaxagoras and the Birth of Physics*. New York: Blaisdell Publishing Company 1964 (references to GG numbers; if the reference is to a page, "p." is added).

Gr D.W. Graham, *The Texts of Early Greek Philosophy*. Cambridge: Cambridge UP 2010 (references to Gr numbers).

KRS G.S. Kirk, J.E. Raven and M. Schofield, *The Presocratic Philosophers*. Cambridge: Cambridge University Press 2009 (references to KRS numbers; if the reference is to a page, "p." is added).

LM A. Laks and G.W. Most, *Early Greek Philosophy*. Cambridge, Massachusetts. 2016 (references to LM numbers).

LSJ H.G. Liddell, R. Scott, H.S. Jones and R. McKenzie, *Greek-English Dictionary*. Oxford: Clarendon Press 1996.

MR J. Mansfeld, and D.T. Runia, *Aëtiana II.2. The Method and Intellectual Context of a Doxographer*. Leiden: Brill 2009 (references to page numbers; MR numbers added when different from P).

P Aëtius in pseudo-Plutarch, *Placita* (numbering according to Dox).

S Aëtius in Stobaeus, *Anthologium* (numbering according to Wachsmuth and Hense). ($\approx$ S) indicates that S's text differs from that of P.

TP1 G. Wöhrle, *Die Milesier: Thales*. Berlin: Walter de Gruyter 2009 (references to TP1 numbers).

TP2 G. Wöhrle, *Die Milesier: Anaximander und Anaximenes*. Berlin: Walter de Gruyter 2012 (references to TP2 numbers).

List of Figures

List of Tables

Chapter 1
Introduction

Contents

Something About the Origin of the Book

Except for members of The Flat Earth Society, some orthodox religious and a few Indians in the Amazon forests, we all know or should know for sure that the earth is a sphere. Yet, in daily life we act as if the earth is flat. When we travel from one city to another, or from one country to another, we do not use a globe to find our way, but look on a two-dimensional map as if the earth were flat. The idea of a spherical earth is as counter-intuitive as the idea of the earth orbiting around the sun. Perhaps this is why we tend to think there cannot be a problem in understanding the world-picture of ancient people who did not know better than that they lived on a flat earth. However, when I ask people who consider themselves as educated, to place themselves in the position of cosmologists living in the time of Aristotle and to mention arguments for the sphericity of the earth without taking refuge to modern knowledge such as the circumnavigation of the earth and photographs of the earth in space taken by astronauts, they are most of the time unable to deliver more than just the argument of an approaching ship at sea, of which we first see the mast and only later the hull. When asked if they have seen this themselves, they often must confess that they have it from hearsay. Perhaps such experiences can make us somewhat more modest towards the ancients who thought that the earth was flat.

The main subject of my book *Heaven and Earth in Ancient Greek Cosmology* (2011) was the transition from the archaic cosmology of a flat earth with the cupola of the heavens above it, to the new world picture introduced by Anaximander, and its reception until the introduction of the concept of a spherical earth. In the years that followed, I realized that there was much more to say about the consequences of one main feature that most of the Presocratic cosmologists shared with archaic

D. L. Couprie, *When the Earth Was Flat*, Historical & Cultural Astronomy,
https://doi.org/10.1007/978-3-319-97052-3_1

cosmology, namely the conception of an earth that is flat. So, the idea matured to write a sequel to *Heaven and earth*, entitled *When the Earth Was Flat*, in which I could also incorporate answers to some of my critics and add some elaborations of subjects that deserved a more extensive treatment. From conversations with other scholars, I learned that I had too easily passed over significant features of the cosmologies of some early thinkers such as Anaximenes, Xenophanes, and Anaxagoras, following what I saw as the standard interpretation of their cosmological ideas. This has resulted in chapters in which the most difficult interpretive puzzles of their flat earth cosmologies are thoroughly discussed: Anaximander's numbers, traditionally associated with the celestial bodies, Anaximenes' cap simile, Xenophanes' cosmological model in general, and the conflicting issues of the Milky Way and lunar eclipses in Anaxagoras. Reading Daniel Graham's *Science Before Socrates* provoked my interpretation of Anaxagoras. If Anaximander was the hero of *Heaven and Earth in Ancient Geek Cosmology*, then Anaxagoras is the hero of the first part of *When the Earth Was Flat*, albeit a tragic hero. At several congresses, especially the yearly London Ancient Science Conferences, the Forth and Sixth Biennial Conferences of Presocratics (Thessaloniki 2014 an Delphi 2018), the Congress "Ex Ionia Scientia" in Athens (2016), and the World Congress "Aristotle 2400 years" (Thessaloniki 2016), I had the opportunity to present early versions of ideas that have finally found their place in my book. It is not my intention, however, to discuss all Presocratic thinkers who can be considered to be among those who believed that the earth is flat, but only those about whom I think I have something new to contribute. Empedocles, for example, is mentioned only incidentally. Not only is it difficult to make sense of several of his cosmological ideas, but I have also become hesitant as to whether or not he believed in a flat earth. The only evidence is to be found in a report on the existence of two suns, which is very hard to understand, and in which his earth is called "round" (κυκλοτερής).[1] This seems to indicate a flat, disk-shaped earth, because otherwise the word στρογγύλος, or even σφαιροειδής, would have been available. In a genuine fragment, κυκλοτερής is used of the moon, which is also called a κύκλος, just like the sun.[2] On the other hand, in another genuine fragment, de word κυκλοτερής is used as an adjective to Σφαῖρος,[3] and can therefore also have the connotation of "spherical" for Empedocles. As for the atomists, in *Heaven and Earth* I have already dealt with some of the most important issues: the alleged tilt of the earth and the unlimited worlds.[4] Where I thought it necessary, these topics will return in this book.

[1]P 2.20.13 = DK 31A56 = S 1.25.3 = LM EMP.D126a = Gr Emp77 = MR 515 and 517, not in KRS. On the use of these abbreviations, see the next chapter.

[2]Achilles Tatius, *Introductio in Aratum* 16 = DK 31B45 = LM EMP.D139 = Gr Emp81, not in KRS.

[3]Simplicius, *In Aristotelis Physicorum Libros Commentaria*. 1183.28–1184.1 = DK 31B27 = LM EMP.D89 = Gr Emp55 = KRS358. Idem: S 1.15.2 = DK 31B28 = LM EMP.D90 = Gr Emp56, not in KRS.

[4]See Couprie (2011, 74–76, 222–224).

In *Heaven and Earth*, I already referred to the ancient Chinese flat earth cosmology. During the work on my new book, I became aware that the ancient Chinese so-called *gai tian* system offered an ingenious version of flat earth cosmology, and I decided to insert a study of it in my book. The contrast of two completely different versions of a cosmology based on the conception of a flat earth makes the present book a kind of comparative study. Several themes and topics return in a curious revival of flat earth cosmology in modern times and turn out to be of interest in understanding the ancient ways of thinking about a cosmos with a flat earth. As a result, this book consists of two parts. Part One focuses on ancient Greek flat earth cosmologies, Part Two on the Chinese *gai tian* cosmology.

An Overview of the Book

The publication of Laks and Most's *Early Greek Philosophy* motivated me to reflect on the confusing variety of textbooks on the Presocratics. Graham's criticism of my methodology inspired me to formulate my methodological principles. Both subjects have found their place in Chap. 2. Chapter 3 is a kind of introduction to the peculiar features of ancient Greek flat earth cosmology. Chapters 4 and 5 contain an extended version of a paper read at the "Ex Ionia Scientia" conference, Athens, December 2016. In Chap. 4, some of Anaximander's main cosmological images are placed in a broader cultural background. In Chap. 5, it is argued that several features of Anaximander's cosmology which at first glance look weird, are in fact rational interpretations of observed phenomena. Chapter 6 was triggered by a recent article by Philip Thibodeau on the unsolved problem of Anaximander's numbers, of which I was able to read preliminary versions years before its publication. Chapter 7 contains my interpretation of Anaximenes' cosmology, in confrontation with McKirahan's and Bicknell's interpretation of the "cap simile." In Chap. 8, I confront my interpretation of Xenophanes' cosmology, including a dissident reading of the word ἄπειρος, with those of Mourelatos and Graham. Chapters 9, 10, and 11 form so to speak the heart of Part One. Chapter 9 reveals and discusses a fundamental incompatibility between what has been handed down about Anaxagoras' opinions regarding the Milky Way and the explanation of lunar eclipses. The interpretation offered in this chapter is a confrontation with Graham's recent publications. Chapter 10 is the sequel of Chap. 9 and discusses the interpretation of Anaxagoras' opinions on the moon's light and phases, based on the findings in the previous chapter. Chapter 11 opposes Graham's suggestion about how Anaxagoras' could have measured the sizes of the sun and moon with the help of a solar eclipse and offers a simple version of another explanation. *Heaven and Earth* already contained a chapter on Aristotle and the sphericity of the earth. Chapter 12 is meant to thoroughly rethink and re-evaluate his arguments for the sphericity of the earth, which heralded the end of the flat earth cosmology in the Western world and much later in the East. Chapters 13 and 14 deal with the ancient Chinese *gai tian* flat earth cosmology, which is completely different from the Greek conception. In Chap. 13, the *gai tian* system is explained in broad outlines. Chapter 14 offers several calculations, based on the calculation of the sun's distance. These two chapters are an augmented version of an article published in 2015 in a Chinese journal (*Tsinghua Studies in Western Philosophy*). In Chap. 15, the differences and similarities of both systems and the supposed influences of the Greek on the Chinese conceptions are discussed. The book ends with a Chap. 16 on the recurrence of several features of the two ancient flat earth cosmologies and their arguments in flat earth theories of the sixth and nineteenth centuries, which may also shed some light on ancient flat earth cosmology. Additional to a general Bibliography, I added a list of References after each chapter. The reader can find a list of Quotations from Ancient Greek and Roman Authors and a list of Quotations from the *Zhou bi* and from Ancient Chinese Authors at the end of the book.

References[5]

Couprie, D.L., *Heaven and Earth in Ancient Greek Cosmology*. New York: Springer 2011.
—, An Ancient Chinese Model of a Flat Earth and a Flat Heaven. *Tsinghua Studies in Western Philosophy*, Vol. 1, No. 2 (2015) 427–483.
Diels, H., *Doxographi Graeci*. Berlin: Walter de Gruyter 1879 (=Dox).
Diels, H. and W. Kranz, *Die Fragmente der Vorsokratiker*. Zürich/Hildesheim: Weidmann 1951[6] (=DK).
Graham, D.W., *The Texts of Early Greek Philosophy*. Cambridge: Cambridge University Press 2010 (=Gr).
—, *Science Before Socrates*. Oxford: University Press 2013.
Kirk, G.S., J.E. Raven and M. Schofield, *The Presocratic Philosophers*. Cambridge: Cambridge University Press 2009 (=KRS).
Laks, A. and G.W. Most, *Early Greek Philosophy I–IX*, Cambridge, MA 2016 (=LM).
Mansfeld, J. and D.T. Runia, Aëtiana. *The Method and Intellectual Context of a Doxographer, Volume Two, Part Two*. Leiden: Brill 2009 (=MR).
Wachsmuth C, and O. Hense, *Ioannis Stobaei Anthologium* 1–5. Berlin: Weidmann 1884–1912.

[5]P = Aëtius in pseudo-Plutarch, *Placita* (numbering according to Dox).
S = Aëtius in Stobaeus, *Anthologium* (numbering according to Wachsmuth and Hense).

Chapter 2
Preliminaries on Sources and Methodology

Contents

Sources

Students of Presocratic philosophy have an abundance of textbooks available. Next to Diels' *Doxographi Graeci* and Diels/Kranz's *Fragmente der Vorsokratiker*, I frequently used Graham's *The Texts of Early Greek Philosophy*, Laks and Most's *Early Greek Philosophy*, and incidentally Dumont's *Les Présocratiques*, Reale's *I Presocratici*, Mansfeld's *Die Vorsokratiker*, and Gemelli Marciano's *Die Vorsokratiker*. Additionally, for Thales, Anaximander, and Anaximenes I used the two volumes of Wöhrle's *Traditio Praesocratica* published so far, for Xenophanes Lesher's *Xenophanes of Colophon* and Untersteiner's *Senofane, Testimonianze e frammenti*, for Anaxagoras Gershenson and Greenberg's *Anaxagoras and the Birth of Physics* and Curd's *Anaxagoras of Clazomenae*. For the most relevant chapter of Aëtius' *Placita* we have now at our disposal Mansfeld and Runia's *Aëtiana* II.2, and for the confrontation of pseudo-Plutarch's and Stobaeus' versions of Aëtius in general Bottler's *pseudo-Plutarch und Stobaeus*. We must be very grateful to the scholars who have provided us with this essential material.

Nevertheless, the user is faced with several serious problems. One of them is the numbering of the texts. One specific problem goes back to an inconvenience in Diels' numbering in *Doxographi Graeci*, where pseudo-Plutarch's and Stobaeus' versions of Aëtius are printed in two columns. Within those columns, Diels put the pieces of text of both authors in the order he thought best reflected Aëtius' original text and numbered them accordingly. For instance: AETII PLAC. II 21 goes as follows: II 21.1 is a text of both pseudo-Plutarch and Stobaeus on Anaximander, II 21.2 is a text of Stobaeus on Empedocles, II 21.3 is pseudo-Plutarch on Anaxagoras, II 21.4 is pseudo-Plutarch and Stobaeus on Heraclitus, and II 21.5 pseudo-Plutarch

D. L. Couprie, *When the Earth Was Flat*, Historical & Cultural Astronomy,
https://doi.org/10.1007/978-3-319-97052-3_2

and Stobaeus on Epicurus. In *Vorsokratiker*, however, it is no longer visible whether a text is handed down by pseudo-Plutarch or by Stobaeus, or by both, and in the latter case, whether their versions are different or not. *Doxographi Graeci* II 21.2, for example, appears as 31A56 in *Vorsokratiker*, but there it is not indicated that this is a text of Stobaeus and not of pseudo-Plutarch. The other way around, II 21.3 can be found in *Vorsokratiker* as 59A72, but it is not indicated there that this is a text of pseudo-Plutarch and not of Stobaeus. The text of *Doxographi Graeci* II 29.6 and 7 is rendered in *Vorsokratiker* under 59A77 in Stobaeus' version, but there is no indication that it is substantially different from pseudo-Plutarch's version. This implies that, always when Aëtius is quoted in *Vorsokratiker*, one should also consult the *Doxographi Graeci*. A special warning is needed, because most other compilations of texts simply follow the choices of the *Vorsokratiker*. Thus, for instance, Laks and Most copy the text of DK 59A77 in their ANAXAGORAS D45a and b. In the reconstruction of Aëtius' text in Mansfeld and Runia's *Aëtiana* II.2, 622(7), the two versions of pseudo-Plutarch and Stobaeus are combined and one must look at an earlier page (614 and, S7 and P2.29.5 respectively) for the original texts.

If authors follow the numbering of *Vorsokratiker*, as, for instance, in Bottler, in Untersteiner's *Senofane*, and in Lesher's *Xenophanes of Colophon*, there is no problem with quoting texts, but several authors introduce their own numbering. In that case, a double concordance (from and to Diels/Kranz) is required, as in Mansfeld's *Vorsokratiker*. The absence of such a concordance is a serious omission in Gershenson and Greenberg. Several times, authors add texts that are not found in *Vorsokratiker*, for instance abundantly in Gershenson and Greenberg and in Wöhrle. Mansfeld and Runia's *Aëtiana* II.2 offers several parallel texts in addition to those of pseudo-Plutarch's and Stobaeus' version of Aëtius. Other compilators leave out texts, either by omission or because they had their own reasons for making a choice. Nowhere, except for *Doxographi Graeci* 627 and Wöhrle TP2 Ar224, one will find, for example, that pseudo-Galen mentions the number 16 for Anaximander's moon circle (pseudo-Galen, *History of Philosophy* 67.1), probably because this is considered to be a mistake. In Laks/Most (LM ANAXIMAND. D24), it is not indicated that the number 27 for the sun is not in Stobaeus' version of Aëtius. Graham's choice to include only "Selected Testimonies" implies that fundamental texts of Aristotle and Aëtius on Anaxagoras and the Milky Way (both in DK 59A80) are missing. The same is the case in Mansfeld's *Vorsokratiker*, because of his "strenge Auswahl" of testimonies, as well as in other collections of texts. Aëtius' testimony on this item is also missing in Laks/Most. Graham's further choice to include only "Major Presocratics" implies that important texts like DK 60A4(4) on Archelaus and the inclination of the heavens are not found in his book. If one wants to know where to find a certain text, or whether it is included in the volume at hand, an index of sources as in the third volume of Diels/Kranz is necessary. Absence of this is a serious omission in Laks and Most's *Early Greek Philosophy*. In *Aëtiana* II.2, Mansfeld and Runia number pseudo-Plutarch's texts consecutively and Stobaeus' texts according to its own numbering. An example of the results is rendered in Table 2.1, which show the difficulties that have resulted especially in quoting texts of pseudo-Plutarch (see the numbers printed in bold).

Table 2.1 A problem of numbering

DK	Dox	MR	MR reconstructed text
Aëtius **2.25.3** (on Parmenides)	**2.25.3** = S 26.1 (not in P)	S 26.1b	Aëtius **2.25.2**
Aëtius **2.25.4** (on Xenophanes)	P **2.25.4** (= S 26.1)	P **2.25.2**	Aëtius **2.25.3**
Aëtius **2.25.8** (on Thales)	**2.25.8** = S 26.1 (not in P)	S 26.1e	Aëtius **2.25.9**
Aëtius **2.25.9** (on Anaxagoras and Democritus)	P **2.25.9** (= S 26.1)	P **2.25.5**	Aëtius **2.25.10**

To avoid this kind of confusion, I have decided in this book to use the numbering according to Diels' *Doxographi Graeci* for pseudo-Plutarch's version of Aëtius (as in the left column of Table 2.1), although Mansfeld/Runia's numbering (as in the second column of Table 2.1) is more logical, and for Stobaeus the numbering of the *Anthologium* (as in the second column of Table 2.1).

Books like Robinson's *Introduction to Early Greek Philosophy* and Kirk, Raven, and Schofield's *Presocratic Philosophers*, which contain extensive commentaries on the texts that are considered to be the most important, also have their own numbering system, but without a concordance to look up the corresponding number of Diels/Kranz, although Kirk c.s. sometimes refer to the DK number.

Most collections of texts are based on the selection made by Diels/Kranz's *Vorsokratiker*, occasionally supplemented with texts that Diels was not yet acquainted with or which the authors found important for some reason. It is Unavoidable that this procedure suffers from an interpretive bias, as was already the case in Diels/Kranz' volumes, despite its incontestable merits. Any kind of arrangement of texts implies a certain degree of interpretation. We are used to Diels/Kranz's distinction in *Vorsokratiker* between A (testimonies) and B (verbal fragments), although one sometimes wonders why a text is put under A instead of B and *vice versa*. The testimonies are subdivided (e.g. in the case of Anaxagoras) in Leben, Apophthegmatik, Lehre, and the last in not specified subcategories. Graham ignores the distinction between A and B texts to distinguish instead between Life, Works, Philosophy, and Reception. Within the category Philosophy, he creates subcategories like (in the case of Anaxagoras) Principles, Mind and Motion, Physical Theory, and Perception, which are occasionally subdivided further (and further). Laks and Most make a main distinction between P (information regarding the philosopher as a person), D (doctrine, including both verbal fragments and testimonies), and R (history of reception). Within these categories, they differentiate between subcategories, which are sometimes subdivided further (and further). Such arrangements of texts according to subjects, which are meant as helpful, can cause inconveniences. In Laks and Most, for instance, Diogenes Laërtius' testimony on Anaxagoras is cut into fifteen different pieces, placed under different headings. Unfortunately, in the process of cutting, relevant information about the size of the sun relative to the Peloponnesus [DK 59A1(8)], about the inclination of the celestial pole [DK 59A1(9)], and about the stone that fell from heaven in Aegospotami [DK 59A1(10)], has been lost. Curiously, Aëtius' item on the size of the sun in relation to the Peloponnesus (cf. DK 59A72) is also omitted, giving that the reader the impression that on this subject Hippolytus (LM ANAXAG. D4) is the only source. Moreover, Hippolytus' text on the size of the sun according to Anaxagoras cannot be found where one would expect it, under the head "Astronomy," subdivision "Sun and Moon." The reason is that the testimony of Hippolytus is not cut into pieces, but placed as a whole under the head "Three Summaries Going Back Ultimately to Theophrastus." Such omissions are not incidental in LM. Another example is the relevant passage of the Suda on Anaximander (DK 12A2), where only the titles of chapters of his book are given (D3) but the information on his occupations with the equinoxes and solstices, the earth, and the gnomon is missing. Similar problems arise in all collections of texts that group them under different headings, although not always so bad as in these examples.

In Wöhrle's *Traditio Praesocratica*, and in Gershenson's and Greenberg's *Anaxagoras*, quite another principle of arrangement of texts is used. They simply place the texts in a chronological order as well as possible. The great advantage of this procedure is that it is as free as possible from interpretation. The problem of finding items on the same subject in texts of different authors or elsewhere in texts of the same author is elegantly dealt with in Wöhrle's *Traditio Praesocratica* by placing cross-references under the texts. It is a pity that until now only two volumes of the intended series, comprising Thales, Anaximander, and Anaximenes, have appeared. Gershenson and Greenberg refer to the relevant texts when they discuss several topics in Part Three of their book, after the texts. Another feature that underscores the interpretation-free approach of these authors is that they try to be as complete as possible, which now and then results in a repetition of similar or almost similar texts of different compilators of Aëtius.

A problem that concerns all these books, regardless of the principle of ordering, is that of texts that are not attributed to a specific thinker. Two examples of this phenomenon will be discussed in Chap. 7. The first is Aristotle's *Meteorologica* 354a28–32, which Diels put into the doxography on Anaximenes (DK 13A14):

2.1 Many of the ancient meteorologists are convinced that the sun does not travel under (ὑπὸ) the earth, but rather around (περὶ) the earth and that (northern) region, and it disappears and causes night because the earth is high toward the north.

A similar text is Epicurus, *On Nature* ΙΑ̅ [33] Arrighetti, from Herculaneum Papyri 1042.8.vi, which Graham placed between the testimonies on Anaximenes:

2.2 They construct walls in a circle [around the earth] so that they may screen us against the vortex, as it whirls around outside the earth, and for all those who drive the heavenly bodies around in a circle overhead.

It is a point of discussion whether or not these texts belong there. On the one hand, we should be grateful that they have found their way in the collections of texts of the Presocratics, but on the other we should beware of accepting too easily the suggestion that these texts are about Anaximenes.

Another example, to be discussed in Chap. 10, is a text of Aëtius (*Placita* 2.29.4), which sounds in pseudo-Plutarch's version:

2.3 The youngers [say that the phases of the moon appear] in accordance with the dissemination of a flame that kindles little by little in an orderly manner, until it produces the complete full moon, and analogously diminishes again until the conjunction [of the sun and the moon], when it is completely quenched.

Diels places this text between those of the Pythagoreans (DK 58B36), while I will argue that it is better understood as belonging to Anaxagoras (and his school). Another and even more serious example of how the shadow of the great Diels still haunts the most recent handbooks and interpretations, and which concerns the paths of the heavenly bodies according to Anaximenes, will be discussed in Chap. 7.

Sometimes, a text that is relevant for a specific thinker is tucked away under the heading of another Presocratic. For instance, Agathemerus, *Geography* 1.1.2, on the earliest maps of the earth, is included in the manuals of the texts of the ancient Greek philosophers under the heading of Democritus (DK 68 A15). However, Agathemerus clearly speaks of "the ancients" in general, and in the previous sentence (DK 12A6) he mentions Anaximander as the first who made an earth map.

Several authors indicate what they consider to be true quotations, either by simply following Diels' distinction between A and B texts, or by somehow marking them, for instance by printing them boldface. Wöhrle, and Gershenson and Greenberg refrain from labeling texts as "authentic fragments" (except incidentally, as in Wöhrle's quotation marks around Anaximander's so-called "fragment" in AR 163). For a text-critical apparatus, one still needs to consult Diels' *Doxographi Graeci*, or for individual thinkers monographs and articles such as Untersteiner's *Senofane*.

The available translations are a great help but can also be a source of nuisance. Each translation is also an interpretation, but as long as we can compare it with the original text, we are in the situation to make our own comparative assessments and choices. The only complete translations of Diels/Kranz are, as far as I know, Dumont's *Les Présocratiques* and Reale's *I Presocratici* (although Dumont starts with Thales and leaves the previous ten chapters, including Pherecydes, untranslated). Gershenson and Greenberg present their (very) free and sometimes wrong translations without the original texts, so that we cannot check which version of a text they have used. McKirahan has taken on the task of translating Wöhrle's volumes into English, of which recently the first volume has been published. I regret that Laks and Most still follow Diels' demonstrably wrong translation "nozzle of a bellows" for πρηστῆρος αὐλός. By way of comparison, see, Graham's *Texts*, Axr22 and 25, and Wöhrle's TP2 Ar57 (plus note 6, continued on p. 55) and Ar88.

In the first part of this book, I have borrowed freely from available translations, but mostly from Graham. For the convenience of the reader, references to several source-books are made in the footnotes in the form of a concordance, for which see also the List of Abbreviations. For instance: "Hippolytus, *Refutatio Omnium Haeresium* 1.7.6 = DK 13A7 = LM ANAXIMEN. D3(6) = Gr Axs12 = TP2

As56 = KRS 156." As already said, for pseudo-Plutarch's version (P) of Aëtius, I follow Diels' numbering in *Doxographi Graeci.* Since Bottler follows the same numbering throughout, the quoted texts of pseudo-Plutarch, Book I and II, are easy to find in her book. This is especially recommended to those who want to read the original texts of the several versions (except that of Qusṭā ibn Lūqā, which is in German).

Methodology

In his *Science Before Socrates*, Daniel Graham reproaches me for trying "to impeach the sources," which he calls "dubious methodology at best, since the sources provide the only ground we have to stand on."[1] Graham's criticism concerns my interpretation of Anaxagoras' measurement of the sun and moon. This specific topic will be elaborated further in Chaps. 3 and 11, but in fact, the issue of the height of the sun and how to measure it on a flat earth is a kind of common thread in the present book and returns in almost every chapter. I guess Graham would make a similar objection to my interpretation of Anaxagoras' ideas about lunar eclipses and phases, or my interpretation of Anaximenes' and Xenophanes' cosmologies in this book. His critical remarks may even have a wider bearing and affect the way I treat the subject of the flat earth as such. In a sense, the first part of this book is a continuous dialogue with Graham's interpretations of ancient Greek cosmology. Therefore, it seems appropriate to make some methodological remarks in advance to allow the reader to decide whether my methods to study Presocratic flat earth cosmology are dubious or not.

I do not have the slightest intention to diminish the relevance of the available textual sources. Without them, the study of ancient flat earth cosmology would not even be possible. The big question, however, is how to evaluate and deal with the texts that have come to us. To begin with, we must always keep in mind that the number of verbal fragments on ancient Greek cosmology is next to nothing. Our knowledge of ancient Greek flat earth cosmology is mainly based on Aristotle and the testimonies of authors who lived centuries later and who had their own reasons for compiling or describing the opinions of the ancient thinkers. Moreover, they wrote within their intellectual contexts, the most relevant characteristic of which was that they were familiar with the fact that the earth is not flat but spherical. This brings me to a second point: when studying the texts on ancient Greek flat earth cosmology we must realize that the sources are *not* the only ground to stand on. The cosmological conceptions of the early Greek philosophers were not just abstract ideas, but ideas about the earth and the heavenly bodies with their movements. This means that, in the case of ancient cosmology, another source of understanding is available,

[1] Graham (2013a, 147).

namely the possibility of imagining how these ancient people would have observed these phenomena.

The most striking feature of the world picture of the ancient cosmologists studied in this book is that they believed the earth to be flat. The conviction that the earth is flat yields—sometimes surprising—consequences for cosmology, map-making, climatology and time measuring, which will be discussed in Chap. 3. It is not so easy to really appreciate the true impact of this ancient world-picture. To understand what it must have been like for those who believed that they lived on a flat earth to develop a cosmology, we need what I once called "creative imagination" or more casually, "mental gymnastics."[2] The creative imagination needed for the understanding of flat earth cosmology is a kind of retrograde paradigm switch. The need for such a paradigm switch is illustrated in the three examples at the beginning of this book under the heading "Spherical versus Flat." They demonstrate how from the same data or from the same experimental arrangement completely different conclusions can be drawn, depending on whether one believes that the earth is spherical or flat.

The fact that a retrograde paradigm switch is both needed and not so easy to achieve makes that there is hardly any other area than the study of archaic cosmology, which to such an extent is subject to anachronistic misunderstandings and misinterpretations. An anachronism is nothing more than a manifestation of our inability to put ourselves in the position of those early thinkers. The most obvious case of the anachronistic trap is the pitfall of reading notions belonging to a spherical earth into a cosmology of a flat earth, into which many an author on early Greek cosmology, accustomed to the concept of a spherical earth, has fallen. I call this inability to make the necessary paradigm switch "the spherical earth bias." An example of how this bias pursues modern interpretations is discussed in Chap. 11 and concerns a typical mistake, made by Graham and Hintz when they try to argue that Anaxagoras' measurement of the sizes of the sun and moon was made with the help of a solar eclipse.

The methodological tool against the anachronistic trap is to always be aware of our tendency to interpret the ancient records in terms of notions to which we are accustomed. In Chap. 10 we will meet a typical example in expressions such as "the moon receives its light from the sun." A special kind of this mistake, which the Greek doxographers were fond of, is to accredit the ancient Greek philosophers as the first to have offered a given theory. I think this attitude is still not absent in the interpretive work of some modern scholars. Take, for example, the recent claims that Parmenides and Anaxagoras were the first advocates of "heliophotism"—the idea that the moon is illuminated by the sun—and that Anaxagoras was the discoverer of the true cause of lunar eclipses, namely that the moon is eclipsed when the earth blocks the sun's light. The danger of such interpretations is that they may easily tend to disregard data that do not concur with them. I must confess that I made this kind of mistake in what I wrote some years ago about Anaxagoras, eclipses and the moon's light. This means that I must withdraw most of what I wrote on page 177 of my

[2]Cf. Couprie (2011, xxxi), see also Couprie and Pott (2002, 58).

Heaven and Earth in Ancient Greek Cosmology, New York (2011). Chapters 9 and 10 offer my current ideas on these subjects.

The doxographic reports on ancient Greek cosmology must be studied in the awareness that they too may contain anachronistic traits. In the following chapters, we will see amply that not only modern interpreters, but already the ancient doxographers, who handed down the teachings of the Presocratic cosmologists were not always sufficiently aware of the necessity for a paradigm switch. Supposing that something has gone wrong in the doxographic tradition is, of course, a last resort in the interpretation of ancient texts. When it can be shown, however, that similar errors occur frequently and systematically, that they are akin to mistakes made by modern authors, and that they are due to a confusion between how things are on a flat earth and how things are on a spherical earth, then it is allowed to suppose that the tradition of ancient cosmology is not always free from anachronisms. When we have the choice between thinking that an ancient cosmologist was confused or that a doxographer was subject to a form a spherical earth bias, it may be helpful to seriously consider the second option. It can be argued that this bias was present, for example, in the tradition of Anaxagoras' contradictory explanations of the Milky Way on the one hand and his explanation of lunar eclipses on the other hand, as will be explained in Chap. 9. The interpretive tool of creative imagination enables us to re-create the ancient world picture and thus to understand the available cosmological texts, to recognize anachronisms in the doxography and to avoid the pitfalls of anachronism in interpreting these texts. Of course, in using creative imagination as a tool, one must be cautious to avoid that other pitfall, fantasy. The remedy against this is to see whether the chosen interpretation makes sense within the context of what we further know about the flat-earth cosmology of the thinker in question.

An additional but equally important methodological tool in avoiding the anachronistic trap is to keep in mind that some ancient ideas that may seem strange to our eyes may nonetheless make sense within their contemporaneous context, the most important of which is that the ancients observed the heavenly phenomena with the conviction that the earth is flat. In other words, the apparent strangeness or even weirdness of an ancient cosmological idea is not as such a reason to distrust it. A good example is Anaximander's explanation of eclipses and lunar phases, as discussed in Chap. 5. A main presupposition of my methodology when studying the testimonies of ancient flat earth cosmologists is the conviction that the ideas of any Presocratic thinker form a consistent whole. As a rule, their ideas are not a mere collection of notions that might be overtly contradictory. Therefore, when we find contradictory statements in the doxography, my method is not to start with blaming the Presocratic cosmologist or ignoring one of the two opinions, but rather to look at whether and how these inconsistencies can be reconciled, or whether and how we can make a reasonable choice between the two conflicting views. Let me take as an example the way I studied Anaxagoras' explanation of lunar eclipses in Chap. 9. My method of investigation is to start with the most reliably documented aspects of Anaxagoras' astronomy and to see if it is possible to interpret the rest of the relevant doxography from that basis, and to obtain a coherent overall understanding of his astronomical thoughts.

The paradigm switch required will be even more demanding when we study the ancient Chinese *gai tian* system, because it asks us to overcome the more general Western bias of a (hemi)spherical firmament. Our mental gymnastics will be challenged to the utmost when we realize that some inconvenient features of the ancient Greek conception of a flat earth, and especially that of the absence of time differences, found an ingenious solution in this ancient Chinese conception of the flat earth. The ancient Greeks and the ancient Chinese share a prejudice resulting from the historical fact that all great civilizations inhabited the northern hemisphere of the earth. I call this "the northern hemisphere bias." We will meet this form of bias, for example, in a curious omission in Aristotle's arguments for the sphericity of the earth (Chap. 12), and in the revival of flat earth conceptions in modern times (Chap. 16). A reminiscence of this bias is the modern nickname "down under" for Australia. A good remedy against this bias is to upside down your globe.

Another interpretive pitfall is that we tend to think that those who provide the right solution to a problem will have the most rational arguments, more than those who fail and stick to false solutions. We tend to think that those who defend the wrong position are less clever. Both tendencies, for which we can coin the term "right solution bias," can be questioned, and the history of the battle on the shape of the earth is a witness, as shown in Chap. 12 on Aristotle's arguments for the sphericity of the earth. Another example of this "right solution bias," which will be discussed in Chaps. 5, 9, and 10, is the explanation of solar and lunar eclipses and the phases of the moon. One of the objectives of this book is to try to make students of ancient cosmology somewhat more aware of all these pitfalls.

For Part Two of this book, I was not able to use the original sources, for the simple reason that I cannot read Chinese. The reasons why I felt free to write on the ancient Chinese *gai tian* system are the following. I used as my main source Christopher Cullen's magnificent introduction to and translation of the *Zhou bi*, knowing that, as director emeritus of the Needham Research Institute, he is an outstanding sinologist and the main western authority in the field. I used his translations and learned many things from his introductory chapters and annotations. And again, the above-described use of creative imagination, trying to place myself in the position of those ancient Chinese cosmologists who thought that both the earth and the heavens were flat, was an important methodological tool. A special feature of their activities, namely the numerous calculations inherent in it, made it possible to check whether I was on the right track or not. For example, #B29 in Cullen's translation of the *Zhou bi* goes as follows:

2.4 On the day of the summer solstice, if one sights due east and west of Zhou then from the subsolar points directly due east and west of Zhou it is 59 595½ *li* to Zhou. On the day of the winter solstices the sun is not visible in the regions due east and west, [however] by calculation we find that from the subsolar points it is 214 557½ *li* to Zhou.

The calculations behind these intriguing figures do not appear in the texts. Based on the numerical information from several parts of the *Zhou bi*, I was able to reproduce the underlying calculations and offer interpretive drawings that show the same outcomes (in this case, Chap. 14, Figs. 14.9 and 14.10). I was encouraged in my belief that my work made sense by the fact that an earlier version of the Chaps. 13 and 14 of this book was considered worth publishing in a distinguished Chinese journal. A final incentive to write on the *gai tian* system was that some eminent scholars offered an interpretation of Presocratic cosmologists that has some aspects in common with it (Dmitri Panchenko and Radim Kočandrle[3] on Anaximenes, Alexander Mourelatos and Daniel Graham on Xenophanes) Mourelatos' and Graham's interpretations of Xenophanes will be dealt with in Chap. 6. Since Panchenko is more explicit on Greek-Chinese parallels, and even suggests influential lines from these Presocratic thinkers to the *gai tian*, his interpretation will be discussed in Chap. 15 in connection with an evaluation, based on acquaintance with both types of flat earth cosmology.

References[4]

Bottler, H., *Pseudo-Plutarch und Stobaios: Eine synoptische Untersuchung*. Göttingen: Vandenhoeck & Ruprecht 2014.

Couprie, D.L., *Heaven and Earth in Ancient Greek Cosmology*. New York: Springer 2011.

Couprie, D.L. and H.J. Pott, Imagining the Universe. *Apeiron* 35 (2002), 47–59.

Cullen, C., *Astronomy and Mathematics in Ancient China: the Zhou bi suan jing*. Cambridge: Cambridge University Press 1996.

Curd, P., *Anaxagoras of Clazomenae*. Toronto: University of Toronto Press 2007 (repr. 2010).

Diels, H., *Doxographi Graeci*. Berlin: Walter de Gruyter 1879 (=Dox).

Diels, H. and W. Kranz, *Die Fragmente der Vorsokratiker*. Zürich/Hildesheim: Weidmann 1951[6] (=DK).

Dumont, J.-P., *Les Présocratiques*. Paris: Gallimard 1988.

Gershenson, D.E. and D.A. Greenberg, *Anaxagoras and the Birth of Physics*. New York: Blaisdell Publishing Company 1964 (=GG).

Graham, D.W., *The Texts of Early Greek Philosophy*. Cambridge: Cambridge University Press 2010 (=Gr).

—, *Science Before Socrates*. Oxford: University Press 2013.

Graham, D.W. and E. Hintz, Anaxagoras and the Solar Eclipse of 478 BC. *Apeiron* 40 (2007), 319–344.

Kirk, G.S., J.E. Raven and M. Schofield, *The Presocratic Philosophers*. Cambridge: Cambridge University Press 2009 (=KRS).

Kočandrle, R., Charakteristické aspekty Anaximenovy kosmologie [Characteristic Aspects of Anaximenes' Cosmology]. *Reflexe* 54 (2018), 19–42.

Laks, A. and G.W. Most, *Early Greek Philosophy I–IX*, Cambridge MA 2016 (=LM).

Lesher, J.H., *Xenophanes of Colophon*. Toronto: University of Toronto Press 1992 (repr. 2001).

Mansfeld, J., *Die Vorsokratiker I*. Stuttgart: Reclam 1983.

—, *Die Vorsokratiker II*. Stuttgart: Reclam 1986.

[3]Kočandrle's article has not yet been published in English at the time I finished the manuscript of this book, but I had the opportunity to look at his manuscript. It is available in Czech: Kočandrle (2018).

[4]P = Aëtius in pseudo-Plutarch, *Placita* (numbering according to Dox).
S = Aëtius in Stobaeus, *Anthologium* (numbering according to Wachsmuth and Hense).

Mansfeld, J. and D.T. Runia, *Aëtiana. The Method and Intellectual Context of a Doxographer, Volume Two, Part Two*. Leiden: Brill 2009 (=MR).

Mourelatos, A.P.D., La Terre et les étoiles dans la cosmologie de Xénophane. In: A. Laks and C. Louguet (eds.), *Que'st-ce que la philosophie Présocratique?* Villeneuve d'Ascq: Presses Universitaires du Septentrion 2002, 331–350.

—, The Cloud-Astrophysics of Xenophanes and Ionian Material Monism. In: P. Curd and D.W. Graham, *The Oxford Handbook of Presocratic Philosophy*. Oxford: University Press 2008, 134–168.

Panchenko, D., Anaximenean Astronomy in the Light of Chinese Parallels. *Tsinghua Studies in Western Philosophy* 1(2) (2015), 412–426.

Reale, G. (ed.), *I Presocratici. Prima traduzione integrale con testi originali a fronte delle testimonianze e dei frammenti nella raccolta di Hermann Diels e Walther Kranz*. Milano: Bompiani, Il pensiero occidentale, 2017.

Robinson, J.M., *Introduction to Early Greek Philosophy*. Boston: Houghton Mifflin Company 1968.

Untersteiner, M., *Senofane: Testimonianze e frammenti*. Milano: Bompiani 2008 (repr. of Florence: La Nuova Italia 1956).

Wachsmuth, C. and O. Hense, *Ioannis Stobaei Anthologium* 1–5. Berlin: Weidmann 1884–1912.

Wöhrle, G., *Traditio Praesocratica. Die Milesier: Thales*. Berlin: Walter de Gruyter 2009 (=TP1).

—, *Traditio Praesocratica. Die Milesier: Anaximander und Anaximenes*. Berlin: Walter de Gruyter 2012 (=TP2).

Wöhrle, G. and McKirahan, R., *The Milesians: Thales. Translation and additional material by Richard McKirahan. Traditio Praesocratica, 1*. Berlin; Boston: De Gruyter, 2014.

Part I
Ancient Greece

Chapter 3
Peculiarities of Presocratic Flat Earth Cosmology

Contents

The issues discussed in this chapter concern peculiarities of the ancient Greek conception of a flat earth. As we will see in Part Two of this book, in the ancient Chinese *gai tian* system, everything, except the meaning of "falling," is different. In the present chapter, there is necessarily some overlap with my earlier publications, but this is inevitable to show the differences between the two systems.

The Shape of the Earth

When the ancient Greeks said that the earth is flat, they did not mean it literally, because the surface of the earth has mountains and valleys, lakes and seas. More importantly, in Presocratic cosmology the flat earth was usually thought to be somewhat concave. This idea can be traced back to Anaximander's drum-shaped earth. Column drums used to be made slightly concave by a technique called ἀναθύρωσις.[1] According to Anaximander's successor Anaximenes, the setting sun

[1]Hahn (2001), 169ff and 195–196.

D. L. Couprie, *When the Earth Was Flat*, Historical & Cultural Astronomy,
https://doi.org/10.1007/978-3-319-97052-3_3

is hidden behind the higher parts of the earth,[2] which implies that the central parts are lower. For Archelaus, the earth is "lofty around the edge and hollow in the middle,"[3] for Democritus "a flat disk, hollow in the middle,"[4] and for Leucippus "a flat drum, hollow in the middle."[5] In one text, the same thing is expressed in a general way, without mentioning names:

3.1 They postulate walls in a circle [around the earth].[6]

The idea of a somewhat concave (κοίλη) earth is understandable to the Greeks, who observed that the inhabitable world (the οἰκυμένη) was situated around the basin of the Mediterranean Sea. In ancient conceptions, the Ocean is thought to encircle the earth, and understandably the water needs a container to keep it inside. Two other reasons are mentioned in the remainder of the text just quoted:

3.2 so that they may screen us against the vortex, as it whirls around outside the earth, and for all those who drive the heavenly bodies around in a circle overhead ([ὕ]π[ερ κε]φά[λ]ης).[7]

To the last-mentioned reason, I will return in Chap. 7, section *Anonymous Texts and Kirk's Interpretation*. The image of a higher periphery of the earth survives even in Plato. In the *Timaeus*, when he tells the story of isle of Atlantis, he also mentions the land that surrounds the Ocean, which enables the Ocean to contain its water:

3.3 the whole of the continent (ἤπειρος) that is around (περὶ) that genuine Ocean.[8]

This presupposes of course the concept of a flat earth, as is fit for a story about a vanished mythical age. In the *Phaedo*, Plato talks about the place where we live as one of the many hollows in the spherical or dodecahedron-shaped earth and compares us, living around the Mediterranean Sea, with ants or frogs around a pond.[9]

---

[2]Hippolytus, *Refutatio Omnium Haeresium* 1.7.6 = DK 13A7(6) = LM ANAXIMEN. D3(6) = Gr Axs12 = TP2 As56 = KRS 156.

[3]Hippolytus, *Refutatio Omnium Haeresium* 1.9.4 = DK 60A4(4), = LM ARCH. D2(4), not in Gr and KRS.

[4]P 3.10.4–5 (not in S) = DK 68A94 = LM ATOM. D111 = Gr Dmc72, not in KRS.

[5]Pseudo-Galen, *History of Philosophy* 82, not in DK, LM, Gr, and KRS, but see Dox. 633. The suggestion in Panchenko (1999), that Archelaus, Leucippus and Democritus meant that the earth is shaped like a shield, concave at its lower side and convex at its upper side, is ingenious but not very convincing.

[6]Epicurus, *On Nature* ΙĀ [33] Arighetti, from Herculaneum Papyri 1042.8.vi = Gr Axs 20, cf. TP2 243, n.2; not in DK, LM, and KRS.

[7]Ibidem.

[8]Plato, *Timaeus* 24E–25A.

[9]Plato, *Phaedo* 109A9–B4.

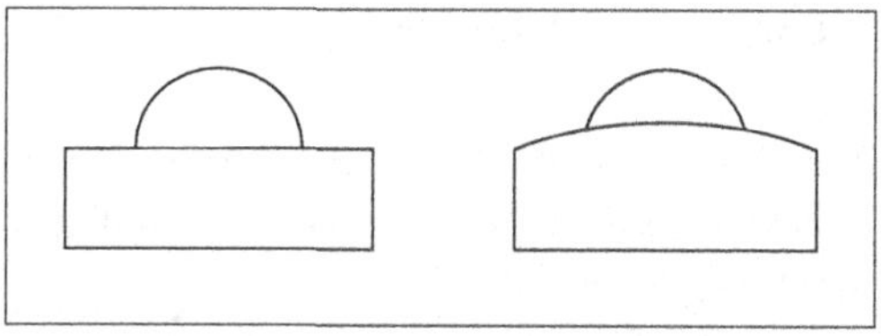

Fig. 3.1 Anaxagoras' proof that the earth is flat, redrawn after Graham (2013a), 129

Arguments Concerning the Shape of the Earth

Originally, when everyone was convinced that the earth is flat, there was no need to argue for its shape, but it was only necessary to suggest some modifications, such being slightly concave. But as soon as some cosmologists put forward the idea of a spherical earth, the believers in a flat earth needed arguments of their own. The debate on the shape of the earth must have already started before Plato, because it is reported that Parmenides taught that the earth is spherical, while others, such as Anaxagoras and Democritus, persisted in the belief that the earth is flat. In the *Phaedo*, Socrates says that someone has convinced him that the earth is a sphere, but he does not tell who this "someone" was and he refrains from producing proofs. Because of his comparison of the earth's sphere with a dodecahedron, however, we can surmise that his arguments were essentially metaphysical and based on the cosmological role he ascribes to the regular polyhedrons.[10] Aristotle is the first of whom we know his arguments for the sphericity of the earth, but even more interesting, he fights arguments that were brought forward for a flat earth being. These arguments will be discussed at length in Chap. 12.

The most obvious argument for a flat earth stems from common sense: if the earth were a sphere we would fall off the earth, and the earth would not be stable. Anaxagoras, apparently confronted with the arguments of some who pleaded for the earth's rotundity, offered the rather ingenious argument that the rising and setting sun or moon are cut off at the horizon by a straight line, while it would be curved if the earth were spherical.[11] The left part of Fig. 3.1 shows the horizon as we see it; on the right side, the horizon as we would see it, according to Anaxagoras, if the earth would be spherical.

Whatever the worth of this argument, Aristotle clearly has difficulty to refute it, as we shall see in Chap. 12. Remarkably, Anaxagoras' argument is repeated several times, in the more general form of the horizon as a straight line, by recent defenders of the flat earth, who obviously are ignorant its ancient origin (see Chap. 16).

[10]For a discussion of Plato on the shape of the earth, see Couprie (2011), 201–212.

[11]Aristotle mentions the argument in *De Caelo* 294a1–4. 4. I discussed it at length in Couprie (2008) and in Couprie (2011), Chap. 15.

Geographical Issues

The geography of a flat earth is different from that of a spherical earth. The ancient Greeks imagined their flat earth as circular in shape, as Herodotus tells us when he describes the ancient maps:

3.4 I laugh to see how many have ere now drawn maps of the earth (. . .). They draw the earth round as if fashioned by compasses, encircled by the river Ocean, and Asia and Europe of like size.[12]

Agathemerus recounts that the ancients described the inhabited earth (οἰκουμένη) as round and adds that Democritus was the first to describe the earth (γῆ) as oblong.[13] Here the words οἰκουμένη and γῆ have probably changed places, because the first is the ancient Greek word for the inhabited earth, which was considered oblong, and the second is the word for the earth which the ancients considered as round.[14]

I take Herodotus' words "Asia and Europe of like size" to mean that the oldest maps showed only two continents, "Europe occupying the northern and Asia the southern segment," as Heidel says.[15] In this conception, Africa, which the ancients called Libya, is the western part of Asia. The grey circle on the periphery of the earth in Fig. 3.2 is meant to represent the edge of mountains that keep the water of the Ocean inside. On a spherical earth, the word "poles" indicates the points where the celestial axis intersects the earth, but this word is not applicable to a flat earth. On a spherical earth, the celestial axis with its north and south pole coincides with the axis of the earth, but this is not the case on a flat earth. Accordingly, the doxography of the Presocratic flat earth cosmologies does not speak of the poles of the earth, but only of the celestial pole or poles. The subpolar point, where the celestial pole is in the zenith, was probably thought to fall near the rim of or outside the disk of the earth, as can be seen in Figs. 3.4 and 3.5b.

Heidel has argued that we might speak of the "Ionian equator and tropics" which run elsewhere than we are used to on a spherical earth. On a flat earth, the "Ionian equator" is the line from where the sun rises to where it sets during the equinoxes, both seen from Greece, or more precisely, Delphi the earth's navel. This line runs approximately through the Pillars of Hercules in the west and Miletus in the east. The

---

[12]Herodotus, *Historiae* 4.36 = Gr Axr8 = KRS 100; not in DK, LM and TP2.

[13]Agathemerus, *Geography* 1.1.2 = DK 68B15 = Gr Dmc152 = LM ATOM. D112 and D113 (but missing several lines); not in KRS.

[14]Cf. Heidel (1937), 12 n. 22: "It is quite certain that the continental mass, not to speak of the οἰκουμένη, was not circular, though the map probably was in the earlier tines."

[15]Heidel (1937), 12. See also Berger (1903), 81 and Pédech (1976), 35. Some scholars divide the surface of the earth on the ancient maps in three parts. For instance, Robinson (1968), 32, and Naddaf (2003), 54, Fig. 1.1. This is, I think, a misunderstanding of another text, in which Herodotus says that the Ionians discerned three parts of the earth: Europe, Asia, and Libya, the river Nile being the border between Asia and Libya. Here, however, Herodotus is not talking about the oldest maps, but about later developments of maps, on which more details of lands were depicted.

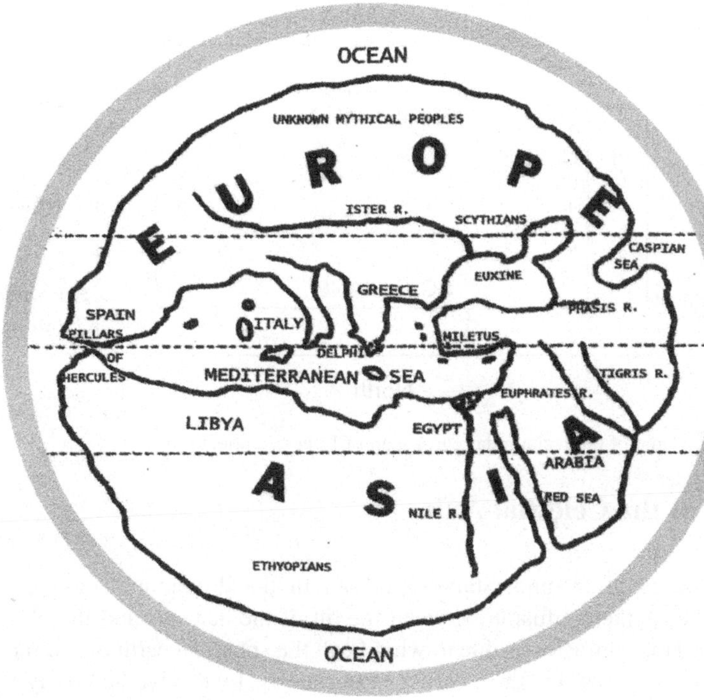

Fig. 3.2 The ancient Greek circular flat earth

points of sunrise and sunset at the summer- and winter solstice, again seen from Greece, determine the "Ionian tropics," as drawn in Fig. 3.2. The area between these "tropics" is roughly the oblong-shaped inhabitable zone, the οἰκουμένη, of the flat earth.[16]

Ephorus, a younger contemporary of Plato, still described the outlines of the οἰκουμένη as a rectangle, whose corners were marked as the sunrises and sunsets of the summer and winter solstices, with the Ethiopians in the south, the Scythians in the north, the Indians in the east and the Celts in the west, which means that Greece must be in the center, on the east-west line of the rectangle. Figure 3.3 shows Heidel's schematic picture of the frame of Ephorus' map.[17]

---

[16]Heidel (1937), 20 and 54. See also Pédech (1976), 35.

[17]Heidel (1937), 17, Fig. 2. See also Pédech (1976), 35–36, and Kominko (2013), 78–79. The data about Ephorus' map are handed down by Strabo, *Geographica*.1.2.28 and Cosmas Indicopleustes, *Topografia Cristiana* 2.79–80. See also Anderson (2013), Plates VIII and IX.

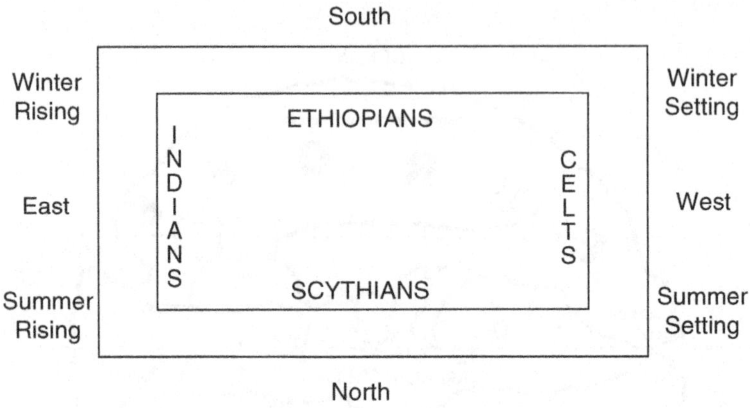

Fig. 3.3 The frame of the rectangular earth-map of Ephorus, after Heidel

The Tilt of the Celestial Axis

A major source of misunderstandings, both in the doxography and in modern commentaries is the confusion between the tilt of the heavens and the obliquity of the ecliptic. This confusion is due to what I call the spherical earth bias. Since I have already discussed it at length in earlier publications, I will give here only the main thrust of the argumentation. The ecliptic is the annual path of the sun among the stars. The Greek texts use the word ζῳδιακός, after the 12 signs of the zodiacal belt through which the sun passes. The Greek word for the obliquity of the ecliptic is λόξωσις.[18] The angle of inclination of the ecliptic is always 23.5° with respect to the celestial equator, regardless of whether the earth is considered flat or as spherical. The tilt of the heavens, on the contrary, is a special feature of the Presocratic conception of the flat earth.[19] The Greek word for the tilt of the heavens is ἔγκλισις.[20] The degree of tilting depends on where one thinks the center of the earth is. According to the ancient Greeks this was in Delphi, the navel of the earth. Instead of the tilt of the heavens, the sources may speak of the tilt of the cosmos, or the tilt of the stars, or the tilt of the pole, or the tilt of the Bears, which all boils down to the same thing. The usual idea was that the celestial axis stood originally perpendicular and that the pole coincided with the zenith of the earth's center, but that the heavens had somehow tilted during the process of cosmogony. In Chap. 4, I

[18]See next note.

[19]Cf. MR 410–411: "In the standard Platonic-Aristotelian cosmological model both the cosmos and the earth cannot be said to be tilted." Nevertheless, in order to visualize the inclination of the ecliptic, modern earth globes are usually positioned with a tilt of 23.5°. Mansfeld and Runia also rightly remark: "It needs some imagination to understand the problem raised in this chapter" (MR 410).

[20]Cf. Kahn (1970), 102: "the term for this general tilting is always ἔγκλισις, whereas λοξὸς (κύκλος) is the technical expression for the obliquity of the ecliptic." See also Dicks (1970), 71.

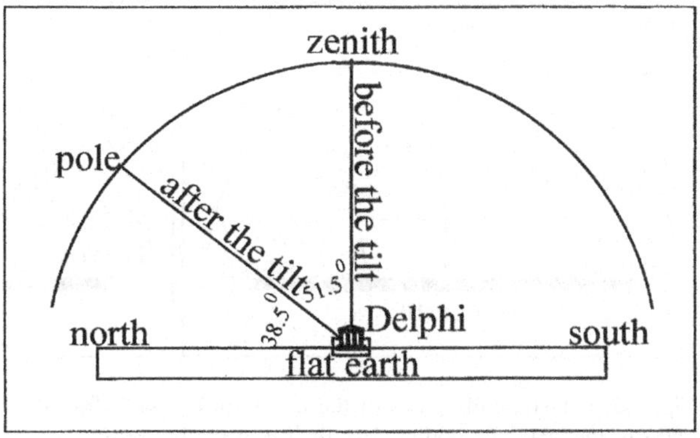

Fig. 3.4 The tilt of the heavens

will argue that these ideas go back as far as Anaximander. At Delphi, the tilt of the celestial axis is considerable, about 51.5° with respect to its former upright position, as shown in Fig. 3.4.

The original situation is described rather graphically in the doxography on Anaxagoras:

3.5 Originally the stars (ἄστρα) moved as in a dome (ϑολοειδῶς) so that the always visible pole (πόλος) was right above (κατὰ κορυφήν) the earth, but later it inclined.[21]

The implication is that in the original situation the daily movements of the heavenly bodies around the pole were in circles parallel to the flat earth's surface. Originally, the sun moved around the edge of the plane of the earth, 6 months below the horizon and the other half of the year above the horizon.[22] We can compare this with what an observer would see at the north pole of the spherical earth. In Fig. 3.5, only the heavens that are visible from a flat earth are shown, because what is rendered is what an observer at the center of the flat earth would see before and after the tilt of the heavens. In the picture on the left, the celestial equator coincides with the horizon, while in the picture on the right, the celestial equator is still at a right angle to the celestial axis, but tilted together with the heavens.

Little is known that the Ionians did not speak of the celestial equator (ὁ ἰσημερινὸς κύκλος). Perhaps they did not even know the concept. Instead, the ancient Ionians spoke of the celestial pole or the celestial axis, around which the heavenly bodies orbit. The expression ὁ ἰσημερινὸς κύκλος is used only once in the doxography on the Presocratics, in a text on Thales that is certainly unreliable, because it is a typical

[21]Diogenes Laërtius, *Vitae Philosophorum* 2.9 = DK 59A1(9) = Gr Axg37(9) = GG 340, not in LM and KRS.

[22]Cf. Heidel (1933), 122: "the early scientists contended that originally the sun had moved round the edge of the earth-plane."

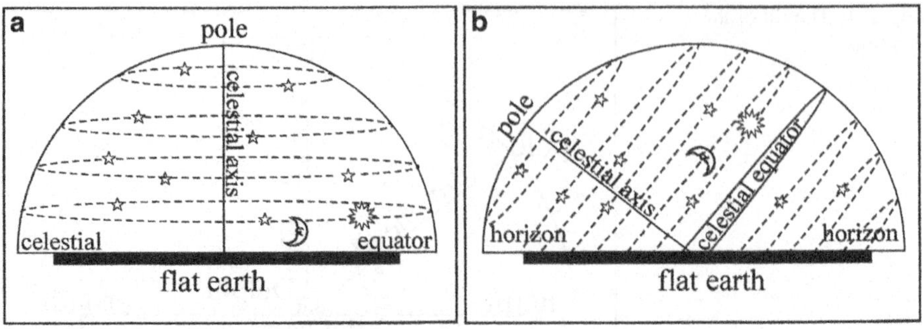

Fig. 3.5 (**a**) The daily paths of the heavenly bodies before the tilt of the heavens. (**b**) The daily paths of the heavenly bodies after the tilt of the heavens

example of the habitude of attributing all kinds of discoveries and knowledge to Thales.[23] The reason for this silence is probably that the concept of the celestial equator, which is a projection of the terrestrial equator into space, is linked to the discovery of the sphericity of the earth. On a flat earth, there is no terrestrial equator in the same sense of the word. That which can be called the "Ionian equator" is not a circle, but the diameter of the flat earth which divides it into a northern and a southern half. Even Plato does not use the expression "celestial equator" when he describes in the *Timaeus* the circle that represents the movement of the Same, although he was familiar with the sphericity of the earth. Aristotle still uses it only once, to indicate the location of a comet.[24] When I used the expression "celestial equator" and drew it in Figs. 3.5 and 3.6, it was mainly to make things clear to the present-day reader.

Because the earth was thought to be surrounded by mountains, Archelaus imagined that in the original situation these mountains were concealing the sun all year around. It was the tilt of the heavens that made the sun shine on earth for the first time, drying up the marsh:

3.6 Archelaus says that the heavens are inclined and this is how the sun came to shine on the earth, made the air transparent, and the earth dry. For in the beginning the earth was a marsh, elevated at its periphery and hollow in the middle.[25]

A typical example of the confusion between the tilt of the heavens and the obliquity of the ecliptic is in Dumont's edition of the Presocratics, in a commentary on a text of Aëtius in which it is said that according to Diogenes and Anaxagoras the cosmos is tilted:

---

[23]P 2.12.1 = S 1.23.3 = DK 11A13c = MR 447; not in Gr, the part on the zodiac and the meridian not in LM THAL. R23 and TP1 156. O'Grady (2002) doesn't even mention this text.

[24]Aristotle, *Meteorologica* 345a3.

[25]Hippolytus, *Refutatio Omnium Haeresium* 1.9.4 = DK 60A4(4) = LM ARCH. D2, not in Gr and KRS.

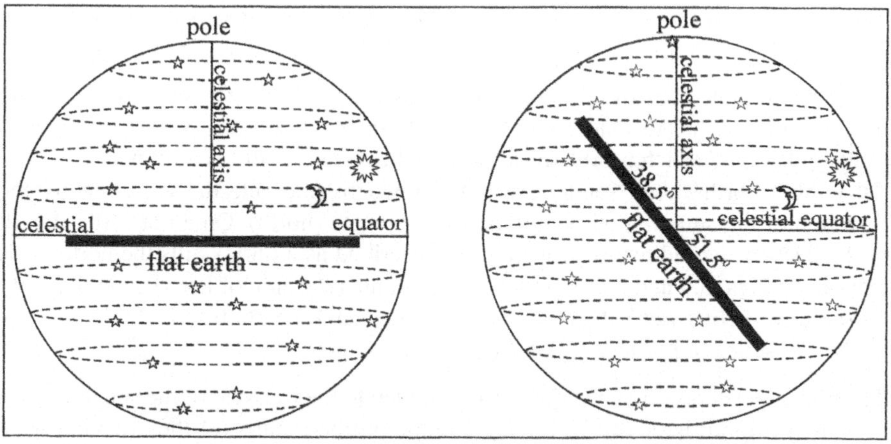

Fig. 3.6 Before and after the alleged tilt of the earth

3.7 Diogenes and Anaxagoras said after the world was formed and brought fourth living things from the earth, the cosmos somehow spontaneously tilted (. . .).[26]

In a note to this text Dumont refers to the discovery of the obliquity of the ecliptic by Oenopides as if it were identical to the tilt of the cosmos.[27] Another example is Dicks, who discusses the tilt of the cosmos and then adds: "this seems to be a recognizable attempt to account for the facts that the plane of the ecliptic is inclined to the plane of the equator (. . .) and that in Mediterranean regions the point in the sky about which all the stars are seen to revolve (i.e., the northern celestial pole) is not at the observer's zenith."[28] The last remark is right, of course, but for the ancient Greeks this was a result of the tilt of the heavens in relation to the surface of the flat earth, which is something quite different from the inclination of the plane of the ecliptic in relation to the plane of the equator. A last example of the same mistake made by several modern authors is in Gershenson and Greenberg. Under the heading "On the inclination of the ecliptic" they mention "the inclination of the cosmos to its southern part."[29]

The same kind of confusion is also not absent in the doxography. The text just quoted continues:

---

[26]P 2.8.1 (≈1.15.6) = DK 59A67 = LM ANAXAG. D32 = Gr Axg42 = MR 365 = GG 165. KRS does not print this text, but on p. 446 it is stated, completely confusing, "that the earth, which is a circle, presumably a round disc, is tilted toward the south (Aëtius II.8.1, DK 59A67) is ascribed also to Anaxagoras and Leucippus." According to DK 59A67 not the earth but the cosmos is tilted. The tilt of the earth is nowhere ascribed to Anaxagoras, but to Leucippus and Democritus (see the next section on the alleged tilt of the earth).

[27]Dumont (1988), 1430, note 3 at page 648.

[28]Dicks (1970), 59.

[29]GG 340–341. For more examples, see Couprie (2011), 77.

3.8 (...) the cosmos somehow spontaneously tilted towards its *southern* portion (...).[30]

Aëtius, to whom this text of pseudo-Plutarch goes back, obviously has in mind the definition of "north" and "south" on a spherical earth, as the opposite directions of the polar axis, which is both the axis of the heavens and the axis of the earth. On a flat earth, "north" and "south" are not defined by the celestial axis, but by the direction of the sun at noon (south) and its opposite (north). On a flat earth, the tilt of the heavens would naturally have been described as a tilt towards the north, as is clear from Figs. 3.4 and 3.5. In other words, in his rendition of Presocratic flat earth cosmology, Aëtius was subjected to a kind of spherical earth bias.

Aëtius writes in his account on Empedocles:

3.9 Empedocles says that because the air yielded to the pressure of the sun, the poles tilted (ἐπικλιθῆναι τὰς ἄρκτους), and the northern parts [of the heavens] were raised, the southern lowered, and accordingly the whole cosmos tilted.[31]

Here again, Aëtius obviously has the concept of a spherical earth within a spherical cosmos in mind. That is why he can speak of the poles in the plural as well as of the northern and southern parts of the cosmos. On a flat earth, it would be natural to speak of the upper part and lower parts of the heavens, which are respectively above the earth's surface and below the earth. On a flat earth, it is natural to speak of the celestial pole, in the singular, which is lowered (meaning "towards the northern part of the flat earth"), due to the tilt of the heavens.

The Alleged Tilt of the Earth

In the doxography on Leucippus and Democritus, probably the last two adherents of a flat earth cosmology in ancient Greece, quite surprisingly not the heavens but the earth is said to be tilted:

3.10 <...> the earth being tilted (κεκλίσθαι) towards the south.[32]

The text is mutilated and Diels has suggested to insert between the angle brackets: τὴν δὲ λόξωσιν τοῦ ζῳδιακοῦ γενέσθαι (the oblique path of the ecliptic is due to), and this emendation has unfortunately been accepted by Graham and others.[33] Heath remarks, however, and I think rightly because in the text the word κεκλίσθαι is used:

[30]My italics.

[31]P 2.8.2 (=S 1.15.6) = DK 31A58 = LM EMP. D 120 = Gr Emp70 = MR 410; not in KRS.

[32]Diogenes Laërtius, *Vitae Philosophorum* 9.33 = DK 67A1(33) = Gr Lcp47(33) = KRS 572; not in LM, but see note 1 at D103.

[33]See also, e.g., Dumont (1988), 730 and 1456; Mansfeld (1986), 250–251; McKirahan (2010), 327; Gemelli Marciano (2013), Band 3, 345.

"But this can hardly be right; the reference must be to the (...) 'inclination of the earth' (ἔγκλισις γῆς), i.e. the angle between the zenith and the pole or between the earth's (flat) surface and the plane of the apparent circular revolution of a star."[34] Just as is the case with a tilt of the heavens, an alleged tilt of the flat earth has nothing to do with the obliquity of the ecliptic. In both cases, the angle between the celestial equator and the ecliptic (or zodiac) remains the same before and after the tilting and is not the result of the tilt. KRS are similarly confused by the spherical earth bias when they write in a commentary to this text: "The tilting of the earth (...) explains both the slant of the zodiac and the differences of climate."[35]

Two other texts also mention a tilt of the earth:

3.11 Leucippus says the earth tilts (παρεκπεσεῖν) towards the south.[36]
3.12 Democritus says because the southern part is weaker than its surroundings, as the earth grew it tilted (ἐγκλιθῆναι) toward the south.[37]

The word παρεκπεσεῖν in the text on Leucippus is a synonym for ἐγκλιθῆναι or κεκλίσθαι, which are used in the other texts. At first sight, we are perhaps inclined to say that the appearances in the case of a tilt of the earth towards the south come down to the same thing as in the case of a tilt of the heavens towards the north. However, when we visualize such a tilt of the earth, as in Fig. 3.6 (right) and make the supposed tilt of the earth in relation to its original position as large as the tilt of the heavens (51.5° in relation to its former position, as discussed in the previous section), something strange appears.

Several scholars have wondered about this enormous dip of the earth. Zeller already expressed his doubts, asking why all the water of the earth does not accumulate in the southern regions.[38] Panchenko voices his bewilderment with the words: "In general, the picture of a tilted earth hanging in the middle of the spherical cosmos is bizarre, to say the least."[39] Wöhrle points to the discrepancy with another report : "Die Neigung der Erdscheibe wäre so beachtlich, daß man sich kaum (...) vorstellen kann, wie sie noch auf dem Luftpolster schwimmen könnte und wieso die Bewohner der Erde von dieser Neigung nicht das geringste verspüren."[40] More precisely, this supposed tilt of the earth is in flagrant contradiction with Aristotle's report, which says that the flat earth does not cleave but covers the air beneath it:

[34] Heath (1913), 122, n. 3.

[35] KRS, p. 420, n. 4.

[36] P 3.12.1 (not in S) = DK 67A27 = LM ATOM. D89(1) = Gr Lcp76; not in KRS, but see note 4 on p. 420.

[37] P 3.12.2 (not in S) = DK 68A96 = LM ATOM. D89(2) = Gr Dmc77; not in KRS, but see note 4 on p. 420.

[38] Zeller and Nestle (1920)^6, 1108 n. 6.

[39] Panchenko (1999), 29.

[40] Wöhrle (1993), 75. See also KRS, p. 157.

3.13 Anaximenes, Anaxagoras, and Democritus say flatness is the cause of [the earth's] staying in place. It does not cleave the air beneath it, but covers it like a lid.[41]

Aristotle's account is repeated, without the crucial reference to the lid, in the doxography:

3.14 (. . .) the earth was formed first, being completely flat. Therefore, it makes sense that it should float on air.[42]
3.15 Anaximenes [says] owing to its flatness the earth floats on air.[43]
3.16 The earth is flat riding on air.[44]

Aristotle's account is probably the most reliable, because he was nearest in time to Democritus and wrote a book about him, which has been lost.[45] He would not have written that the Democritus' flat earth does not cleave the air beneath it if he would have known of such an enormous dip of Democritus' earth. Moreover, he mentions Democritus in one breath with Anaxagoras, of whom we know that he taught the tilt of the heavens. Perhaps the doxographers also had in mind this text by Aristotle:

3.17 Many of the ancient meteorologists are convinced that the sun does not travel under the earth, but rather around the earth and that (northern) region, and it disappears and causes night *because the earth is high toward the north.*[46]

When we read this text in its context, there is no trace of a reference to a dip of the earth. All Aristotle says is "that the earth is high toward the north." Kirk already remarked: "Yet attractive as this interpretation [sc. a dip of the earth ascribed to Anaximenes] is, it is made very doubtful by [the text in the *Meteorologica*]; here Aristotle refers to the theory of higher parts," and Kirk continues: "but his context, which is concerned with showing that the greatest rivers flow from the greatest mountains, in the north, makes it quite clear that he understands 'the earth being high to the north' to refer to its northern mountain ranges," the mythical Rhipaean mountains.[47] I will come back to this text in Chap. 7, section *Anonymous Texts and Kirk's Interpretation*.

---

[41] Aristotle, *De Caelo* 294b13–21 = DK 13A20 = LM ANAXIMEN. D19 = Gr Axs13 = TP2 As3 = KRS 150.

[42] Ps. Plutarch, *Stromata* 3 = DK 13A6 = LM ANAXIMEN. D2 = Gr Axs11 = TP2 As83; not in KRS.

[43] P 3.15.8 (not in S) = DK 13A20 = Gr Axs15 = LM ANAXIMEN. D20b = TP As46; not in KRS, but see p. 153.

[44] Hippolytus, *Refutatio Omnium Haeresium* 1.7.4 = DK 13A7(4) = LM ANAXIMEN. D3(4) = G Axs12(4) = TP2 As56(4) = KRS 151.

[45] Cf. Simplicius, *In Aristotelis De Caelo Commentaria* 294.33 = DK 68A37 = LM ATOM. D29 = Gr Dmc12; not in KRS.

[46] Aristotle, *Meteorologica* 354a28–32 = DK 13A14 = LM ANAXIMEN. D16 = Gr Axs18 = TP2 As4 = KRS 157. Graham and Laks/Most translate "this region," but meant is the northern region mentioned just before.

[47] KRS, p. 157.

Fig. 3.7 The dip of a spherical earth equals the inclination of the ecliptic

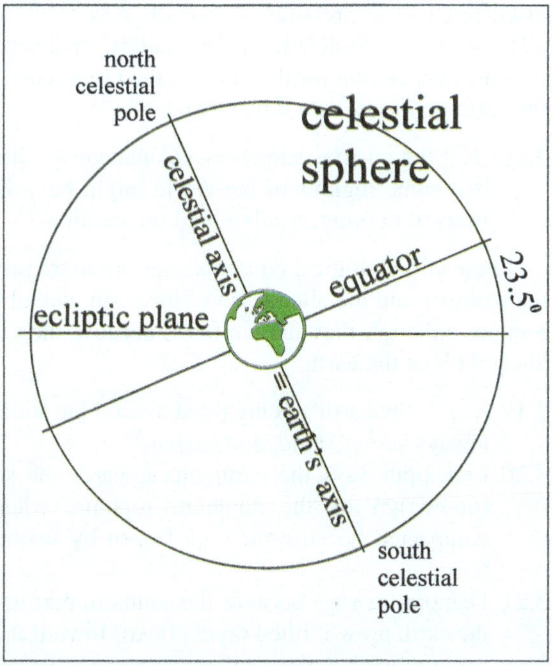

On a *spherical* earth, it is possible to speak of a dip of the earth towards the south, indicating the inclination of the ecliptic, which is also the way our globes are placed (see Fig. 3.7).

Based on these considerations, I dare to state that the obvious conclusion is that the doxography must be mistaken on the alleged dip of the earth in Leucippus and Democritus and that the atomists taught a tilt of the heavens, just like the other Presocratic flat earth cosmologists, the source of confusion being the same as before: the spherical earth bias.

Climatological Issues

On a flat earth, the northern regions are the colder ones, until one enters those lands where it is too cold to live decently. The southern regions, on the contrary, are the warmer lands, until one comes into the lands where people are burnt black by the sun.[48] The inhabitable zone lies in between and consists mainly of the lands around the Mediterranean Sea. This is how the doxography speaks about the geography and climates of the Presocratic flat earth. The cause of these climatic differences from north to south is the tilt of the heavens in relation to the surface of the flat earth, as

[48]Cf. Pédech (1976), 35.

discussed in the previous section. It was even suggested in what is sometimes referred to as a "Stoicizing interpretation," that according to Diogenes and Anaxagoras the very reason for this tilt was the providence of nature, to make some parts of the earth temperate and thus inhabitable:

3.18 (...) the cosmos somehow spontaneously tilted (...), perhaps by providence, that some regions of the world might be uninhabited, some inhabited on the basis of cooling, heating, and moderation.[49]

These climatological consequences are more precisely mentioned in the texts on the atomists and the alleged tilt of the earth, already partially quoted in the previous section, although they are confused because they were thought to be linked to the alleged tilt of the earth:

3.19 <...> the earth's being tilted toward the south. The region toward the north is always snowy, cold, and frozen.[50]

3.20 Leucippus says the earth tilts towards the south because of the porousness (ἀραιότης) of the southern regions, whereas the northern regions are compacted because they are frozen by frosts, while the contrary regions are fiery.[51]

3.21 Democritus says because the southern part is weaker than its surroundings, as the earth grew it tilted (ἐγκλιθῆναι) toward the south. For the northern regions are intemperate, the southern temperate; hence this region is heavy, where there is a greater abundance of flora, as a result of the growth.[52]

In text 3.20, Graham translates "because of the rarity [of the air]," but this is confusing. Gemelli Marciano translates rightly: "Leukip [behauptet,] die Erde neige sich wegen der lockeren Beschaffenheit, die die südlichen Teile aufweisen, nach Süden,"[53] which leaves no doubt that the southern parts of the earth are meant. In text 3.21, Guthrie translates: "the southern part of the surrounding [atmosphere]," but this is again confusing.[54] Laks and Most translate "Democritus: because the southern part of the periphery [sc. of the earth] was weaker (...)," which leaves undetermined whether the southern parts of the earth or the surrounding air of the

[49]The first part of this text was already quoted twice in the previous section; see n. 28 and 32. The qualification "Stoicizing interpretation" in LM, Volume vi, 170; see also MR 412.

[50]Diogenes Laërtius, *Vitae Philosophorum* 9.33 = DK 67A1(33) = Gr Lcp47(33) = KRS 572; not in LM, but see note 1 on D103.

[51]P 3.12.1 (not in S) = LM ATOM. D89(1) = Gr Lcp76 = DK 67A27; not in KRS, but see note 4 on p. 420.

[52]P 3.12.2 (not in S) = DK 68A96 = LM ATOM. D89(2) = Gr Dmc77; not in KRS, but see note 4 on p. 420.

[53]Gemelli Marciano (2013), Band 3, 34. See also Mansfeld (1986), Leukipp 5: "(...) wegen der in den südlichen Teilen anzutreffenden lockeren Beschaffenheit (...), obviously meaning the loose constitution of the southern *earth*."

[54]Gemelli Marciano translates, inconsequently: "Demokrit [behauptet,] die Erde neige sich, wenn sie wachse, nach dieser Seite hin, weil der südliche Teil der umgebenden Luft schwächer sei."

southern countries is meant. However, it is clear from the climatological context and the parallel with the text on Leucippus, that the intention is that the southern part of the *earth* is weaker or more porous than the northern part. The doxographer mistakenly understood these climatological remarks as an explanation for the alleged tilt of the earth. That the climate on a flat earth is getting warmer the farther one goes southward, caused Herodotus and Democritus to doubt about Anaxagoras' explanation of the Nile floods as a result of melting snow from mountain peaks in Ethiopia.[55]

The climatological consequences of the tilt of the heavens over a flat earth are fundamentally different from the climatological consequences of the obliquity of the ecliptic on a spherical earth. This was already recognized by Parmenides, who is said to have been the first to teach the sphericity of the earth, for he is also credited with the division of the earth in five zones, with two inhabitable temperate zones north and south of the torrid zone, and two arctic zones.[56]

Falling on a Flat Earth

For some reason, usually understood as due to the cosmic whirl, the heavenly bodies move in circles around the earth. Apart from this circular movement of the heavenly bodies, the direction of falling things on a flat earth is perpendicular, in parallel lines towards the earth's flat surface, unless some additional force, such as wind or a throwing hand, forces them into another direction (Fig. 3.8). This concerns heavy things, whereas light things, such as air or fire, tend to go upwards.

That falling is perpendicular to the earth's flat surface is not documented as such, but it follows from common sense, because otherwise it would mean that for most people things would fall in oblique lines, which is obviously not the case. Therefore, Rovelli's interpretation of how things fall according to what he calls "centrifocal dynamics" on Anaximander's earth[57] is wrong. "Centrifocal dynamics" is essentially the description of how things fall in Aristotle's conception of a spherical earth in the center of a spherical cosmos.

That all falling is perpendicular also follows from the question of why the earth does not fall, which bothered the ancient cosmologists. Aristotle investigates the answers given by his predecessors to this question and calls the answers more incomprehensible than the question itself: Xenophanes said that the earth extended infinitely downwards, but this is begging the question.[58] Thales supposed that the

[55]For Democritus, see Diodorus Siculus, *Bibliotheca Historica* 1.39.1–3 = DK 68A99 (partially) = LM ATOM. D120B = Gr Dmc84; not in KRS. Also in *Scholium on Apollonius of Rhodes* 4.269–271a = DK 68A99 = LM ATOM. D121, not in Gr and KRS. For Herodotus, see *Historiae* 2.22.1 = DK 59A91 = LM ANAXAG. R3 = Gr Axg55, not in GG and KRS.

[56]Cf. Strabo, *Geography* 2.2.2 = DK28A44a = LM PARM. D38 = Gr Prm43; not in KRS. Also in P 3.11.4 (not in S) = DK28A44a = LM PARM. D39 = Gr Prm44; not in KRS.

[57]Rovelli (2009), 55, Fig. 4.

[58]In Chap. 8, I will express my doubts as to this rendition of Xenophanes' ideas.

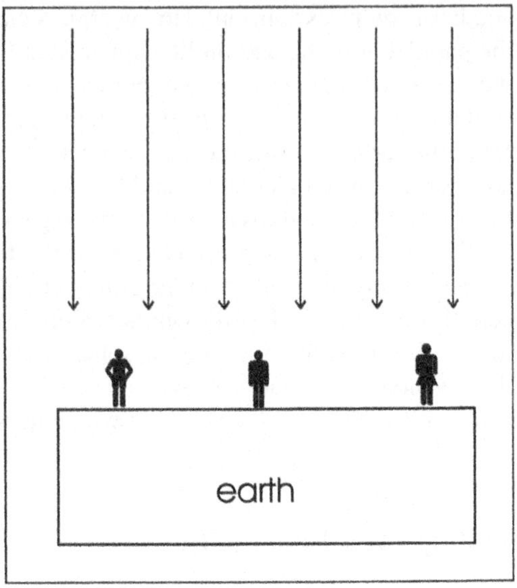

Fig. 3.8 The direction of falling on a flat earth

earth floats on water, but this leads to the question on which the water rests. Anaximenes, Anaxagoras and Democritus thought that the earth settles like a lid on the air under it, but their argument is about the size of the earth rather than its flatness. Empedocles argued that the heavenly vortex prevented the movement of the earth, but the light and heavy existed before the vortex arose. Anaximander held that the earth remains at rest because it is equally related to the extremes and therefore has no impulse to move, but this does not explain why the earth is in the center, and not, for example, fire.[59]

A lot has been written about this discussion, but here I want to emphasize a significant dichotomy in the solutions proposed. According to Aristotle, the Presocratic flat earth cosmologists worried about the question of why the earth does not fall and thus they were looking for something the earth could rest on. Apparently, they were not interested in the questions why the earth does not move sideways or upwards.

The issue of falling has yet another dimension: why do the heavenly bodies not fall on the earth? For Anaximander, they are enclosed within the heavenly wheels, for Anaximenes they are nailed into the firmament, but generally the answer of the Presocratic cosmologists was that the heavenly bodies stayed in place because of the heavenly whirl, the vortex. This was an understandable solution as long as the heavenly bodies were considered as fiery or airy, but since the time that they were regarded as earthy and stony this answer became questionable. The stone that fell in

[59]Cf. Aristotle, *De Caelo* 295b10–20 = DK 12A26 = LM ANAXIMAND. D30 = Gr Axr21 = TP2, Ar6 = KRS 123. As regards Democritus, we must probably distinguish three kinds of motion: the movements of the atoms, which is a random jostling because of collisions, the motion of the heavenly bodies, probably caused by the cosmic whirl, and the motion of heavy things, including the earth, which is downwards unless some other body prevents it. Cf. Furley (1989), 11–12.

Aegospotamoi[60] could be seen as evidence that the heavenly bodies were made of stony matter, but at the same time it undermined the theory of the vortex.

Distance of the Heavens

Quite different from what is the case when the earth is spherical, on an earth conceived as flat the heavenly bodies are relatively close by. The ancient Greek flat earth cosmologists were aware of this feature, or better, they could not even imagine the enormous distances of the heavenly bodies that are inherent in the conception of a spherical earth. The image of the heavens as a kind of felt cap with the stars attached to it, ascribed to Anaximenes, is a good example of this feature, regardless of its precise interpretation.[61] Another example is Xenophanes, according to whom the heavenly bodies are incandescent clouds, to be compared with St. Elmo's fire, and the sun is composed of tiny flares gathered together from the moist evaporation.[62] Xenophanes' cosmology is admittedly difficult to understand, but the comparison of the heavenly bodies with clouds and St. Elmo's fire implies, just like Anaximenes' cap simile, that they cannot be at a great distance. From Empedocles two contradictory reports on the distances of the moon and the sun are handed down:

3.22 Empedocles (held) that the distance of the moon from the sun is the moon is twice that from the earth.
3.23 Empedocles (held) that the distance of the moon from the earth is twice that from the sun.[63]

Whichever of the two versions is right, the obvious implication of both is that the moon and sun are not at a great distance from the earth. Another report on Empedocles clearly states that the distance between earth and heaven is not great:

---

[60] See, e.g., Diogenes Laërtius, *Vitae Philosophorum* 2.10 = DK 59A1(10) = Gr Axg1(10) = GG 340 = KRS 503; not in LM.

[61] See Hippolytus, *Refutatio Omnium Haeresium* 1.7.6 = DK 13A7(6) = LM ANAXIMEN. D3 (6) = Gr Axs12(6) = TP2 As56 (7.6) = KRS 156. For a discussion of Anaximenes' cosmology, see Chap. 7.

[62] See P 2.13.14 (=S 1.24.1) = DK 21A38 = LM XEN. D36a = Gr Xns60 (abusively referring to DK21A8) = MR 454 (2.13.7); not in KRS. Also in P 2.18.1 (≈S 1.24.1) = DK 21A39 = LM XEN. D38 = Gr Xns73 = MR 505 (abusively numbered P2.17.1); not in KRS, but see p. 174. See also P 2.20.3 (≈S 1.25.1) = DK 21A40 = Gr Xns 61 = MR 514 (P2.20.2) = KRS 177. For a discussion of Xenophanes' cosmology, see Chap. 8.

[63] The first is P's version (2.31.1), the second that of S (1.26.5), see Dox 362. Both versions in DK 31A61, followed by a kind of reconstruction of the original; LM EMP. D 136 and Gr Emp86 give P's version, but that of S, in footnotes, without translation; MR give also both versions: 635 (S1.26.5 S1) and 636 (P2.31.1), but in their reconstruction, they follow P; neither version in KRS. According to Mansfeld (2000), 185, the genitive τῆς σελήνης in Stobaeus' text is an obvious mistake.

3.24 Empedocles says that the distance [of the heaven] along the breadth [of the earth] is greater than the height from earth to heaven, which is its altitude above us (. . .).[64]

If the heaven can be identified with the sphere of stars, the moon and the sun must orbit below it and therefore even nearer to the earth. That the sun is not at a great distance from us when the earth is flat is easy to prove with by using the method Thales is already said to have used for measuring the height of a pyramid. I have tried to explain this method of measuring the height of the sun above a flat earth in *Heaven and earth*, referring to version used by ancient Chinese cosmologists.[65] Graham called this method "excessively complicated,"[66] and in a sense, he is right, as will be shown when the Chinese method is discussed more in detail in the second part of this book. I will therefore offer a simpler version and three other, also simple, observations that lead to the same conclusion: on a flat earth the sun must be close by.

For the simpler version, let us start again with Thales' method of measuring the height of a pyramid as described by both Pliny and Plutarch. It is rendered in Fig. 3.9.

Thales compared the two similar triangles AGH and APQ, and since he knew the length of the gnomon HG, the length of its shadow AG, and the length of line AP from the end of the shadow of the pyramid to the point right below the top of the pyramid, he could easily calculate the height of the pyramid QP. We still use this method when we want to estimate the height of a tree or building and compare its shadow with our own shadow. The picture invites us, as it were, to draw the perpendicular SX, which is the height of the sun above the flat earth. Thales, who had visited Egypt, could easily have noticed (if this was not yet common knowledge) that the shadows become shorter the more southward one goes, and thus that there must inevitably be a place where a gnomon at noon at the summer solstice does not cast any shadow at all. Even if the story that Thales was in Egypt to measure the height of a pyramid would not be true, the ancient Greeks certainly knew that there are places on earth, for example in southern Egypt, where the sun stood in the zenith at noon during the summer solstice. For the measurement of the sun's height, the pyramid is no longer needed, but only the gnomon and the measuring method. Since the triangles AGH and AXS are similar, you only need to know or estimate the distance AX from the observer to the point right below the sun in order to calculate the height of the sun SX. The ancient Greeks were not able to measure the distance between Miletus or Athens and the place where the sun stood in the zenith, but they could have made a rough estimate. The result is that the sun must be quite close. And the moon could not be far below the sun, for otherwise, orbiting the earth, the moon would not be able to pass the edge of the earth. At this short distance from the earth,

[64]S 1.26.5 (not in P) = DK 31A50 = LM EMP. D119 = Gr Emp69 = MR 535 (S1.26.5.1 S4); not in KRS. According to the rest of this report, Empedocles concluded that the heavens (the κόσμος) is egg-shaped.

[65]See Couprie (2011), 193–200.

[66]Graham (2013a), 147.

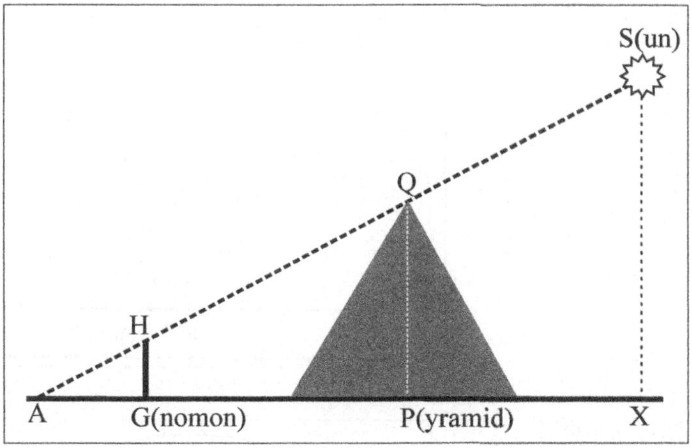

Fig. 3.9 Measuring the height of a pyramid and the height of the sun on a flat earth (drawing not to scale)

the sun cannot be very large. Actually, the sun must be much smaller than the earth itself. It is only a small step from the method illustrated above to calculate the size of the sun. And again, it is Thales, who is said to have measured the visible diameter of the sun, which is about 0.5°, or 1/720th of its full orbit around the earth.[67] The radius of the sun's orbit is the line AS in Fig. 3.9, which can be calculated with Pythagoras' theorem or by a comparison with AH, which can be measured by using a measuring cord. In Chap. 11, this method will be implemented for Athens during the summer solstice.

Virtually the same conclusion can be drawn from of a simple observation and an even simpler drawing, without any further calculations. In Athens at noon at the summer solstice the sun is about 76° above the southern horizon. Figure 3.10 shows that if there exists a place on earth where the sun at noon at the summer solstice stands in the zenith, the sun must be close to the earth and thus much smaller than the earth, because otherwise the perpendicular line from the sun would fall outside the disk of the flat earth. Figures 3.9 and 3.10 are the fundamental pictures of this book: whoever fully understands these drawings will understand what it means to live on a flat earth.

Similarly, it can be argued that not only the sun and the moon, but also the stars must be nearby in the Presocratic conception of the flat earth. We already saw that in Empedocles' estimation of the distance of the heaven. It can easily be illustrated by the way the stars are seen from different places on the flat earth. The celestial pole, for example, is seen under a different angle when an observer goes towards the north or the south, as will be shown in Chap. 11, Figs. 11.8–11.10. The only way to cope

[67]The angular size of the sun was not so easy to measure in those times; two centuries later, Aristarchus calculated with an angular diameter of 2°. With Aristarchus' number, the size of the sun would have been about 275 km.

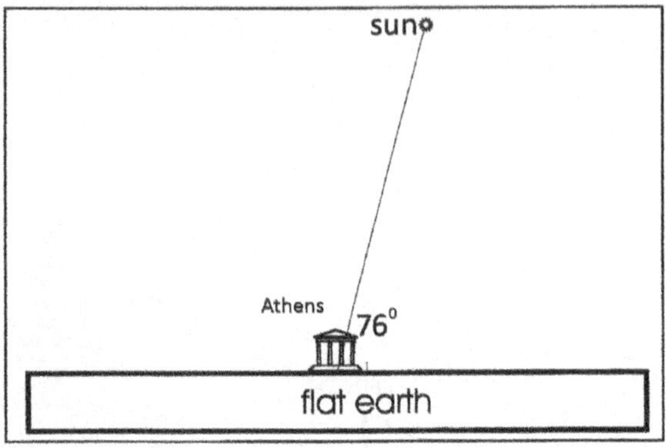

Fig. 3.10 On a flat earth, the sun is nearby and rather small (drawing to scale, except for the temple)

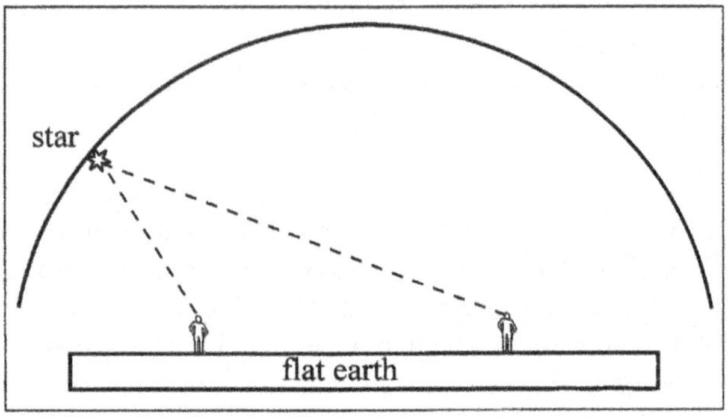

Fig. 3.11 On a flat earth, the stars must be nearby

with this phenomenon on a flat earth is the assumption that the stars are not far away. This is illustrated in Fig. 3.11.

An argument *a contrario* can be built on the idea that, if the sun were far away, its rays would hit the earth parallel to each other. This was Eratosthenes' assumption when he measured the circumference of the earth, showing that a gnomon in Syene did not cast a shadow at noon on the longest day, while a gnomon in Alexandria did cast a shadow (See Fig. 11.14a). On a flat earth, on the other hand, if a gnomon in Syene did not cast a shadow at noon on the longest day, the parallel rays of a distant sun would not cause any shadow either, wherever a gnomon would be placed (see Fig. 3.12). This point can be generalized: the parallel rays of a distant sun would always produce identical shadows of gnomons, placed anywhere on a flat earth (see Fig. 3.13). This is demonstrably not the case, and therefore, on a flat earth, the sun must be close to and smaller than the earth to ensure that gnomons cast different shadows.

In Chap. 11, we will see that Graham and Hintz overlooked this simple fact. In Chap. 16, it will be shown how Cosmas Indicopleustes, a supporter of the flat earth in Late Antiquity, used a version of this argument to prove that the sun must be close to and therefore much smaller than the earth.

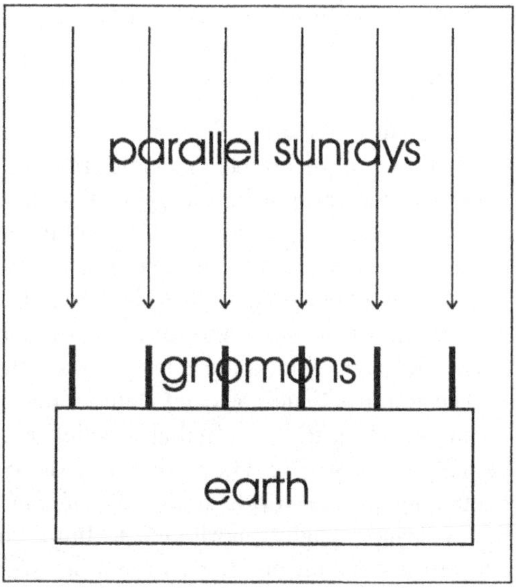

Fig. 3.12 On a flat earth, parallel rays of a far away sun would not produce anywhere a shadow of a gnomon at noon on the longest day

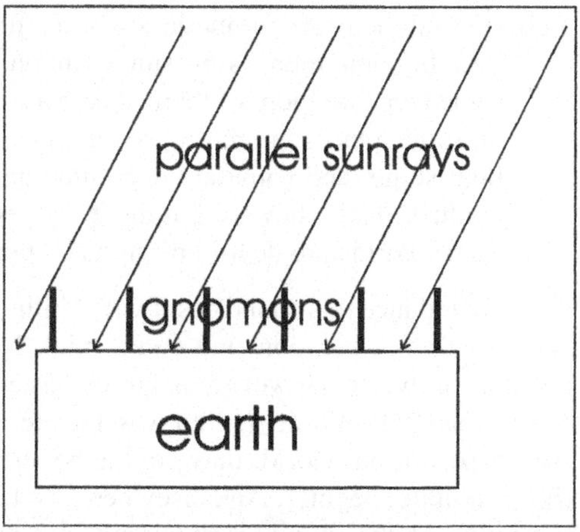

Fig. 3.13 On a flat earth, parallel rays of a far away sun would produce identical shadows everywhere

Aristotle was well aware of the fact that the sphericity of the earth, which he strongly advocated, had enormous consequences for the distances and sizes of the celestial bodies: they became, as it were, catapulted into space and must therefore be much larger than the earth. At the end of the chapter of *De Caelo*, in which he presents his arguments for the sphericity of the earth, he concludes with an obvious reference to the contrary opinions of the flat earth cosmologists:

3.25 From these arguments we must conclude not only that the earth's mass is spherical, but also that it is not large (μὴ μέγαν) in comparison with the size of the other heavenly bodies (ἄστρα).[68]

Apparently, Aristotle does not count here the moon, which is smaller than the earth, as one of the ἄστρα. Aristotle still was in discussion with the defenders of the flatness of the earth, trying to prove that they were wrong. We may read his words about the relative sizes of the earth and the celestial bodies as a direct refutation of the sizes and distances that necessarily follow from a conception of the earth as flat. The doxographers, on the other hand, who lived later than Aristotle, were all used to the concept of the sphericity of the earth and, unlike Aristotle, they sometimes no longer fully understood the implications of the concept of a flat earth.

The atomists Leucippus and Democritus, and perhaps also Anaximander, seem to be an exception to the rule that a belief in the flat earth implies that the heavenly bodies are close by. They were said to believe that the universe was infinite, containing infinite worlds. In the case of Anaximander, some scholars doubt whether this doctrine is rightly attributed to him, while others maintain that this does not concern coexisting but successive worlds.[69] On Democritus, however, the report states quite clear that we must imagine an infinity of coexisting worlds:

3.26 He said there are numberless worlds (ἀπέιρους δὲ εἶναι κόσμους) differing in size. In some there is no sun or moon, in some they are larger than ours, in some there are more of them. There are unequal distances between worlds, and in some places more, in some places less; some are growing, some are flourishing, and some are declining, and in some places they are being born, in other places they are failing. They perish when they collide with each other. Some worlds are devoid of animals, plants, and any moisture.[70]

At first glance this sounds as if with "infinite worlds" Democritus meant the stars. With some exaggeration one could think he was a modern astronomer speaking about the universe. However, as Furley has convincingly explained: "In the cosmology of classical antiquity (. . .) it was a matter of common agreement that the stars we see are part of our world: they are the boundary beyond which the infinite universe (if it is infinite) begins." And elsewhere: "What they saw in the night sky was not the beginning of the infinite universe: it was rather the boundary beyond which the infinite universe began."[71] In other words, the stars we can see belong to our world (κόσμος), making up its boundary, while the "infinite worlds" lie invisibly beyond this boundary. They are not only ἄ-πειρος in the sense of "infinite," but also in the

[68]Aristotle, *De Caelo* 298a18–21; see also 290b19: ἄστρων τὸ μέγεθος.

[69]For a thorough discussion on Anaximander and infinite worlds, see McKirahan (2001), and more recently Kočandrle (2019).

[70]Hippolytus, *Refutatio Omnium Haeresium* 1.13.2–4 = DK 68A40(2–4) = LM ATOM. D81 = Gr Dmc53(2–4) = KRS 565.

[71]Furley (1987), 136. See also Furley (1989), 2.

sense of "beyond our experience."[72] The idea of "infinite worlds" is a metaphysical conclusion drawn by the atomists from their speculation about space, but in principle their existence cannot be verified by observation.

According to the sources, Anaximander indicated specific numbers for the heavenly bodies. Their interpretation, however, is extremely difficult. All interpretations of Anaximander's numbers as indications of the distances from the earth to the heavenly bodies make them much too far away from his flat earth. I will discuss this problem in Chap. 6.

Temporal Issues

On our spherical earth, theoretically speaking, only places on the same meridian have the same time of the day, but apart from this each place on earth has its own time. In fact, it is not too long ago that every city had its own time. Nowadays this is regulated by means of the 24 time-zones, and within each zone it is supposed to be the same time for all people. On the Presocratic flat earth, however, it is always for everyone everywhere the same time. For instance, when the sun rises, it rises all over the flat earth. Ptolemy uses this as his first argument to prove that the earth is a sphere rather than a flat disk, saying:

3.27 It is possible to see that the sun and moon and the other stars do not rise and set at the same time for every observer on the earth, but always earlier for those living towards the orient and later for those living towards the occident. (...) If it were flat, the stars would rise and set for all people together and at the same time.[73]

Similarly, in Cleomedes:

3.28 If the earth were flat in shape (...), the sun's risings and settings, and thus also the beginnings of daytimes and nighttimes, would occur at the same time everywhere. But this does not in fact happen. (...) Daytimes would also turn out to be of equal length for everyone.[74]

In the doxography on the discussion on the shape of the earth in Presocratic times, this argument is mentioned only once, as far as I know. According to Hippolytus, Archelaus used it in a strange attempt to overcome the problem of simultaneity:

---

[72]See LSJ on the two different words ἄπειρος; see also Chap. 8, 155–158.

[73]Ptolemy, *Almagest* I.4.

[74]Cleomedes in Bowen and Todd (2004), 66. Panchenko (1999), 24 mentions several other ancient authors who used this argument.

3.29 He gives as a proof of this concavity the fact that the sun does not rise and set at the same times at all places, which would inevitably be the case if the earth were flat.[75]

If the intention was to overcome the problem of simultaneity in this way, there must have been some confusion. Paul Tannery already wondered how Archelaus could have drawn conclusions which are the exact opposite of those he should have drawn, because on a concave earth the sun would be seen earlier by people in the occidental regions than by those in the oriental regions.[76] Instead of concluding that Archelaus was stupid, I suppose that Hippolytus must have been confused. When we read the argument in its context (quoted already partly in the section on the tilt of the heavens), it looks rather misplaced within a story about cosmogony and the origin of life on earth:

3.30 Archelaus says that the heavens are inclined and this is how the sun came to shine on the earth, made the air transparent, and the earth dry. For in the beginning the earth was a marsh, elevated at its periphery and hollow in the middle. He gives as a proof of this concavity the fact that the sun does not rise and set at the same times at all places, which would inevitably be the case if the earth were flat. Concerning animals, he says that when the earth grew warm, first of all in its lower part, where hot and cold were blended, many kinds of animals appeared, including man.[77]

The line of reasoning is clear: originally, the sun was not shining on earth but circled around and behind its lofty edge; due to the tilt of the heavens, the sun began to shine on the earth and when the earth warmed up, animal life was could originate. The differences in time on earth have nothing to do with this story. Probably, Hippolytus, or whoever wrote the book that is traditionally ascribed to him, felt obliged to comment on the tilt of the heavens, inserting something he had heard or read about time differences. In the *Turba Philosophorum*, the argument is made right by simply replacing the concave earth by a convex earth:

3.31 Arisleus (=Archelaus) said: Know that the earth is a hill and not flat, which is why the sun does not rise above the regions of the earth at one time. For if [the earth] were flat, [the sun] would rise in one moment above the whole earth.[78]

If I am right that Hippolytus here does not render Archelaus truthfully, there is not a single text left in the doxography in which the question of time differences is discussed. Apparently, the argument was not or hardly present in the Presocratic

---

[75]Hippolytus, *Refutatio Omnium Haeresium* 1.9.4 = DK 60A4(4) = LM ARCH. D2(4), KRS 515 (4); not in Gr.

[76]Tannery (1887), 288. See Panchenko (1999), 23 for more scholars on this subject.

[77]Hippolytus, *Refutatio Omnium Haeresium* 1.9.4–5 = DK 60A4(4–5) = LM ARCH. D2(4–5), KRS 515(4–5); not in Gr.

[78]*Turba Philosophorum*, Sermo 5 = LM ARCH. R7b [referring to Plessner (1975), 57, 1–58.9], not in DK, Gr, and KRS; cf. Ruska (1931), 178 [5].

discussions about the shape of the earth. The absence of the argument of time differences in Presocratic cosmology would also explain why Plato in the *Phaedo* can describe us as living in a hole in the earth, without bothering about temporal consequences, and why Aristotle in the *De Caelo* does not mention the argument when he seeks to prove the spherical shape of the earth. After Aristotle, it became a common argument for the earth's sphericity, for instance in Ptolemy and Cleomedes, as we have seen. In retrospect, these quotations demonstrate that the problem of time differences on a flat earth could not be solved within the framework of the ancient Greek conception of the flat earth, for otherwise Ptolemy, Cleomedes and others would not have used it as an argument against it.

The non-existence of the argument of time differences in Presocratic cosmology needs not be surprising, for it is not so easy to recognize time differences on earth when you are not able to make a phone call to people in the far east or west and ask them what time it is. Hence, the ancient proof of time differences came about in a rather sophisticated way that is not recorded for Presocratic times, namely with regard to the observation of eclipses. To quote Ptolemy again:

3.32 For we find that the phenomena of eclipses taking place at the same time, especially those of the moon, are not recorded at the same hours for everyone (...); but we always find later hours recorded for observers towards the orient than for those towards the occident.[79]

Cleomedes also bases the notion of time differences on the observation of eclipses:

3.33 The heavenly body that is eclipsed among the Iberians in the 1st hour is detected as undergoing an eclipse among the Persians in the 5th hour.[80]

Another phenomenon that could have triggered a discussion about the problem of simultaneity on a flat earth, namely the differences in the duration of daylight at different latitudes, is also not mentioned in the doxography. Apparently, the ancient Greek flat earth cosmologists did not notice it or did not recognize it as a problem in relation to the shape of the earth. Perhaps the reason was that the differences in the duration of daylight at the latitudes between which the ancient Greeks traveled (let us say between northern Egypt and the Crimea), were not considerable enough (no more than 1 h and a half) to be clearly recognized. Herodotus recounts that it is told that in the far north people live who sleep 6 months of the year. This is the oldest indication of the polar regions where the sun remains for half a year below the horizon, but Herodotus underlines twice that he does not believe a word of it,[81] obviously because it does not fit in with his conception of a flat earth.

[79]Ptolemy, *Almagest* I.4.

[80]Cleomedes in Bowen and Todd (2004), I.5.

[81]Herodotus, *Historiae* 4.25.

In a text that is difficult to understand, Xenophanes is said to have discerned between "regions, sections and zones (κλίματα, ἀποτομάι, ζώναι) of the earth," each having their own sun.[82] It is hard to say whether this was mentioned as an allusion to time differences. In the doxography on Parmenides, he is said to have divided the spherical earth into five zones (ζώναι).[83] The main purpose of these zones was to make a climatological differentiation between uninhabitable (cold and torrid) and inhabitable (temperate) parts of the earth. The topic of differences in the duration of daylight is also not mentioned in Aristotle's *Meteorologica*, where the zones on earth are differentiated by means of their being habitable or uninhabitable because of the temperature, but not in relation to differences in the duration of daylight.[84] The argument is also not used by Aristotle in *De Caelo*[85] as evidence of the earth's sphericity, which can be seen as an indication that it did not play a role in the discussion at that time. In Ptolemy's *Almagest*, the argument does not only occur in the chapter with arguments for the sphericity of the earth, named *That also the Earth, Taken as a Whole, Is Sensibly Spherical*, but also elsewhere, in the Expos*ition of the Properties of Each Parallel*, where he records the time of daylight on the longest day for all parallels.[86]

Perhaps we must even say that not coincidentally, in the doxography on the Presocratic flat earth cosmologists, the problem of differences in time at different places on earth is not mentioned. One could argue that they were blind to it, precisely because it was a problem that could not be solved within their flat earth cosmology. In other words, acknowledging the existence of time differences would have undermined the very conception of a flat earth and a (hemi)spherical heaven as developed in ancient Greece. On the contrary, as we will see in Part Two of this book, it can be argued that the solution of the problem of time differences was the *raison d'être* of the ancient Chinese *gai tian* system.

References[87]

Anderson, J.C., *The Christian Topography of Cosmas Indikopleustes*. Roma: Edizioni di Storia e Letteratura 2013.

Berger, H., Geschichte der wissenschaftlichen Erdkunde der Griechen: Abteilung 1. *Die Geographie der Ionier*. Leipzig: Verlag Von Veit & Comp. 1903.

[82]P 2.24.9 (=S 1.25.3) = DK 21A41a = LM XEN. D35 = MR 563 (reading κλίμα instead of κλίματα) = Gr Xns66 = KRS 179. Untersteiner (2008), 83 (125), n., considers the word ζώναι not authentic, but inserted by the doxographer.

[83]Strabo, *Geographica* 2.2.2 = DK 28A44a = LM PARM. D38 = Gr Prm43; not in KRS.

[84]Aristotle, *Meteorologica* 391a31ff.

[85]Aristotle, *De Caelo* 297b23ff.

[86]Ptolemy, *Almagest* II.6.

[87]P = Aëtius in pseudo-Plutarch, *Placita* (numbering according to Dox).
S = Aëtius in Stobaeus, *Anthologium* (numbering according to Wachsmuth and Hense).

Bowen, A.C. and R.B. Todd: *Cleomedes' Lectures on Astronomy*. Berkeley: University of California Press 2004.

Cosmas Indicopleustes, *Topografia Cristiana*. Greek text of Chapter 6: http://www.hs-augsburg.de/~harsch/graeca/Chronologia/S_post06/Cosmas/cos_i060.html.

Couprie, D.L., Anaxagoras and Aristotle on the Sun at the Horizon. *Hyperboreus* 14 (2008), 39–49.

—, *Heaven and Earth in Ancient Greek Cosmology*. New York: Springer 2011.

Dicks, D.R., *Early Greek Astronomy to Aristotle*. Ithaca N.Y.: Thames and Hudson 1970.

Diels, H., *Doxographi Graeci*. Berlin: Walter de Gruyter 1879 (=Dox).

Diels, H. and W. Kranz, *Die Fragmente der Vorsokratiker*. Zürich/Hildesheim: Weidmann 1951[6] (=DK).

Dumont, J.-P., *Les Présocratiques*. Paris: Gallimard 1988.

Fehling, D., Das Problem der Geschichte des griechischen Weltmodells vor Aristoteles. *Rheinisches Museum für Philologie* 128 (1985), 195–231.

Furley, D., *The Greek Cosmologists. Volume I: The Formation of the Atomic Theory and Its Earliest Critics*. Cambridge: Cambridge University Press 1987.

—, *Cosmic Problems. Essays on Greek and Roman Philosophy of Nature*. Cambridge: Cambridge University Press 1989.

Gemelli Marciano, M.L., *Die Vorsokratiker* (3 Bände). Düsseldorf: Artemis & Winkler. 2007–2013.

Gershenson, D.E. and D.A. Greenberg, *Anaxagoras and the Birth of Physics*. New York: Blaisdell Publishing Company 1964 (=GG).

Graham, D.W., *The Texts of Early Greek Philosophy*. Cambridge: Cambridge UP 2010 (=Gr).

—, *Science Before Socrates*. Oxford: University Press 2013.

Hahn, R., *Anaximander and the Architects*. Albany: SUNY 2001.

Heath, T., *Aristarchus of Samos. The Ancient Copernicus*. Oxford: Clarendon Press 1913.

Heidel, W.A., *The Heroic Age of Science*. Baltimore: Williams & Wilkins Company 1933.

—, *The Frame of the Ancient Greek Maps. With a Discussion of the Discovery of the Sphericity of the earth*. New York: American Geographical Society 1937.

Kirk, G.S., J.E. Raven and M. Schofield, *The Presocratic Philosophers*. Cambridge: Cambridge University Press 2009 (=KRS).

Kahn, C.H., On Early Greek Astronomy. *Journal of Hellenic Studies* 90 (1970), 99–116.

Kočandrle, R., Infinite Worlds in the Thought of Anaximander. *The Classical Quarterly* (2019, in print).

Kominko, M., The Science of the Flat Earth: The Cosmography of Cosmas, 67–81 in. J. C. Anderson, *The Christian Topography of Cosmas Indikopleustes*. Roma: Edizioni di Storia e Letteratura 2013.

Laks, A. and G.W. Most, *Early Greek Philosophy I–IX*, Cambridge MA. 2016 (= LM).

Liddell, H.G., R. Scott H.S. Jones and R. McKenzie, *Greek-English Dictionary*. Oxford: Clarendon Press 1996. Also available on the internet, e.g.: http://stephanus.tlg.uci.edu/lsj/#eid=1714&context=search (=LSJ).

Mansfeld, J., *Die Vorsokratiker II*. Stuttgart: Reclam 1986.

—, Cosmic Distances. Aëtius 2.31 Diels and Some Related Texts. *Phronesis* 45.3 (2000), 175–204.

Mansfeld, J. and D.T. Runia, Aëtiana. *The Method and Intellectual Context of a Doxographer, Volume Two, Part Two*. Leiden: Brill 2009 (=MR).

McKirahan, R.D., Anaximander's Infinite Worlds, In A. Preus, ed., *Before Plato. Essays in Ancient Greek Philosophy VI*. Albany: State University of New York Press 2001, 40–65.

—, *Philosophy Before Socrates*. Indianapolis/Cambridge: Hackett Publishing Company, Inc. 2010.

Naddaf, G., Anthropogony and Politogony in Anaximander of Miletus. In: Couprie, D.L., R. Hahn and G. Naddaf, *Anaximander in Context New Studies in the Origins of Greek Philosophy*. Albany: State University of New York Press 2003, 7–69.

O'Grady, P. *Thales of Miletus: The Beginnings of Western Science and Philosophy*, Aldershot: Ashgate 2002.

Panchenko, D., The Shape of the Earth in Archelaus, Democritus and Leucippus. *Hyperboreus* 5 (1999), 22–39.

Pédech, P., *La géographie des Grecs*. Paris: Presses Universitaires de France 1976.
Plessner, M., *Vorsokratische Philosophie und griechische Alchemie in ararbisch-lateinischer Überlieferung. Studien zu Text und Inhalt der Turba philosophorum*. Wiesbaden: Franz Steiner Verlag 1975.
Robinson, J.M., *Introduction to Early Greek Philosophy*. Boston: Houghton Mifflin Company 1968.
Rovelli, C., Anaximander's Legacy. *Collapse: Philosophical Research and Development* 5 (2009), 50–71.
Ruska, J., *Turba Philosophorum: Ein Beitrag zur Geschichte der Alchemie*. Berlin: Springer-Verlag 1931.
Tannery, P., *Pour l'histoire de science Hellène : de Thalès à Empédocle*. Paris : Félix Alcan 1887.
Untersteiner, M., *Senofane: Testimonianze e frammenti*. Milano: Bompiani 2008 (repr. of Florence: La Nuova Italia 1956).
Wöhrle, G., *Anaximenes aus Milet. Die Fragmente zu seiner Lehre*. Stuttgart: Franz Steiner Verlag 1993.
—, *Traditio Praesocratica. Die Milesier: Anaximander und Anaximenes*. Berlin: Walter de Gruyter 2012 (=TP2).
Zeller E. and W. A. Nestle, Die Philosophie der Griechen in ihrer geschichtliche Entwicklung. 1. Teil, 2. Hälfte: Allgemeine Einleitung, Vorsokratische Philosophie Leipzig: 1920^{6}.

Chapter 4
Anaximander's Images

Contents

Introduction

Anaximander's cosmology is often described as a bizarre concoction of strange images that has nothing to do with observational data. One example out of many is Dicks, who writes: "The tertiary sources (. . .) attribute to him a fantastic theory (. . .). The unsatisfactory nature of the evidence, which is garbled and contradictory (. . .) makes it highly doubtful whether it has any historical worth."[1] Another example is Daniel Graham, who writes: "Anaximander's imaginative model accounts for the apparently circular orbits of the heavenly bodies at the cost of making them radically different from their manifestations."[2] In Chap. 4, on the contrary, I will show that the images in Anaximander's cosmology are not the result of a bizarre fantasy and that they are never chosen arbitrarily. They refer to things and events from everyday life, but at the same time they reflect deep-rooted archetypes. In Chap. 5, I will show that Anaximander's account of the astronomical phenomena remains as close as possible to what we can observe directly in the heavens. Therefore, I would call it a phenomenological astronomy. More specifically, I will show that his explanation of the phases of the moon and solar and

[1]Dicks (1970, 45–46).

[2]Graham (2013, 60).

D. L. Couprie, *When the Earth Was Flat*, Historical & Cultural Astronomy,
https://doi.org/10.1007/978-3-319-97052-3_4

lunar eclipses is not weird but purely rational and describes precisely what we see happening in the heavens. In Chap. 6, the problem of Anaximander's numbers for the heavenly bodies, which has intrigued me since I began studying his cosmology, is dealt with again. Referring to a recently published paper, I argue that a new interpretive approach is needed.

The Cosmic Tree

In the doxography on Anaximander's cosmogony we read about the origin of the cosmos:

4.1 He says that at the coming to be of this cosmos that which since days of old is generative (τὸ ἐκ τοῦ ἀιδίου γόνιμον) of hot and cold was secreted (ἀποκριθῆναι)' and from this (ἐκ τούτου) a sort of sphere of flame (τινα φλογὸς σφαῖραν) grew around the air that was around the earth like bark around a tree (ὡς τῷ δένδρῳ φλοιόν). This subsequently broke off (ἥστινος ἀπορραγείσης) and was closed into individual circles (καὶ εἴς τινας ἀποκλεισθείσης κύκλους) to form the sun, the moon, and the stars.[3]

Since Baldry's influential article[4] it has become common practice to emphasize the embryological image inherent in this text, mainly expressed by the words γόνιμον, ἀποκριθῆναι and more specifically φλοιός, which is also used in the doxography on Anaximander's account of the origin of humans.[5] The suggestion is that the biological image of a germ goes back to that of the world-egg that plays a role in many ancient cosmogonies. It is beyond doubt, though, that the expression (περὶ) τῷ δένδρῳ φλοιός, "the bark around a tree," as such does not contain any reference at all to an egg or the shell of an embryo or whatever kind of animal. West offers a strange suggestion for the interpretation of the bark-simile: "It would be reasonable to imagine that the original unit was something smaller than the present cosmos, and that it grew like a tree-trunk, leaving rings behind it."[6] The image, however, speaks about the bark of the tree and not about the rings of the tree trunk.

In the discussion of Anaximander's cosmogony, the importance of the image of the cosmic tree has been almost completely downplayed or overlooked. I disagree with Baldry when he remarks that the image of the tree is not to be stressed.[7] Guthrie

---

[3]Pseudo-Plutarch, *Stromata* 2 = DK 12A10 = LM ANAXIMAND. D8 = Gr Axr19 = TP2 Ar101 = KRS 121. I translate ἐκ τοῦ ἀιδίου as "since days of old," following a suggestion of Heidel (1912, 229, n. 2). See also KRS, p. 132, n. 1, TP2, 85 n. 2, and Kočandrle and Couprie (2017, 64–66).

[4]Baldry (1932).

[5]Cf. P 5.19.4 (not in S) = DK 12A30 = LM ANAXIMAND. D38 = Gr Axr 37 = TP2 Ar67 = KRS 133. For two recent studies on this text, see Gregory (2009) and Kočandrle and Kleisner (2013).

[6]West (1971, 95).

[7]Baldry (1932, 30).

even suggests that "the reference to trees (. . .) may have been added by Theophrastus or even later," and therefore "is obviously not intended to be pressed."[8] Andrew Gregory in his book on ancient Greek cosmogony has nothing more to say about it than that, although Anaximander's cosmogony is modeled on biological processes, we must differentiate between passive cosmological process and what living beings are capable of.[9] And in his recent book on Anaximander he explicitly remarks: "Nor do I see that much can be made of Anaximander theorising (. . .) something like the growth of a tree in the early phase of cosmogony."[10]

I think, on the contrary, that the simile of a tree and its bark is typical for Anaximander's style with its abundance of vivid images. When we look somewhat closer to the text, the suspicious word is not 'tree' (δένδρον), but 'sphere' (σφαῖρα), which is strange in two respects. In the first place, a sphere does not seem to be the right word for the bark of a tree. An indication might be that the text speaks of "a sort of sphere" (τινα σφαῖρα), as if the author himself already feels that the words "sphere" and "bark" do not go very well together. In the second place, spheres play no role in Anaximander's cosmology, which consists of a cylinder (the earth) and rings or wheels (the heavenly bodies). Heidel rightly remarks: "it is obvious that the bark of a tree is annular and not spherical," and he speaks of "the *envelope* of fire" and even "the *circle* of flame," referring to Diels' translation "Waberlohe."[11] In German mythology, "Waberlohe" is the impenetrable firewall that surrounds the castle where Brunhilde is lying in enchanted sleep until Siegfried manages to cross the wall of fire and awakens her. West's casual suggestion goes even further, namely to discount the word σφαῖρα.[12] I tend to regard σφαῖρα as an anachronistic addition to φλόξ by pseudo-Plutarch, who was presupposing a spherical cosmos.

I think there is more to say about the image of the tree in ancient Greek cosmology, and especially that of Anaximander, than just this. In several cultures the world-tree is thought to somehow support the heavens and is associated with the axis of the heavens.[13] This tree is usually thought to be standing upright. The cosmic tree stands in the center of the earth and spreads its leaves that represent the heavens. West begins a section called "*The tree*" with the remark that the conception of the Cosmic Tree is unfamiliar in Greece."[14] Still, he mentions Pherecydes' winged tree, the roots of the earth of which Hesiod and others speak, and the tree of the Hesperides with its golden apples, "but these are only traces, echoes of a forgotten

---

[8]Guthrie (1962, 91).

[9]Gregory (2007, 45). See also Gregory (2009, 51).

[10]Gregory (2016, 216).

[11]Heidel (1913, 688, 689, my italics). Cf. Diels (1897, 229): "Denn ursprünglich umgab die Erde eine zusammenhängende Waberlohe, welche alles umfasste wie die Rinde den Baum, nach seinem Ausdrucke."

[12]West (1971, 85).

[13]See, e.g., Beck (2008) and Holmberg (1923).

[14]West (1971, 55).

cosmography."[15] In addition to the instances mentioned by West, I would draw attention to some lines in the cosmology of Anaximenes:

4.2 According to some, [Anaximenes says] the heavenly bodies are fiery leaves.[16]
4.3 Anaximenes says the sun is flat like a leaf.[17]

The strange looking combination of the stars compared to both leaves and paintings can become clearer when we look upon it as a reference to the world mantle which is often associated with the world tree. This combination also occurs in the doxography on Pherecydes as:

4.4 (. . .) the winged oak and the embroidered robe upon it.[18]

Just like Anaximenes, Pherecydes was a contemporary of Anaximander. In spite of his just quoted skeptical words, West remarks with regard to Pherecydes' winged tree and the embroidered robe: "it was clearly in a sense a 'world tree'."[19] I think it is not a far cry to see also in Anaximander's tree of fire and air at the beginning of the cosmogony a reminiscence of the erected cosmic tree.[20] Anaximander, however, did not use the image of the tree in the ancient mythological way but in a profane context, in which he intended to develop a cosmogony in natural terms. Accordingly, his heavenly tree, which can be interpreted as the column of the axis of heaven, consists of air (the trunk) and fire (the bark). There is an intimate connection between the flame or fire and the air: the flame is said to have "grown around" the air. Just like the bark belongs intrinsically to the tree, so the fire and the air together make up the heavenly tree.

The word "tree," rather than the misplaced word "sphere," suggests a column, which is, in cosmological terms, the celestial axis. The association between tree and column is not so strange at all: originally, the columns of the early Greek temples were made of trees, which since the seventh century BC were replaced by stone

[15] West (1971, 59).

[16] P 2.14.4 (not in S) = DK 13A14 = LM ANAXIMEN. D14 = GrAxs17 = MR 470 = KRS 154. MR 475 reads: "But some (philosophers declare that they) are fiery leaves, like pictures." I agree, however, with Bottler 375–376, who refers to Aëtius 2.22.1 (here text 4.3: "Anaximenes says the sun is flat like a leaf") and concludes that this lemma fits better with Anaximenes.

[17] P 2.22.1 (not in S) = DK 13A15 = LM ANAXIMEN. D15 = Gr Axs 22 = MR 547 = TP2 As40 = KRS 155.

[18] Clemens, *Stromata* 6.53.35 = DK 7B2 = LM PHER. D10, KRS 55; not in Gr. And similarly: "But consider also the man of Syros (. . .) and the tree and the robe" (Maximus Tyrius, *Dissertations* 4.4.5 = DK7 A11 = LM PHER. D3, KRS 56; not in Gr). Elsewhere, the robe is described more fully: "Pherecydes of Samos says: Zas makes a mantle large and beautiful and decorates on it Ge and Ogenos and the houses of Ogenos" (Grenfell-Hunt, *Greek Papyri*, Ser. II. N.11, p. 23 = DK 7B2 = LM PHER. D9; not in Gr and KRS). LM put behind "Ogenos" between square brackets "[i.e. Ocean]," but according to Eisler's intriguing interpretation, this mantle is the mantle of the heavens, Ogenos is the celestial river (the Milky Way), and the houses of Ogenos are the twelve signs of the Zodiac. See Eisler (1910, 203–209, 566, 596).

[19] West (1971, 27).

[20] Cf. Classen (1986, 62): "(. . .) der φλοιός-Vergleich, der an den Weltenbaum anknüpfen mag."

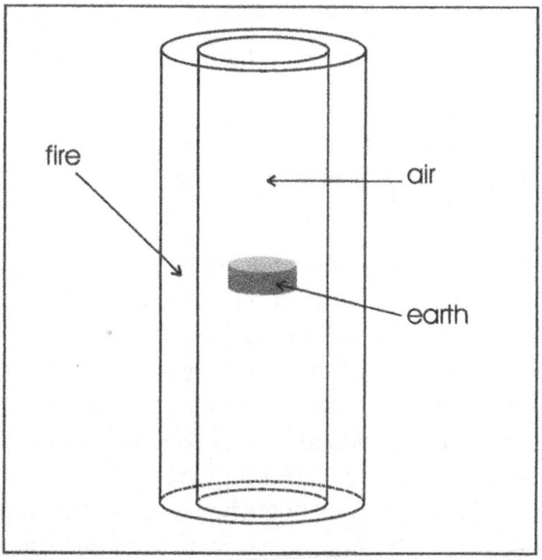

Fig. 4.1 Anaximander's cosmic tree/column/axis

columns.[21] And, as Hahn writes: "the column has cosmic significance; it symbolically separates, joins, or interpenetrates the cosmos as its axis."[22] An essential feature of Anaximander's account is that, from the very beginning of the origin of the cosmos, the earth is an integral and even central element of the heavenly column of air and fire, which are said to have grown around the earth. According to the doxography, Anaximander compared the earth with a column drum.[23] Usually, a column drum is part of a column, and the image expresses that in a similar way, at the origin of the cosmos, the earth was part of the heavenly tree or column. The three times repeated "around" (περί + dative case) suggests that from the very beginning the earth was in the center, where it was protected from the fire by the surrounding air (Fig. 4.1).

A curious parallel can be found in Exodus 13: 21:

4.5 And the LORD went before them by day in a pillar of a cloud (עַמּוּד עָנָן), to lead them the way; and by night in a pillar of fire (עַמּוּד אֵשׁ) to give them light; to go by day and night.

We might suppose that the God who led the Jewish people out of Egypt, and who was the same God who had created the world, marked his presence with a column of

[21]Cf. Hahn (2001, 71): "The first temples (. . .) were constructed out of wood and mud brick, the columns were made of large trees."

[22]Hahn (2001, 87).

[23]See pseudo-Plutarch, *Stromata* 2 = DK 12A10 = LM ANAXIMAND. D8 = Gr Axr19 = TP2 Ar101 = KRS 121. Hippolytus, *Refutatio Omnium Haeresium* 1.6.3 = DK 12A11 = LM ANAXIMAND. D7[3] = Gr Axr20 = TP2 Ar75 = KRS 122(B); P 3.10.2 (not in S) = DK 12A25 = LM ANAXIMAND. D29 = TP2 Ar65; not in Gr and KRS, but see p. 133, n. 1.

cloud and fire like an *axis mundi* that moved with Him and his chosen people until it would have found its final place in his temple in the promised land.

The Tilted Tree

Heidel already remarked: "the words καὶ εἴς τινας ἀποκλεισθείσης κύκλους refer not specifically to σφαῖρα but to φλόξ."[24] I would add the suggestion that, despite the feminine word ending, the last sentence as a whole not only refers to φλόξ as the fiery bark (φλοιός) of the heavenly tree, but also to the entire tree (δένδρον), which was broken off. Just like the temple columns, the cosmic tree evidently stands upright. But Anaximander added to the image of the cosmic tree as *axis mundi* an element typical for ancient Greek cosmology: the tilt of the axis of the heavens. The ancients observed that the daily orbits of the heavenly bodies, as they turn around an axis that ends in the celestial pole, all make the same, and fairly arbitrary, angle in relation to the surface of the flat earth. As explained in Chap. 2, the usual idea of the ancient Greek cosmologists was that originally this axis stood perpendicular to the surface of the flat earth, but that somehow during the process of cosmogony the heavens had tilted. This is reported in the doxography on Anaxagoras, Diogenes, Archelaus and Empedocles, where it is expressed as the tilt of the cosmos, of the heavens, of the stars, of the pole, or even of the Bears.[25] Elsewhere I have argued that the same holds true for the atomists Leucippus and Democritus.[26] Alleged cap simile attributed to Anaximenes implies the same tilt of the heavens, as will be argued in Chap. 7.

Although it is not said in so many words in the doxography, we may assume that the breaking off of the heavenly tree in Anaximander's cosmology resulted in its tumbling down, or cosmologically speaking, in the tilt of the celestial axis.[27] This is confirmed by the reports that Anaximander's heavenly wheels of the sun and the moon lie aslant. This will be discussed below in the section on the tilted wheels. One could therefore say that the idea of the tilted celestial axis goes as far back as to Anaximander, who probably was its originator.

---

[24]Heidel (1913, 689).

[25]Diogenes and Anaxagoras (tilt of the κόσμος): P 2.8.1 (≈S 1.15.6) = DK 59A67 = LM ANAXAG. D32 = Gr Axg42 = MR 409 = GG 165 = not in KRS, but see p. 446. Anaxagoras (tilt of the ἄστρα and the πόλος): Diogenes Laërtius, *Vitae Philosophorum* 2.9 = DK59A1(9) = Gr Axg37 = GG 340; not in LM and KRS. Archelaus (tilt of the οὐρανός): Hippolytus, *Refutatio Omnium Haeresium* 1.9.4 = DK 60A4(4) = LM ARCH. D2 = KRS 515(4); not in Gr. Empedocles (tilt of τὰ ἄρκτα and the κόσμος): P 2.8.2 (= S 1.15.6) = DK 31A58 = LM EMP. D120 = Gr Emp70 = MR 409; not in KRS.

[26]For Leucippus and Democritus, see also: Couprie (2011, 74–76).

[27]This idea has been suggested earlier by Robinson (1968, 44–45): "Anaximander (...) plac[ed] the rings bearing the sun and moon obliquely to the plane of the earth, but he did not explain (as far as we know) how this state of affairs came about, unless *he thought of it simply as a chance result of the breaking up of the original sphere of flame*" (my italics).

Fig. 4.2 (**a** and **b**) A tree stump and a column drum (Thessaloniki)

When a tree is broken off, all that remains is a tree stump. Anaximander compares the earth with a column drum, another vivid image from daily life that plays with the association of temple columns with (heavenly) trees. When the heavenly tree, the column of fire and air which is the celestial axis, is broken off and tilted, what remains upright is the earth in the shape of a column drum (Fig. 4.2).

The Reversal in the Relationship Between Air and Fire

Somehow, during the breaking off of the heavenly tree, the initial cosmic structure of the column of fire and air changed completely: the bark of fire was ripped off and transformed in the circles of sun, moon and stars. This process involved a complete reversal of fire and air, the essential features of the original cosmos. Hippolytus provides this essential further information:

4.6 The heavenly bodies came to be as circles of fire, surrounded by air.[28]

Originally the air, as we saw, was surrounded by fire. But after the fall of the heavenly tree, when the heavenly bodies were formed, the relationship has become the other way around: now the fire is enclosed in air. The reversed relationship between fire and air is confirmed by other sources, in which the heavenly bodies are described as wheels of air, their felloes filled with fire.[29] The word "wheels" was probably Anaximander's term, rather than the anachronistic "circles." I will return to the image of wheels further on in this chapter. But first there is something more to say about the heavenly fire and its relationship to the air. Unfortunately, our sources

[28]Hippolytus, *Refutatio Omnium Haeresium* 1.6.4 = DK 12A11(4) = LM ANAXIMAND. D7 [4] = Gr Axr20(4) = TP2 Ar75(4) = KRS 125.

[29]Cf. P 2.20.1 (≈S 1.25.1) = DK 12A21 = LM ANAXIMAND. D23 = Gr Axr22 = MR 514 = TP2 Ar57 and 150 = KRS 126. P 2.25.1 (≈S 1.26.1) = DK 12A22 = LM ANAXIMAND. D26 = Gr Axr25 = MR 572 = TP2 Ar60 and 151; not in KRS.

only tell us that this conversion of the mutual relationship of fire and air, which ended in the entrapment of fire within air, took place, but not why and how this took place. Anyway, we may conclude that at the end of the cosmogony, somehow the tables are turned, and the original all-embracing fire has become tamed or disciplined by being enclosed within circular shapes of air, like the shrew in Shakespeare's play is tamed by becoming obedient to her husband. As far as I know, this essential reversal has been recognized only once, in an interesting article by Nie Minli, in which he says that this text "seems to provide a different positional relationship between the Hot and the Cold, i.e. that the sphere of flame is enclosing the air."[30]

Tamed Fire

The natural behavior of fire is to spread and devour everything in its neighborhood, unless it is either extinguished or confined and thus tamed. Anaximander knew from everyday life that fire can only be tamed and made useful by somehow enclosing it, for instance in a candle or lamp, or in a fireplace or stove. The taming of fire lies at the cradle of human civilization.[31] The taming of cosmological fire by becoming trapped within air marks the origin of the regulated cosmos as we see it. In Anaximander's phenomenological astronomy, the image of the stars, the sun, and the moon as tamed specimens of fire, somehow enclosed within air, is an adequate expression of what we see when we look at the heavens. The appearances of the heavenly bodies are similar to those of controlled fire in the cultivated society, and cosmogony can be regarded as a process similar to civilization. Just as the earthly fire is tamed by the ingenuity of mankind, so in the heavenly fire Anaximander's discourse is tamed by the power of nature. With his story of the changing relationship between fire and air that took place in the prehistoric era of cosmogony, Anaximander also intends to emphasize that it was a completely natural process, in which no gods played a role but only the power of nature. There is more to be said about the tamed cosmic fire, but for that we first need to look at another of Anaximander's images.

Turning Wheels

As stated above, some sources use the plastic image of wheels. Aëtius' texts describe the formation of the celestial wheels more precisely: according to Anaximander, as a result of his cosmogony, the containers of the heavenly fire took the shapes of the rims of gigantic wheels of air that hide the fire inside.

[30]Nie Minli (2016, 85).

[31]For the essential role of fire for civilization, see Goudsblom (1995).

4.7 Anaximander says the sun is a circle twenty-eight times the size of the earth, similar to a chariot wheel (ἁρματείῳ τροχῷ παραπλήσιον), having its rim hollow, full of fire (...).[32]

4.8 Anaximander says the moon is a circle nineteen times the size of the earth, like a chariot wheel (ὅμοιον ἁρματείῳ <τροχῷ>) having a hollow rim full of fire, like that of the sun (...).[33]

the stars (or: the heavenly bodies, τὰ ἄστρα) are wheel-shaped (τροχοειδῆ) compressions of air (πιλήματα ἀέρος).[34]

The image of the hollow rim of a wheel is perfectly suited to visualize the fire trapped in airy shapes that account for their regular circular movements. In Stobaeus' text, the probably more accurate expression "compressed air" or literally "feltings of air" (πιλήματα ἀέρος) is used for the material the heavenly wheels are made of. The celestial wheels of condensed air are, so to speak, the materialization of the way in which the heavenly fire is disciplined into steady circular movements. The image of circling wheels explains why sun, moon and stars do not err randomly in the heavens but in a regular, predictable way. In the heavenly wheels not only the fire is tamed, trapped within air, but also the movement is tamed and forced into circular motion. This makes that we can predict where a star or the sun or the moon will rise at the eastern horizon after having set at the western horizon.

And again, it was an analogy with observable facts from everyday life that provided the appropriate picture. The word "wheels," rather than the anachronistic word "circles," indicates motion. Usually, moving objects as falling or thrown stones, or strolling animals do not move in circles. The only kind of bodies that are forced to move accurately in circles are round, man-made objects like wheels. Therefore, Anaximander used the image of wheels to explain the regular circular movements of the heavenly bodies. Moreover, the image of the wheel is another archetypal image, in addition to that of the captured fire. The invention of the wheel, which forced the natural movement into a circular shape, was another crucial step in the development of human civilization. Just as in the course of human civilization wheels were made to get a circular turning movement, in Anaximander's cosmogony the power of nature imposed a circular motion upon the heavenly bodies.

The implication of the image of heavenly bodies as wheels revolving around the earth is that they continue their arcs along the heavens as full circles, also under the earth, where we are unable to see them. The imagination of what happens under the horizon and even under the earth provides the rational supplement to the

[32]P 2.20.1 (≈S 1.25.1) = DK 12A21= LM ANAXIMAND. D23 = Gr Axr 22 = MR 514 = TP2 Ar57 and 150 = KRS 126.

[33]P 2.25.1 (≈S 1.26.1) = DK 12A22 = LM ANAXIMAND. D26 = Gr Axr25 = MR 572 = TP2 Ar60 and 151; not in KRS, but see 126.

[34]S 1.24.1 (not in P) = DK 12A18 = LM ANAXIMAND. D20 = MR 455 (S1.24.1 S7) = TP2 Ar148; not in Gr and KRS.

observed phenomena of the heavenly bodies above the horizon. In other words: Anaximander's rational explanation of the movements of the heavenly bodies needs the help of imagination of the unseen. The consequences were tremendous: creative imagination guided by reason led to the inevitable conclusion that the heavenly bodies go under the earth and that thus the earth floats unsupported in space. Here we see our Western world picture in the making. The difference with all earlier answers to the question "what happens to the celestial bodies when they have set?" is that Anaximander's explanation, although it made use of imagination and analogies, was completely rational. It is typical for Anaximander's way of arguing that in his explanation of the orbits of the heavenly bodies he does not take resort to the metaphysical conception of the circle as the most perfect shape and the circular movement as the most perfect movement, as in later Greek cosmology, for instance in Aristotle, but to the mundane example of the man-made wheel. I think it is even allowed to consider Anaximander's image of the heavenly wheels as a kind of pun, directed against the current mythological explanation: it is not Helios who drives the chariot of the sun across the sky, but the sun itself is a chariot wheel.

Two Images for Escaping Fire

The very reason why we do not perceive these wheels in the heavens is that they are made of—compressed—air. In other words, they are of the same substance as the medium in which they move. This means that we cannot see the fire that is hidden inside the wheels either. Therefore, a final feature completes Anaximander's explanation of the heavenly bodies: we cannot see the fire that is trapped inside, except at one aperture in the wheel, and this is what we call the sun, the moon, or a star. To explain this, Anaximander seems to have made use of two different images. As the descriptions of the heavenly wheels in the sources say:

4.9 (of the sun:) in one part emitting its fire through an opening (διὰ στομίου).[35]
4.10 (of the moon:) with one outbreathing (ἔχοντα μίαν ἐκπνοὴν).[36]
4.11 (of the stars:) there are some outbreathing spots, airy passages (ἐκπνοὰς … τόπους τινὰς ἀερώδεις), through which the heavenly bodies shine.[37]

In the last quoted text, I read τόπους τινὰς ἀερώδεις as in the manuscripts. Diels' emendation πόρους τινὰς αὐλώδεις is unnecessary and confusing.[38] In Anaximander's

[35]P 2.20.1 (=S 1.25.1) = DK 12A21 = LM ANAXIMAND. D23 = Gr Axr22 = MR 514 = TP2 Ar57 and 150 = KRS 126.

[36]P 2.25.1 (=S 1.26.1) = DK 12A22 = LM ANAXIMAND. D26 = Gr Axr25 = MR 572 = TP2Ar60 and 151; not in KRS, but see 126.

[37]Hippolytus, *Refutatio Omnium Haeresium* 1.6.4 = DK 12A11(4) = LM ANAXIMAND. D7 [4] = Gr Axr20 = TP2 Ar75(4) = KRS 125.

[38]See Conche (1991, 192 n. 1), and Couprie (2001, 198): "There is no need for tube-like gadgets in order to understand this text."

Fig. 4.3 Bronze founding; the "mouth" or "breathing hole" is clearly visible. Photograph by Steve Hoerner

vivid image, the fire is said to permanently escape through openings, as if it were mouths spitting fire. Perhaps there were already human fire spitters in Anaximander's time, using what the ancients called naphtha (νάφθα), a kind of petroleum.[39] The combination of the heterogeneous images of wheel and mouth looks rather strange to us. Workman, however, has put forward that the words ἐκπνοή (outbreathing) and στόμιον (mouth) were technical terms in bronze founding (Fig. 4.3).[40] There too, two heterogeneous things, casting mold and exhaling mouth, are combined to describe how a stream of hot air escapes from an object. A more specific reason why Anaximander used the image of mouths for the openings in the celestial wheels could have been that a mouth is easily associated with opening and closing, which Anaximander needed for his explanation of solar and lunar eclipses and the phases of the moon. I will return to the rationality of this explanation in the next chapter.

From the sources it appears that Anaximander also used another image in this context. According to him:

4.12 (the light of the sun comes to us) ὥσπερ διὰ πρηστῆρος αὐλοῦ.[41]
4.13 (the light of the moon comes to us) οἷον πρηστῆρος αὐλοῦ.[42]
4.14 (the light of the stars comes to us) ὥσπερ πρηστῆρας.[43]

---

[39]The use of naphtha is not documented from Anaximander's time. It is mentioned in the apocryphal *Second Maccabees* 1:36 as a miraculously flammable liquid, and Strabo mentions it (*Geography* 16.1.15). He calls it, after Eratosthenes, "liquid asphaltus," notes that it was known by the Babylonians, and mentions a spring of it near the Euphrates.

[40]See Workman (1953, at 46). Nowadays, "foundrymen call the hole a 'pop up,' as you are pouring you watch for it to pop up the smaller hole, indicating it is full. A pressure relief for the mold." (Information by Steve Hoerner, bronze founder).

[41]P 2.20.1 (≈S 1.25.1) = DK 12A21 = LM ANAXIMAND. D23 = Gr Axr22 = MR 514 = TP2 Ar57 and 150 = KRS 126.

[42]P 2.25.1 (≈S 1.26.1) = DK 12A22 = LM ANAXIMAND. D26 = Gr Axr25 = MR 572 = TP2 Ar60 and 151; not in KRS, but see 126.

[43]Achilles Tatius, *Introductio in Aratum* 19 = DK 12A21 = TP2 Ar88, not in LM, Gr and KRS.

Ever since Hermann Diels wrote the ominous words "immo πρηστήρ est *follis fabrorum*,"[44] almost everyone slavishly followed this great scholar and translated "as (through) the nozzle of a bellows." However, as I have extensively shown elsewhere,[45] there is no scratch of evidence for this translation, either in the contemporaneous literature, or in the image as such. A bellows blows air into the fire, while the openings in the celestial wheels are supposed to blow fire into the air. The Greek word for "bellows" is not πρηστήρ but φῦσα, and in the doxography of the Presocratics, including Anaximander, πρηστήρ is the word for a vehement meteorological event, associated with windstorm and lightning.[46] The word αὐλός does not only mean "pipe" or "flute," but also since Homer "stream" or "jet" (of blood),[47] which fits better into a cosmological context. The expressions οἷον πρηστῆρος αὐλοῦ and ὥσπερ πρηστῆρας probably reflect Anaximander's usage better than ὥσπερ διὰ πρηστῆρος αὐλοῦ, in which the word διά plus genitive obviously is inserted by a doxographer because of the misunderstanding of αὐλός as "pipe." Without Diels' unfortunate suggestion, nobody would ever have thought of the translation "nozzle of a bellows." So, I suggested that Anaximander meant to compare the light of the sun, the moon and the stars with that of lightning: whereas lightning is a momentary flash of fire, that of the heavenly bodies is like a permanent jet of fire. I was pleased to see that my translation was followed by Daniel Graham in his volume with texts of the Presocratics.[48] The reason for using this image could have been that Anaximander wanted to underline that in both cases, the light of celestial bodies and lightning, we are dealing with a completely natural process, a kind of meteorological phenomenon in which no gods played a role but only the power of nature. With this he stood at the cradle of a tradition in which the heavenly bodies and their lights were associated with meteorological phenomena.[49]

Tilted Wheels

An important feature of Anaximander's conception of the heavenly bodies as wheels brings us back to the broken heavenly tree. In the doxography on Anaximander's cosmology, which describes the final state of the cosmos, we read:

---

[44]Dox (1879, 26–27).

[45]See Couprie (2001, 2011), Chap. 11.

[46]Cf. The text under the head Περὶ βροντῶν ἀστραπῶν κεραυνῶν πρηστήρων τε καὶ τυφώνων: P 3.3.1 (=S 1.29.1) = DK 12A23 = LM ANAXIMAND. D33a = Gr Axr30 = KRS 130.

[47]Homer, *Odyssea* 22.18. In Euripides, fragment 384, πρηστήρ is used in the meaning of "stream" (of blood).

[48]Gr, 59 and 68; see also TP2, 53–55, n. 6 and 77 n1.

[49]For instance, in Xenophanes, who considers the heavenly bodies as originating every day from incandescent clouds. See P 2.13.14 (=S 1.24.1) = DK 21A38 = LM XEN. D36 = Gr Xns60 (abusively referring to DK 21A8) = MR 454; not in KRS. Also in P 2.20.3 (≈S 1.25.1) = DK 21A40 = LM XEN. D28 = Gr Xns61= MR 514 = KRS 177.

4.15 Anaximander says the moon is a circle 19 times the size of the earth, like a wagon wheel having a hollow rim full of fire, like that of the sun, *lying aslant* (κείμενον λοξόν) [the plane of the earth] *as does the sun* (. . .).[50]

The last words, printed in italics here, are handed down by Stobaeus and not in the parallel text of pseudo-Plutarch.[51] The doxographer clearly had the obliquity (λόξωσις) of the ecliptic in mind. Yet most scholars agree that this is an anachronism, due to what I call the spherical earth bias. On this issue, I completely agree with Dicks: "the words κείμενον λοξόν are a late addition in the doxographic tradition, inserted by someone who was so familiar with the slanting ecliptic of late Greek astronomy that he could not conceive of its not being a well-known concept in this early period."[52] The concepts of the ecliptic and its obliquity belong to the model of a spherical earth at the center of a celestial sphere. The word "obliquity" is used in a report on Oenopides which says that according to Eudemus it was he who discovered the obliquity of the ecliptic (εὗρε πρῶτος τὴν τοῦ ζῳδιακοῦ λόξωσιν) with regard to the celestial equator.[53] In the context of Anaximander's cosmology, however, the words κείμενον λοξόν must be understood as "lying aslant *the plane of the earth*," as Graham rightly adds,[54] just like the wheels of the stars. The slanting position of the heavenly wheels is just another expression for the tilt of the celestial axis, as we saw in Chap. 3 (Fig. 3.5).

When we compare this text with the text quoted above on the heavenly tree generating the circles that form the sun, the moon, and the stars, the conclusion must be that the idea of the tilt of the heavens is already present in Anaximander. During the process of the generation of the world, not only the relationship between air and fire reversed (from air around fire to fire inside airy wheels), but also the axis of the heavens, symbolized by the cosmic tree, tilted. The only remnant of the cosmic tree in its upright position is, as we saw, the tree stump of the flat earth.

---

[50] S 1.26.1 (italicized words not in P 2.25.1) = DK 12A22 = LM ANAXIMAND. D26 = Gr Axr25 = MR 574 = TP2 Ar151; not in KRS.

[51] See Dox 355. In textbooks, it is usually not clearly indicated that this part of the text has no parallel in P. An exception is Gemelli Marciano (2007), *Anaximander* 14.

[52] Dicks (1966, 35–36).

[53] Theon of Smyrna, *Expositio* 198.14 = DK 11A17 = Gr Ths24 = TP1 Th167, = KRS 76; this part of the text not in LM THAL. R16). Pliny's report that Anaximander discovered the obliquity of the Zodiac (*Natural History* 2.30–31 = DK 12A5 = LM ANAXIMAND. R16 = TP2 Ar40; not in Gr and KRS, but see p. 103, n. 3) is evidently anachronistic.

[54] Gr, in his translation of Axr25.

References[55]

Baldry, H.C., Embryological Analogies in Pre-Socratic Cosmogony. *The Classical Quarterly* 26 (1932), 27–34.
Beck, W. Der Weltenbaum. In: Müller U. and W. Wunderlich (Eds.): *Mittelalter-Mythen 5 – Burgen, Länder, Orte.* Konstanz 2008, 965–979.
Classen, C.J., *Ansätze: Beiträge zum Verständnis der frühgriechischen Philosophie.* Amsterdam: Rodopi B.V. 1986.
Conche, M., *Anaximandre. Fragments et Témoignages.* Paris: Presses Universitaires de France 1991.
Couprie, D.L., πρηστῆρος αὐλός Revisited. *Apeiron* 34 (2001), 193–202.
—, *Heaven and Earth in Ancient Greek Cosmology.* New York: Springer 2011.
Dicks, D.R., Solstices, Equinoxes, and the Presocratics. *The Journal of Hellenic Studies* 86 (1966), 26–40.
Diels, H., *Doxographi Graeci.* Berlin: Walter de Gruyter 1879 (=Dox).
—, Über Anaximander's Kosmos. *Archiv für Geschichte der Philosophie* 10 (1897), 228–237.
Diels, H. and W. Kranz, *Die Fragmente der Vorsokratiker.* Zürich/Hildesheim: Weidmann 1951[6] (=DK).
Eisler, R., *Weltenmantel und Himmelzelt.* München: Beck 1910.
Gemelli Marciano, M.L., *Die Vorsokratiker* (3 Bände). Düsseldorf: Artemis & Winkler 2007–2013.
Goudsblom, J., *Fire and Civilization.* London: Penguin Books 1995.
Graham, D.W., *Science Before Socrates.* Oxford: University Press 2013.
Gregory, A., *Ancient Greek Cosmogony.* London: Duckworth 2007.
—, Anaximander's Zoogony. In: M. Rossetto, M. Tsianikas, G. Couvalis and M. Palaktsoglou (Eds.), *Greek Research in Australia: Proceedings of the Eight Biennial International Conference of Greek Studies. Flinders University June 2009.* Adelaide: Flinders Department of Languages – Modern Greek 2009. (Repr. in Gregory 2016a, 29–50).
—, *Anaximander, A Re-assessment.* London: Bloomsbury 2016.
Guthrie, W.K.C., *A History of Greek Philosophy. Volume I. The Earlier Presocratics and Pythagoras.* Cambridge: Cambridge University Press 1962.
Hahn, R., *Anaximander and the Architects.* Albany: SUNY 2001.
Heidel, W.A., On Anaximander. *Classical Philology* 7 (1912), 212–234.
—, On Certain Fragments of the Pre-Socratics: Critical Notes and Emendations. *Proceedings of the American Academy of Arts and Sciences* 48 (1913), 681–734.
Holmberg, U., *Der Baum des Lebens.* Helsinki: Suomalainen Tiedeakatemia 1922–1923.
Kirk, G.S., J.E. Raven and M. Schofield, *The Presocratic Philosophers.* Cambridge: Cambridge University Press 2009 (=KRS).
Kočandrle, R. and D.L. Couprie, *Apeiron: Anaximander on Generation and Destruction.* New York: Springer 2017.
Kočandrle, R. and K. Kleisner, Evolution Born of Moisture: Analogies and Parallels Between Anaximander's Ideas on Origin of Life and Man and Later Pre-Darwinian and Darwinian Evolutionary Concepts. *Journal of the History of Biology* 46 (2013), 103–124.
Laks, A. and G.W. Most, *Early Greek Philosophy VI,* Cambridge, MA 2016.
Mansfeld, J. and D.T. Runia, Aëtiana. *The Method and Intellectual Context of a Doxographer, Volume Two, Part Two.* Leiden: Brill 2009 (=MR).
Nie Minli, Yin and *Yang* and the Hot and the Cold. *Frontiers of Philosophy in China* 11 (2016), 73–87.
Robinson, J.M., *Introduction to Early Greek Philosophy.* Boston: Houghton Mifflin Company 1968.

[55]P = Aëtius in pseudo-Plutarch, *Placita* (numbering according to Dox).
S = Aëtius in Stobaeus, *Anthologium* (numbering according to Wachsmuth and Hense).

West, M.L., *Early Greek Philosophy and the Orient*. Oxford: Clarendon Press 1971.

Wöhrle, G., *Traditio Praesocratica. Die Milesier: Anaximander und Anaximenes*. Berlin: Walter de Gruyter 2012 (=TP2).

Workman, A., "La terminologie sculpturale dans la philosophie présocratique," *Actes du XIème congrès de la philosophie: méthodologie Antiquité et Moyen Âge*, 45–50. Amsterdam: North-Holland Publishing Company 1953.

Chapter 5
Anaximander's Phenomenological Cosmology

Contents

Closing Fire Spots

Anaximander's model of the heavens stays as close as possible to what we actually see when we look at the sky. When we look at the stars with an unbiased eye, they give the impression of some kind of fire contained within the surrounding air, which fixes their positions relative to each other and makes them move in circles around the polar axis. Anaximander's picture of the heavens is, so to speak, the direct translation of this experience: the heavenly bodies are regularly circling wheels of air with fire hidden inside that we see through openings. The star wheels circle always in the same orbit, but the wheels of the sun and moon have a second movement up and down the celestial axis to cause the seasons. At first sight, the weirdest part of Anaximander's conception of the heavenly bodies is his explanation of the phases of the moon and lunar and solar eclipses. They are said to result from the partially or totally closing of the apertures in their wheels:

5.1 There are some airy breathing places (ἐκπνοὰς τόπους τινὰς ἀερώδεις), through which the heavenly bodies appear. Accordingly, when the holes are blocked there are eclipses. The moon appears to be waxing or waning by turns according to whether the passages are blocked or opened.[1]

[1]Hippolytus, *Refutatio Omnium Haeresium* 1.6.4–5 = DK 12A11(4–5) = LM ANAXIMAND. D7 [4] and [5] = Gr Axr20(4–5) = TP2 Ar75(4–5) = KRS 125.

D. L. Couprie, *When the Earth Was Flat*, Historical & Cultural Astronomy,
https://doi.org/10.1007/978-3-319-97052-3_5

5.2 Anaximander says an eclipse of the sun occurs when the mouth of the fire-hole is closed off.[2]
5.3 Anaximander says an eclipse of the moon occurs when the opening on the wheel is blocked.[3]

The simplest and generally accepted interpretation of these texts is that in the case of solar and lunar eclipses or phases of the moon the aperture in the celestial wheel is closed off by the airy envelope that surrounds and hides the fire inside. In the first quoted text, I read τόπους τινὰς ἀερώδεις, as in the manuscripts.[4] I fully agree with Conche, when he remarks that "la correction de Diels, πόρους τινὰς αὐλώδεις, 'sortes de trous de flute', est inutile." Conche translates "il y a des orifices respiratoires—certains lieux de l'air," and he explicates: "les ouvertures sont de nature aérienne, étant pratiqués à travers l'enveloppe d'air."[5] The only reason for Diels' emendation πόρους τινὰς αὐλώδεις seems to be his obsession with bellows. In the *Doxographi Graeci*, Diels defends his emendation by saying "ut πρηστῆρος αὐλοὶ quodammodo significarentur."[6] Unfortunately, all textbooks still follow Diels' emendation, mostly without mentioning the text of the manuscripts.[7] Curiously, in Reale's *I Presocratici*, Salvatore Obinu follows Diels' emendation πόρους τινὰς αὐλώδεις, but he translates "certi passagi d'aria," which is the translation of τόπους τινὰς ἀερώδεις."[8]

Later thinkers came up with the right explanations of eclipses and the moon's phases, and we are inclined to think that those who offer the right explanations are the only ones who have rational arguments. In Chap. 2, I called this "the right solution bias." Sometimes, however, the right explanation can be obtained with insufficient or even invalid arguments. In the case of Anaximander at stake here, I will show that he was able to solve a problem that the early adherents of the right explanations were unable to solve and that therefore his explanation of phases and eclipses was, in a sense, as rational as later (and right) ones.

---

[2]P 2.24.2 (=S 1.25.1) = DK 12A21 = LM ANAXIMAND. D25 = Gr Axr24 = MR 562 = TP2 AR59 and 150; not in KRS.

[3]P 2.29.1 (=S 1.26.3) = DK 12A22 = LM ANAXIMAND. D28 = Gr Axr27 = MR 614 = TP2 Ar62 and 153; not in KRS.

[4]See the Greek text in Miller (1851), 11.

[5]Conche (1991), 192, n. 1.

[6]Diels, Dox *Prologomena* 156 (in Dox 560 abusively referred to as 204), refers to Cedrenus 277.2–3, who has πόρους τινὰς αὐρώδεις, which Bekker (1838, 277) translates as "expirationes esse quasdam aurae permeatus." See also the cross-reference to DK 12A11 in DK, note at 84, lines 10f.

[7]In his recent edition of Hippolytus' *Refutatio Omnium Haeresium*, Litwa (2016, 30 and 31) also reads πόρους τινὰς αὐλώδεις, without indication that this is an emendation, and translates "tube-like passages.

[8]Reale (2017), 184 and 185. In one of the front pages, Salvatore Obinu is mentioned as the translator of the texts on Anaximander and other thinkers, although on the next page his name does not appear as one of the collaborators.

Phases of the Moon

Parmenides is usually considered as the first to give the right explanation of the moon's phases. Graham translates the most relevant texts, which are direct quotations of Parmenides' poem, as follows:

5.4 [moon] shining by night (νυκτιφαές), wandering around earth with borrowed light.[9]
5.5 Ever (αἰεί) peeking towards the rays of the sun.[10]

Although the phases of the moon are not mentioned in these lines, it has been argued that their explication is implied. The word νυκτιφαές poses an interpretive problem, because one of the essential features of the moon is that it shines both at night and by day, unlike the sun that only shines during the day. The meaning of the word 'ever' (αἰεί) in the second line should be, Mourelatos argues, that the moon is always illuminated by the sun, both during the day and at night, even when the sun is below the horizon.[11] We may paraphrase: the light part of the moon is that which is illuminated by the sun and the dark part of the moon is that which does not receive the sun's light. This should explain what we call the phases of the moon during the month from new moon to full moon. This explanation of the phases of the moon, to which we still adhere to, replaced earlier explanations such as those of Anaximander and Xenophanes. It may be questioned whether this is the only and necessary explanation of Parmenides' texts. Here, however, I will confine myself to Anaximander's explanation of the phases of the moon and argue that it is at least as rational as the explanation that Parmenides is said to have first proposed or implied.

The best way to foster understanding for the rationality of Anaximander's explanation is to point out an interesting fundamental problem in the theory attributed to Parmenides. The phenomena of the moon's phases are incomparable with what we usually see when the sun shines on an object. Compare, in Fig. 5.1a, b, the pictures of the moon and my moon globe, both lit by the sun.

Usually, the side of an object that is not lit by the sun does not disappear, as is the case with the dark part of the moon, but remains visible, like the part of my globe that is not lit by the sun. Nowadays, we know that this is due to the scattering of light being reflected by air molecules or other atmospheric particles, but this knowledge was not available to the ancients. Neither is the disappearance of the unilluminated part of the moon is due to its dark surroundings. The moon is often visible during the day (see Fig. 5.2), when its unilluminated side looks blue, just like the surrounding sky, which is not the case with any other object that we see around us and that is illuminated by the sun.

---

[9]Plutarch, *Against Colotes* 1116a = DK 28B14 = LM PARM. D27 = Gr Prm32 = KRS 308.

[10]Plutarch, *On the Face in the Moon* 929b = DK 28B15 = LM PARM. D28 = Gr Prm33; this part of the text not in KRS.

[11]Cf. Mourelatos (2012 and 2013), 98–104. See also Graham (2013), 91.

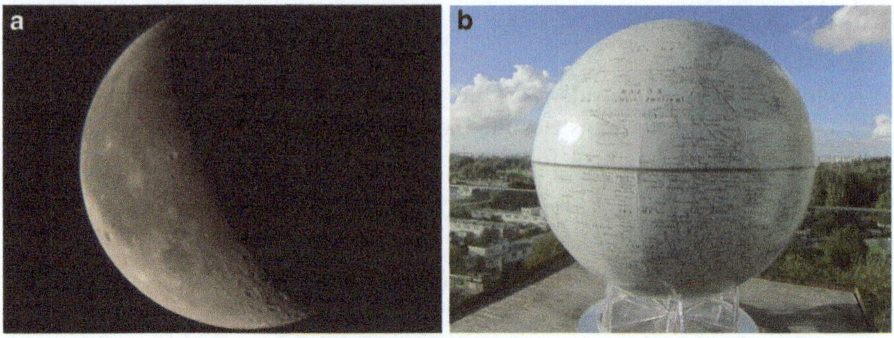

Fig. 5.1 (**a** and **b**) The moon and a moon globe, both lit by the sun. Left photograph by Keith Schengili-Roberts, July 12, 2012

Fig. 5.2 The unlit part of the moon is blue, just like the surrounding air. Photograph June 13, 2016, CC0 Creative Commons, free for commercial use

Therefore, schematic pictures of the shapes of the moon like that in Fig. 5.3, which can be found in many books on astronomy are in a sense misleading, since they show the unilluminated side of the moon as a black part of a sphere against a white background. However, what we see is not a full sphere, partly illuminated and partly black, but only the illuminated part of the moon against the black or blue background of the sky.

We must conclude that the moon behaves quite differently from other objects that are lit by the sun. Where usually the unilluminated side of an object remains visible, the unilluminated side of the moon always takes the color of its surroundings. Parmenides' theory of the phases of the moon does not explain this strange behavior of the moon. This behavior appears even stranger when we consider that sometimes during a crescent moon the unilluminated part of the moon is vaguely visible.

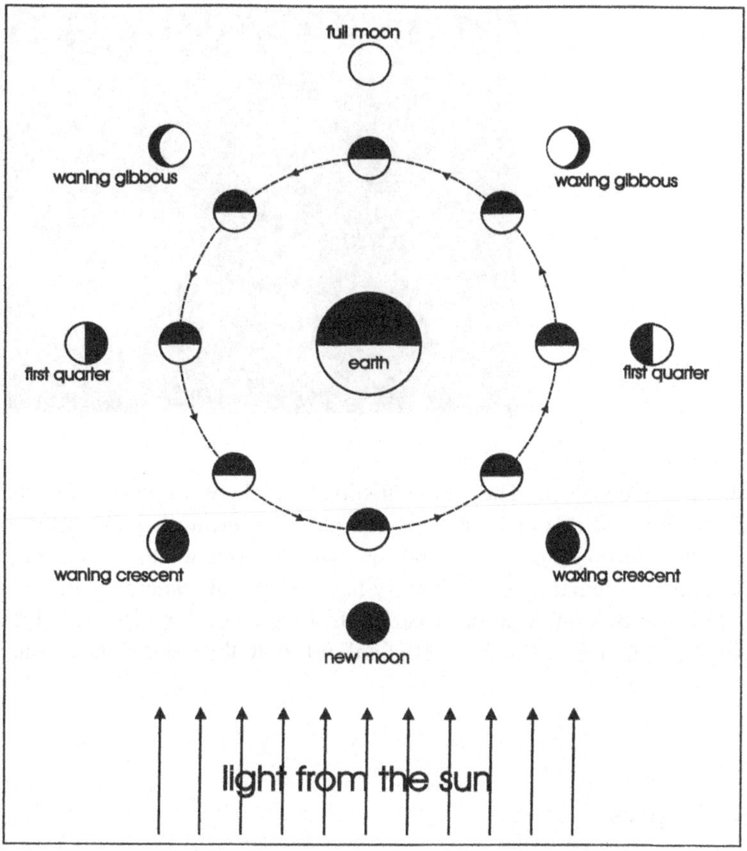

Fig. 5.3 The phases of the moon

Nowadays we know that this phenomenon, called "earth-shine" (also called "the moon's ashen glow," or "the old moon in the new moon's arms") originates from sunlight reflected by the earth on the moon, but this explanation was not available to the ancients (Fig. 5.4).[12]

It could be argued that Parmenides, in an attempt to provide an explanation of this phenomenon, held a double theory of the moon's light, not only a "borrowed light" but also a light of itself. Anaximander's explanation, on the other hand, enabled him to explain these phenomena very well. Anaximander's idea was completely rational: the air of the wheel that hides the fire inside can also partially or totally cover the fire that shines through the opening in the wheel. At night, the air of the wheel looks black, just like the rest of the sky at night, and by day the air looks blue, just like the rest of the sky by day. As was argued in Chap. 4 in the section *Two Images for Escaping Fire*, we are not able to see Anaximander's celestial wheels, because they

[12]As far as I know, Galilei (1632, Day one, 92) was the first to suggest this explanation.

Fig. 5.4 Earth-shine. Photograph by Sebastien Lebrigand, February 2, 2014, 18:50

are made of condensed air. The blocking slide is made of the same air as that of the heavenly wheel, and this explains why the closed aperture has the color, black or blue, of the surrounding sky. And as for the phenomenon of earth-shine, Anaximander's explanation could simply have been that sometimes the slide is not fully opaque but of somewhat transparent air, like a cloud, which still allows us to vaguely see the fire inside. In Chap. 10, I will return to the issue of the explanation of the moon's phases.

Lunar Eclipses

Let us now turn to eclipses of the moon. The right explanation of this phenomenon—the earth blocks the light of the sun—is reportedly first given by Anaxagoras. In Chap. 9, I will challenge this claim. But whoever it was who discovered that the shadow of the earth causes lunar eclipses, this theory was unable to explain an everyday phenomenon. Normally, a shadow cast on an object does not make that object invisible. Let us see what happens when I imitate a lunar eclipse with the help of an earth globe and a moon globe.

An earth globe like the one in Fig. 5.5, throwing a shadow on a moon globe does not make the shadowed part of the moon globe invisible. During a real eclipse of the moon, on the contrary, its eclipsed part is invisible and colored black or (dark) blue, depending on whether it is night or twilight (Fig. 5.6). Again, If Anaxagoras had been the one who discovered the real cause of lunar eclipses, it would have been difficult for him to explain why the shadow of the earth adapts to the color of the air, whereas this is evident in Anaximander's theory.

Sometimes, a phenomenon similar to earthshine can be seen during a full eclipse of the moon, in which a reddish moon remains visible although it is fully eclipsed, the so called "blood moon" (Fig. 5.7). Nowadays, this is explained by the atmosphere that

Fig. 5.5 Imitation of an eclipse of the moon

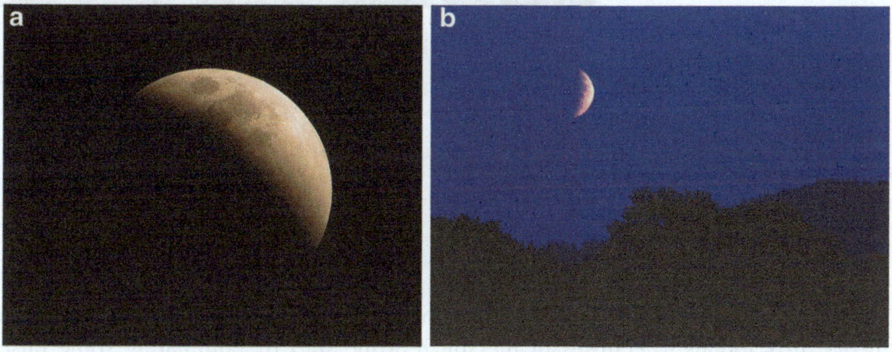

Fig. 5.6 (**a**) Lunar eclipse at night. Photograph CC0 Public Domain. (**b**) Lunar eclipse in twilight. Photograph by Rolf Maier (Reference: 003-067), Picture: Edenpics.com

filters out the blue light of the sun so that only the red parts of the spectrum can reach the moon. This explanation was, of course, not available to the ancients.

In an attempt to clarify the theory of Anaxagoras and Democritus about the Milky Way, which we shall discuss more extensively in Chap. 9, Olympiodorus suggests that the moon has a twofold light:

5.6 And the case of the moon makes this clear. For this has one kind of light of its own and another from the sun. Its own light is coal-like, which the moon's eclipse shows us.[13]

---

[13]Cf. Olympiodorus, *In Aristotelis Meteora* 67.32 = Gr Axg46 = GG 607; not in DK, LM, and KRS.

Fig. 5.7 Blood moon. Lunar eclipse September 27, 2015. Photograph Alfredo Garcia Jr

Anaximander's explanation could have simply been, again, that sometimes the airy slide is not fully opaque but semi-transparent, like a cloud.

Solar Eclipses

Until now, we have only discussed the moon, its phases and eclipses. Anaximander's theory also applies to eclipses of the sun. Let me start by eliminating some false information on what we see at a solar eclipse. Usually, a partial solar eclipse passes without being noticed at all by non-professionals and by people who have not been warned that something will happen to the sun. This is because the light of the sun is so bright that even if a large part of it is blocked, hardly any diminution in sunlight is noticeable. That is why you always use eye protection when looking at a solar eclipse, or observe it indirectly, for instance on the surface of a fluid.

In the second place, most photographs of eclipses of the sun are deceptive, because they show the sun against a black background (Fig. 5.8).

This is again caused by the brightness of the sun, which makes photographers use special filters. During partial eclipses, the sky remains blue as usual, and even during complete eclipses the sky is not black. Thirdly, pictures of the sky during a total eclipse are sometimes highly exaggerated, showing a sky full of stars, as in a recent article by a distinguished scholar, obviously the result of an incorrect adjustment of a computer program (see Fig. 11.6). This issue will be discussed in more detail in Chap. 11. During a solar eclipse, what we see is the sun missing a larger or smaller

Fig. 5.8 Deceptive photograph of a solar eclipse. Photo by Dirk Rabe, March 20, 2015

Fig. 5.9 The missing part of the sun has the same color as the surrounding air. Photo by Staff Sgt. Jarad A. Denton (released), March 20, 2015

part of its disk. And what is important, the missing part of the sun has the same color as the surrounding sky, as in Fig. 5.9.

Within the theory that the blocking moon causes a solar eclipse this is hard to explain. Usually, when a light is blocked, the blocking object remains visible and does not take the color of its surroundings. This is demonstrated by my moon globe blocking a light, but remaining visible and not taking the color of its surroundings (Fig. 5.10).

The moon behaves completely different: when there is a solar eclipse we do not see the moon but instead the sun missing a smaller or larger part that has the same color as the surrounding sky. Or to put it negatively: from the immediate appearances there is no reason at all why we should believe that it is the moon that blocks the sunlight. In Anaximander's explanation, however, everything is clear: a part of

Fig. 5.10 Simulation of a solar eclipse. The respective angular diameters of the sun and moon are not correct in this photograph, because my moon globe was larger than the available light source. This does not affect the argument

the surrounding air of the solar wheel slides partially or completely before the fire inside, and that is why it has the same color as the surrounding sky.

Perhaps Anaximander had some difficulty in explaining a total or an annular eclipse, because then the light of the sun protrudes on all sides of the sun. Probably, however, Anaximander never saw a total solar eclipse (even the eclipse of Thales, May 28, 558 BC was not total in Miletus), and certainly not an annular eclipse. My conclusion is that, given the available knowledge and actual observations, Anaximander's solution was at least as rational as the correct explanations.

References[14]

Bekker, I. (ed. and transl), *Georgius Cedrenus Ioannis Scylitzae ope*. Tomus Prior. Bonn: Weber 1838. Available on the internet: https://archive.org/stream/georgiuscedrenu00scylgoog#page/n0/mode/2up.

Conche, M., *Anaximandre. Fragments et Témoignages*. Paris: Presses Universitaires de France 1991.

Diels, H., *Doxographi Graeci*. Berlin: Walter de Gruyter 1879 (=Dox).

[14]P = Aëtius in pseudo-Plutarch, *Placita* (numbering according to Dox).
S = Aëtius in Stobaeus, *Anthologium* (numbering according to Wachsmuth and Hense).

Diels, H. and W. Kranz, *Die Fragmente der Vorsokratiker*. Zürich/Hildesheim: Weidmann 1951[6] (=DK).

Galilei, G., *Dialogo sopra i due massimi sistemi del mondo*. Fiorenza: Landini 1632.

Graham, D.W., *The Texts of Early Greek Philosophy*. Cambridge: Cambridge University of Press 2010 (=Gr).

—, *Science Before Socrates*. Oxford: University Press 2013.

Kirk, G.S., J.E. Raven and M. Schofield, *The Presocratic Philosophers*. Cambridge: Cambridge University Press 2009 (=KRS).

Laks, A. and G.W. Most, *Early Greek Philosophy VI*, Cambridge MA 2016.

Litwa, M.D., *Refutation of All Heresies*. Atlanta: SBL Press 2016.

Mansfeld, J. and D.T. Runia, Aëtiana. *The Method and Intellectual Context of a Doxographer, Volume Two, Part Two*. Leiden: Brill 2009 (=MR).

Miller, E., *Origenis Philosophumena sive Omnium Haeresium Refutatio*. Oxonii: E Typographio Academico 1851. Available on the Internet: https://archive.org/details/origenisphiloso00origgoog.

Mourelatos, A.P.D., 'The Light of Day by Night': *nukti phaos*, Said of the Moon in Parmenides B14. In: R. Patterson et al. (ed.), *Presocratics and Plato: Festschrift at Delphi in Honor of Charles Kahn*, Las Vegas: Parmenides Publishing 2012, 25–58.

—, Parmenides, Early Greek Astronomy, and Modern Scientific Realism. In: J. McCoy, ed., *Early Greek Philosophy*, Washington, D.C.: The Catholic University of America Press 2013, 91–112.

Reale, G. (ed.), *I Presocratici. Prima traduzione integrale con testi originali a fronte delle testimonianze e dei frammenti nella raccolta di Hermann Diels e Walther Kranz*. Milano: Bompiani, Il pensiero occidentale, 2017.

Wöhrle, G., *Traditio Praesocratica. Die Milesier: Thales*. Berlin: Walter de Gruyter 2009 (=TP1).

—, *Traditio Praesocratica. Die Milesier: Anaximander und Anaximenes*. Berlin: Walter de Gruyter 2012 (=TP2).

Chapter 6
Anaximander's Numbers

Contents

Introduction

In 1995, when I discussed several visualizations of Anaximander's world picture and added my own attempt,[1] I took the interpretation of Anaximander's numbers by Tannery and Diels for granted.[2] Later, I got my doubts about the right understanding of the doxography on the distances of the heavenly bodies. In 2001, I wrote about my own interpretation: "Even if some parts of this reconstruction might be wrong. I think the conclusion still stands that Anaximander is the originator of the Western world picture or, in other words, the discoverer of space."[3] In 2003 I wrote: "Any interpretation entailing unacceptable observational consequences that were easy for Anaximander himself to observe must be wrong. In other words, Anaximander's numbers cannot be in flagrant discrepancy with observational data, for otherwise he

[1]Couprie (1995).

[2]See Tannery (1887, 94–98) and Diels (1897).

[3]Couprie (2001, 47), see also Couprie (2003, 238).

D. L. Couprie, *When the Earth Was Flat*, Historical & Cultural Astronomy,
https://doi.org/10.1007/978-3-319-97052-3_6

would have noticed it."[4] In 2009, in an article, entitled *Problems with Anaximander's Numbers*, I discussed a number of interpretive problems and proposed solutions for understanding them. I concluded that even a symbolical interpretation of the numbers such as "far," "farther," "farthest" would not help, because on a flat earth the heavenly bodies are not far away. I tried to save the standard interpretation of the numbers by suggesting that Anaximander did not make a three-dimensional model of his conception of the cosmos. In 2011, this conclusion was repeated in Chap. 9 of *Heaven and Earth*, although I felt increasingly uneasy about it. Since then, my doubts as to whether Anaximander's numbers as transmitted in the sources are in any of their current interpretations consistent with the conception of the celestial bodies being close by have only become more serious. In this chapter, several interpretations will be discussed and especially one that no longer takes the numbers as indications of distances in the universe but as a kind of calculator for a lunar-solar calendar. At the end of this chapter, I will make up my mind and outline my current position.

An Ordered Universe

Anaximander was convinced that there must be an order in the universe. He studied the movements of the heavenly bodies, following the lead of Thales who had tried to find some regularity in the occurrences of eclipses to be able to predict a solar eclipse. The regular sequences of day and night, the phases of the moon, and the year with its seasons had been studied in Mesopotamia and Egypt for making calendars and for timekeeping. The sources tell us that Anaximander marked the times, and more specifically the solstices (τροπαί), the equinoxes and the seasons with the help of a gnomon.[5]

Our sources clearly attest that Anaximander was also concerned with the distances of the heavenly bodies, but at the same time these reports are a source of confusion and misunderstanding and seem to resist all attempts of consistent interpretation. Simplicius, appealing to the authority of Aristotle's pupil Eudemus, reports:

6.1 Anaximander was the first to find the explanation of the sizes and distances of the planets (τῶν πλανωμένων).[6]

---

[4]Couprie (2003, 217).

[5]Diogenes Laërtius, *Vitae Philosophorum* 2.1.1 = DK 12A1 = LM ANAXIMAND. R14 = Gr Axrl = TP2 Ar92 = KRS 94. Eusebius, *Preparatio Evangelica* 10.14.11 = DK 12A4 = LM ANAXIMAND. R15 = GR Axr2 = TP2 Ar102; not in KRS. Suda, *Lexicon alpha* = DK12A2 = Gr Axr4 = TP2 Ar237 = KRS 95. These lines omitted in LM ANAXIMAND. D3.

[6]Simplicius, *In Aristotelis De Caelo Commentaria* 471.1–11 = DK 12A19 = LM ANAXIMAND. R17 = TP2 Ar185, not in Gr and KRS.

Although this report only mentions planets, it is often quoted as saying that Anaximander has measured the sizes and distances of the celestial bodies in general.[7] Perhaps Simplicius counted also the sun and the moon as planets, as was customary in ancient times, but in a report on Anaximander, Aëtius says Anaximander placed the planets at the same distance as the fixed stars, other than that of the sun and moon. The planets, however, do not return in the rendition of Hippolytus:

6.2 Anaximander and Metrodorus of Chios and Crates say the sun is highest of all, then the moon, and below these the fixed stars and the planets.[8]
6.3 The sun is the highest body, and lowest are the circles of the fixed stars.[9]

At any rate, these reports state that Anaximander differentiated between the distances of the heavenly bodies. This raises the question of how Anaximander could have found out these relative distances between the heavenly bodies. Anaximander positioned the stars below the sun, which seems to conflict with the observation of star occultation. Several attempts have been made to explain this remarkable detail,[10] but at least it does not help to raise our expectations regarding the reliability of Anaximander's alleged calculations or estimates of the distances of the heavenly bodies. We are used to infer from a solar eclipse that the sun must be farther away than the moon, because the moon blocks the light of the sun, but Anaximander's explanation of a solar eclipse forbade him from making this inference:

6.4 Anaximander says an eclipse of the sun occurs when the mouth of the fire hole [in the sun wheel] is closed off.[11]

For this reason, too, Simplicius' suggestion that Anaximander could have discovered the method for measuring the sizes and distances of the sun and the moon by means of eclipses cannot be correct.[12] Anaximander's idea that the heavenly bodies

[7]See, e.g., Kahn (1994, 63): "Anaximander's concern for the sizes and distances *of the heavenly bodies* is cited by S(implicius) not from Theophrastus but from the ἀστρολογικὴ ἱστορία of Eudemus." White (2008, 100): "the sizes and distances *of the sun and moon* (fr. 146)," and 105: "According to Eudemus (fr. 146), he was the first to specify sizes and distances *for the heavenly bodies*." Couprie (2011, 118): "Anaximander was the first to gain insight into the measurements and distances (*of the celestial bodies*), as Eudemus reports," repeated at 128–129. Hahn (2010, 60): "Simplicius reports, not on the authority of Theophrastus but rather Eudemus (...), that Anaximander was the first to describe the sizes and distances (...) *of the heavenly bodies*." (all italics are mine). See also Gemelli Marciano [2007, 43 (12)].

[8]P 2.15.6 (≈S 1.24.1) = DK 12A18 = LM ANAXIMAND. D22 = MR 476 = TP2 Ar55 and 148, not in Gr and KRS.

[9]Hippolytus, *Refutatio Omnium Haeresium* 1.6.5 = DK 12A11(5) = LM ANAXIMAND. D7(6) = GrAxr20(4) = TP2 Ar75(5) = KRS 125.

[10]See Couprie (2011, 116–117).

[11]P 2.24.2 (=S 1.25.1) = DK 12A21 = LM ANAXIMAND. D25 = Gr Axr24 = MR 563 = TP2 Ar59 and 150; not in KRS.

[12]See Simplicius, *In Aristotelis De Caelo Commentaria* 471.1–11 = DK 12A19 = LM ANAXIMAND. R17 = TP2 Ar185, not in Gr and KRS.

are at different distances from the earth was an astonishing case of creative imagination, it was his "discovery of space,"[13] of which we are not able to uncover on what kind of observation or argument it was based. *Mutatis mutandis*, the same impossibility of drawing conclusions from observation applies to the sizes of the heavenly bodies. As we will see in the next section, Anaximander did not put forward absolute figures for the distances and sizes of the heavenly bodies, but numbers that indicated relations between them. The size of the earth, also expressed as a relation between its width and height, seems to have served him as a kind of module, to which all other numbers refer and from which they are somehow derived.

Anaximander's Numbers of the Heavenly Bodies

According to the doxography, Anaximander tried to express the dimensions of the cosmic order in numbers, numerical proportions. The relevant reports on Anaximander's numbers are:

6.5 He says the earth is cylindrical in shape, and has a depth one third its width.[14]
6.6 Anaximander says there is (or: the sun is) a circle 28 times the size of the earth.[15]
6.7 Anaximander says the sun is equal to the earth (ἴσον τῇ γῇ), and the circle on which it has its hole and on which it revolves is 27 times as big as the earth.[16]
6.8 The circle of the sun is 27 times as big as the moon.[17]
6.9 Anaximander says that the circle of the moon is 19 times the size of the earth.[18]
6.10 Anaximander believed that the sun is a circle 18 times as big as the earth.[19]
6.11 Anaximander says that the moon is a circle, 16 times the earth.[20]

When we try to better understand the notion of order in Anaximander's cosmology, we stumble upon some major problems. The numbers as such seem to be a great mess. Various attempts have been made to make sense of them as indications of distances in the cosmos, but there seems to be no observational or a rational way in

[13]This was the title of my contribution to Couprie, Hahn, and Naddaf (2003, 165–254).

[14]Pseudo-Plutarch, *Stromata* 2 = DK 12 A10 = LM ANAXIMAND. D8 = Gr Axr19 = TP2 Ar101 = KRS 122(A).

[15]P 2.20.1 (=S 1.25.1) = DK 12A21 = LM ANAXIMAND. D23 = Gr Axr22 = MR 514 = TP2 Ar57 and 150 = KRS 126.

[16]P 2.21.1 (≈S 1.25.1) = DK 12A21 = LM ANAXIMAND. D24 = Gr Axr23 = MR 534 = TP2 Ar58 and150 = KRS 127.

[17]Hippolytus, *Refutatio Omnium Haeresium* 1.6.5 = DK 12A11(5) = LM ANAXIMAND. D7 (5) = GrAxr20(5) = TP2 Ar75(5) = KRS 125.

[18]P 2.25.1 (=S 1.261) = DK 12A22 = LM ANAXIMAND. D26 = Gr Axr25 = MR 572 = TP2 Ar60 and 151; not in KRS, but see p. 136.

[19]Qusṭā ibn Lūqā in Daiber (1980, 155), not in DK, LM, Gr, TP2, MR, and KRS, but see Bottler 404, and Thibodeau (2017, 95).

[20]Pseudo-Galen, *De Historia Philosopha* 67.1 = TP2 Ar224, not in DK, LM, Gr, and KRS.

which Anaximander could have reached them. It is worth noting that numbers for the circles of the stars are completely absent. The biggest problem is how to understand the numbers of the sun and the moon. Most of the quoted texts relate to the sizes of the circles of sun and moon with respect to the earth, and the numbers 27 and 18 are multiples of 3, the number that expresses the earth's diameter in relation to its height. Hippolytus' text, in which the circle of the sun is said 27 times the moon, is quite different from all the other texts, in which the circles of the sun and the moon are related to the earth. Most scholars, therefore, accept Diels' emendations, with which it reads:

6.12 The circle of the sun is 27 times as big as <the earth, that of> the moon <18 times>.

However, Diels' emendation of the number 18 for the moon is not supported by any other text but results from his idea that the numbers must be multiples of 3 (+1). A simpler emendation would be to suppose a miswriting, ("moon" for "earth"), as already suggested by Gruppe and Röper,[21] so that the text would read, in accordance with Aëtius (P 2.21.1):

6.13 The circle of the sun is 27 times as big as the earth.

Aëtius' number 28 for the sun is mentioned by pseudo-Plutarch and Stobaeus, as well as by Eusebius and pseudo-Galen,[22] the number 27 in pseudo-Plutarch is also in Eusebius and pseudo-Galen, but not in Stobaeus, who refers to the "aforementioned number," which was 28.[23] Pseudo-Galen's number 16 for the moon is usually disregarded as somehow mistaken. Qusṭā ibn Lūqā's number 18 for the sun is a strange anomaly, quoted by Daiber but until recently always ignored.

A further complication is that the words "sun" and "moon" are ambiguous. In some texts, they indicate the openings in the circles or wheels. In other texts, they seem to indicate the whole of the heavenly wheel. Since the words "sun" and "moon" are

[21]Gruppe (1851, 45 n): "Dagegen ist in der Stelle des Origenes (Philos. Cap. 6) [the work, in the mss. ascribed to Origenes, is usually ascribed to Hippolytus, although this ascription is probably also not right, DC], ein offenbarer Fehler, wenn er von die Sonne dieselbe Zahl meldet, aber als Einheit nicht die Erde sondern den Mond nimmt. Röper (1852, 608–609): "(. . .) und das urtheil Gruppe's, kosm. Syst. D. griech. S. 45, dass bei unserem verfasser ein fehler obwalte, indem er als einheit anstatt der erde den mond annehme, ist gewiss richtig." See also Gregory (2016, 261 n. 21), although his explanation of this emendation, "the circle of the sun is now 27 times *that of* the earth," (my italics) is dubious.

[22]For S 1.25.1, see TP2 Ar150, and cf. Dox. 348. For Eusebius, *Preparatio Evangelica* 15.23.1, see TP2 Ar105. For pseudo-Galen, *De Historia Philosopha* 62.1–3, see TP2 Ar221. Both items not in LM, Gr, and KRS.

[23]For S 1.25.1, see TP2 Ar150, and cf. Dox. 351. For Eusebius, *Preparatio Evangelica* 15.24.1, see TP2 Ar106. For pseudo-Galen, *De Historia Philosopha* 63,1–2, see TP2 Ar222, with a miswriting: πόλος instead of κύκλος. Both items not in LM, Gr, and KRS.

ambiguous, the statement "the sun is equal to the earth" (text 6.7) is also ambiguous and can be taken to refer to the opening in the sun wheel or to the width of the sun wheel.[24]

Tannery and the Standard Interpretation

The usual interpretations of Anaximander's numbers are, in one way or another, modifications of Tannery's suggestion[25] that the numbers 9 and 10 for the stars and 18 for the moon should be extrapolated from the numbers 27 and 28 for the sun and 19 for the moon. Tannery and most scholars after him neglected (or were unacquainted with) the divergent numbers 18 for the sun and 16 for the moon. Following Tannery's suggestion, the numbers 9, 18, and 27 are usually assumed to indicate the diameters of the inner perimeters of the celestial wheels of respectively the stars, the moon and the sun, expressed as multiples of the earth's diameter, while the numbers 10, 19, and 28 should indicate the diameters of the outer perimeters of their wheels. An important argument for this option is that it compares like with like, diameters with diameters. Others have argued that the numbers refer to the radii of the heavenly wheels, according to the idea that Anaximander would have measured the distances of the heavenly bodies and that the radii of the wheels indicate their distances to the earth. In Fig. 6.1 these two versions are rendered. On the left that of Diels (and others), with a diameter of the solar wheel of 28 earth diameters, on the right that of my former self (and others), with a radius of the solar wheel of 28 earth diameters.[26] The most obvious difference is that the latter option makes Anaximander's universe larger than the former, as can be seen in Fig. 6.1. In the left image, the width of heavenly wheels is one half earth diameter, a feature of which Tannery and Diels were well aware.[27] Kirk saw here a problem here, because the sun does not seem to fit in its own ring, and the same applies to the moon.[28]

I used to visualize my conception of Anaximander's cosmos in three-dimensional drawings like Fig. 6.2.

[24]Cf. Guthrie (1962, 95): "We may assume that the rings are one earth-diameter thick."

[25]Tannery (1887, 91–92, 119). For a survey of the interpretations, based on Tannery's suggestion, see Couprie (2011, 121–136). For a more recent survey, see Gregory (2016, 173–192).

[26]See Diels (1897, 236), Couprie (2003, 213; 2011, 130).

[27]Cf. Tannery (1887, 91): "La double épaisseur du cerceau est ainsi égale au diamètre de la terre;" Diels (1897, 232): "so ist die die Breite dieser Ringe auf einen Erdradius zu veranschlagen."

[28]See KRS, p. 136 n. 1.

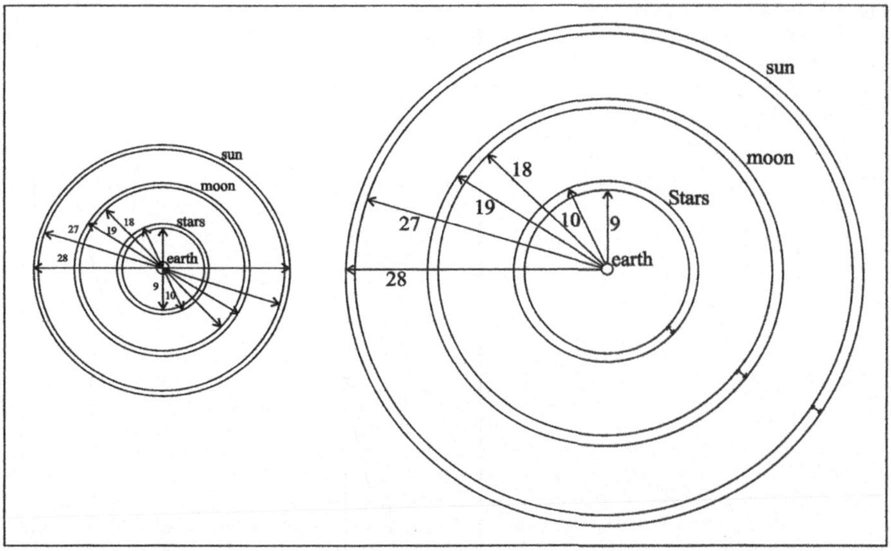

Fig. 6.1 Two versions of the dimensions of Anaximander's universe: with a diameter of 28 earths (left) and with a radius of 28 earths (right)

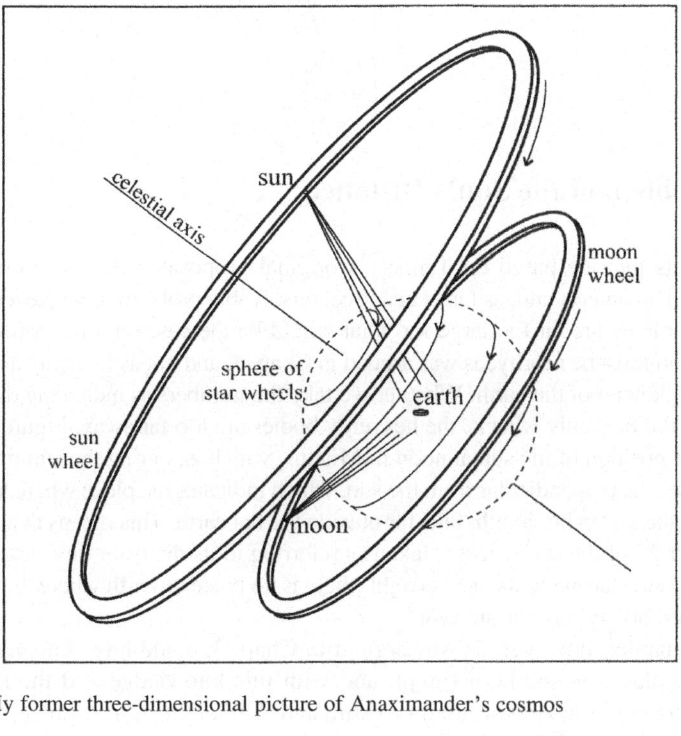

Fig. 6.2 My former three-dimensional picture of Anaximander's cosmos

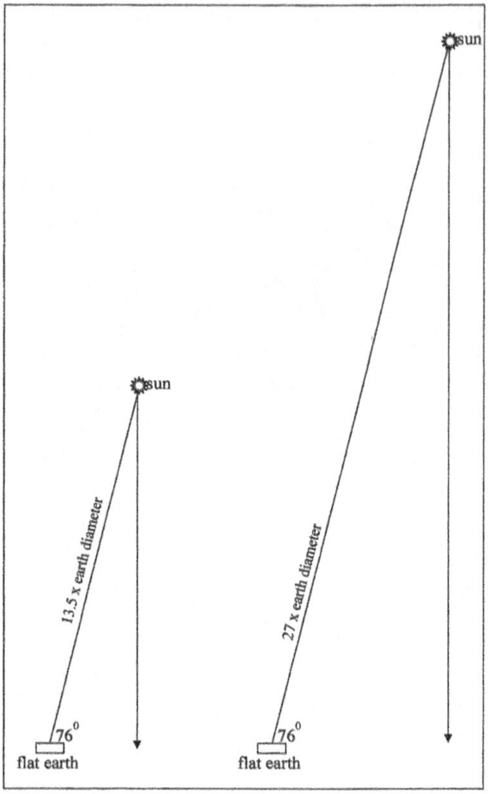

Fig. 6.3 In Miletus, at noon at the summer solstice, the sun is 76° high. The perpendicular from the sun falls far outside the flat earth

The Problem of the Sun's Distance

All attempts that are based on Tannery's original proposal suffer from one or more problems. The biggest and, as far as I can see, unsolvable problem in whatever version is that the numbers are far too large for what would be the case on a flat earth. On a flat earth the sun must be nearby, as was argued in Chap. 3, and not as far away as 27 or 13.5 times the diameter of the earth. Whether we take the numbers as indicating diameters or as radii of the heavenly wheels, the heavenly bodies are too far away. Figure 6.3 shows the highest position of the sun at noon in Miletus, which is, during the summer solstice. At that time, the perpendicular from the sun, which indicates the place where an observer would see the sun in his zenith, falls far outside the flat earth. This means that both when the number 27 of the sun wheel is taken as referring to its diameter (left) and when it is understood as referring to its radius (right) there is no place on earth where the sun stands in the zenith at any time of the year.

Anaximander, however, as was argued in Chap. 3, could have known that there were such places in southern Egypt, and with this knowledge and the help of his gnomon, he could have measured or estimated the height of the sun. It is a typical case of spherical earth bias that this problem has not been recognized before. We are so accustomed to the idea that the heavenly bodies are at great distances, that we take for granted that the ancient flat earth cosmologists shared the same idea. Diels, for example, wrote: "errechnet sind diese Zahlen nicht (*dafür sind sie viel zu klein*)."[29] On the contrary, if the earth is conceived of as flat, they are far too big, as follows from Fig. 6.3. Gregory's recent discussion of several interpretive difficulties related to Anaximander's numbers does not mention this main problem and thus suffers from the same bias.[30]

Anaximander is said to have been at the Black Sea, where he was the leader of a colony,[31] and perhaps he was in Egypt, or he had heard Thales tell of his conversations with Egyptian scholars about the height of a pyramid, or he had heard the stories of the Milesian merchants who visited that country on a regular basis. It is hard to believe that he, who was associated with the introduction and use of the gnomon and who reportedly used it at the solstices,[32] would not have noticed that the sun is higher in the sky in southern countries than in northern countries. If he had not seen it himself or heard it from travelers, he would easily have been able to deduce it from the map of the earth which he is said to have made[33] and for which he must at least have had some idea of the distances included. And even if he had no knowledge of the method of measuring the height of a pyramid with the help of a gnomon, ascribed to Thales, Anaximander could easily have found out how to estimate the height of the sun with the method explained in Fig. 3.10. How could he, with all his experience, skill and understanding of the gnomon, not have known what others, who also thought that the earth is flat, took for granted, namely that the heavenly bodies are not far away from the earth?

Attempts to Explain the Origin of Anaximander's Cosmological Numbers

Several scholars have tried to answer the question of the origin or source of inspiration of Anaximander's numbers, which are handed down in the doxography (and the numbers that could be extrapolated from them). Their efforts yielded amazingly divergent results,

[29]Diels (1923, 72, my italics). See also Diels (1897, 232). Obviously, Diels has in mind the real distance from the earth to the sun, 149,597,870.7 km, which would equal, given the diameter of the spherical earth of 12,756.32 km, to a distance of 11,727 spherical earth diameters.

[30]See Gregory (2016, 173–192).

[31]See Claudius Aelianus, *Varia Historia* 3.17 = DK 12A3 = LM ANAXIMAND. P8 = TP2 Ar 78; not in Gr and KRS, but see p. 105.

[32]Diogenes Laërtius, *Vitae Philosophorum* 2.1.1 = DK 12A1 = LM ANAXIMAND. R14 = Gr Axr1 = TP2 Ar92 = KRS 94: "Anaximander first discovered the gnomon and set one up at the sundials in Sparta, as Favorinus says in his *Miscellaneous Studies*, to mark solstices and equinoxes; and he constructed hour-indicators."

[33]Cf. Strabo, *Geographica* 1.1.11 = DK 12A6 = LM ANAXIMAND. D4 = Gr Axr7 = TP2 Ar32 = KRS 99.

but in one way or another they all take for granted Tannery's suggestion that Anaximander's cosmological numbers indicate the distances of the heavenly bodies.

Some authors have suggested that Anaximander's numbers for the heavenly bodies were derived from sources in other civilizations. Diels saw a parallel with shamanistic rituals that were intended to guide the steps of the soul in its journey to heaven.[34] Burkert pointed out that Anaximander's strange order of the heavenly bodies—stars, moon, sun—occurs already in the *Avesta*, where it is supplemented by the "beginningless lights" or "paradise."[35] The identical order of the heavenly bodies is certainly striking, but the Iranian sources do no mention Anaximander's numbers.

Diels already pointed to a passage in Hesiod for a parallel with Anaximander's numbers.[36] Hesiod describes the distance between the heavens and the earth as enormous:

6.14 For a brazen anvil falling down from heaven nine nights and days would reach the earth upon the tenth.[37]

However, Anaximander could also have read other passages in Hesiod, in which the heavens are obviously conceived of as not so far away. Hesiod describes twice how Atlas bears the heaven.[38] Elsewhere, he tells that from the house of Styx in Hades, silver pillars reach out to heaven.[39] It is hardly conceivable that Atlas or those pillars are as long as the 9 days fall of an anvil. And again, when Hesiod describes the relation between earth and heaven, he says:

6.15 And earth first bore starry Heaven, equal to herself, to cover her on every side,

The verb καλύπτω, to cover, can be said of a woman's headgear or veil, which sounds more like Anaximenes' cap simile than as the image of the falling anvil. After all, it does not seem to be such a good idea to consult Hesiod for the interpretation of Anaximander's numbers.

More generally, Diels argued that in many cultures 3 is a holy number, of which 9 and 27 are multiples.[40] In earlier publications, I suggested as an explanation for Anaximander's choice of numbers that we must understand them as mere symbols, not as indicating real distances but as meaning "far," "farther," and "farthest."[41] I argued that in the Greek counting system, the number 9 ($=3 \times 3$) was used to indicate a long

[34]Cf. Diels (1897, 233).

[35]Cf. Burkert (1963, 106–112, esp. 111).

[36]Cf. Diels (1897, 232).

[37]Hesiod, *Theogonia* 722–723.

[38]Cf. Hesiod, *Theogonia* 517–520 and 746–747.

[39]Cf. Hesiod, *Theogonia* 778–779.

[40]Cf. Diels (1897, 232–233). Diels does not mention here the number 18.

[41]See Couprie (2001, 40–41; 2003, 215; 2011, 136).

distance or a long time. Hesiod's falling anvil could be taken as expressing that the Titans made a big fall indeed, much more than a human being can ever imagine falling.[42] Similarly, Troy was conquered in the 10th year after having been besieged for 9 years, and Odysseus scoured the seas 9 years to return to his homeland in the 10th year. The numbers 18 and 27 could be said to indicate the comparative and superlative degrees of this symbolic "9." It might even be argued that the Pythagoreans followed Anaximander's example by looking for numbers of musical harmonies indicating the dimensions in the cosmos.[43] At first sight, the purely symbolic interpretation of Anaximander's numbers seems to offer an elegant answer to the question of why he took precisely these numbers and not others. But apart from the fact that in my drawings I considered them not only as symbolic, but also as indicating real distances, this suggestion seems not tally with the fundamental idea that on a flat earth the heavenly bodies are not far (farther, farthest) away but, on the contrary, close by.[44]

West offered the speculative theory that the outer οὐρανός in Anaximander's cosmology must be at 36 diameters or radii distance. The doxography, however, does not mention an outer οὐρανός, and the number 36 is West's extrapolation of Tannery's series 9, 18, 27, which was itself an extrapolation of the number 27 for the sun. His reference to the 36 *decans* (specific stars) of Egyptian astronomy is also strange, because the number 36 of the *decans* does not indicate their distance to the earth.[45]

Eggermont has made the suggestion that Anaximander's numbers had to do somehow with the gold/silver ratio, which was, in Croesus' time $13\frac{1}{3}$:1, which equals 360:27. Eggermont's presuppositions are that Anaximander was familiar with the Babylonian (lunar) year of 360 days and a supposed sidereal year of $13\frac{1}{3}$ months of 27 days ($13\frac{1}{3} \times 27 = 360$), and that gold is equal to the sun and silver equal to the moon.[46] Thus "Anaximander selected the figure 27;" the numbers for the moon and the stars he found by giving "$\frac{2}{3}$ and $\frac{1}{3}$ of that amount to the respective diameters of the lunar and stellar circles."[47] The number 18 for the moon "by a lucky coincidence happened to allude to the complement 20," since $18 \times 20 = 360$," and this "alluded to the 20 hemistaters comprised in 1 gold coin."[48] This curious attempt leads to the same problematic observational consequences as Tannery's and Diels' interpretation, as is clear from Eggermont's picture of Anaximander's universe (made by B.M.W. van Gelder),

[42] Cf. Gregory (2016, 203): "I take this as a poetic expression of something exceedingly heavy dropping exceedingly fast."

[43] Cf. KRS, p. 136: "His proportionate distances may have influenced Pythagoras." See also Zhmud (2012, 292).

[44] A somewhat problematic solution would be to accept that Anaximander, when measuring the height of the heaven, made a calculation error similar to that of the Chinese astronomers; see Chap. 13.

[45] Cf. West (1971, 92).

[46] Cf. Eggermont (1973, 124–125)

[47] Eggermont (1973, 128).

[48] Eggermont (1973, 128, 127).

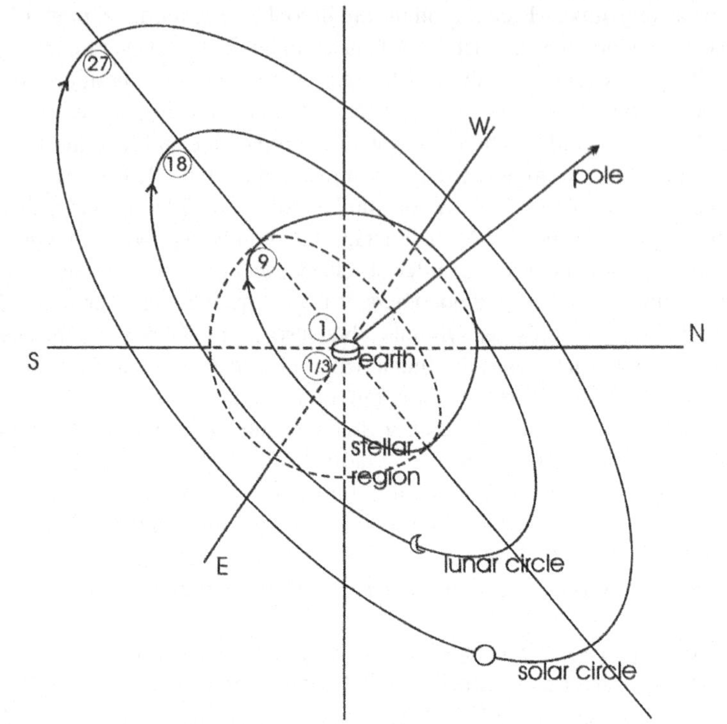

Fig. 6.4 Anaximander's universe according to Eggermont. Redrawn after B.M.W. van Gelder in Eggermont (1973, 120)

reproduced in Fig. 6.4. Perhaps the most intriguing part of Eggermont's article is that he somehow relates Anaximander's numbers not only to Croesus' coinage but also to questions of calendar making.

Hahn has argued that Anaximander has been inspired by the ratios used by contemporaneous architects. However, he was not able to explain satisfactorily the architectural origin of the 3:1 ratio of Anaximander's column-drum shaped earth, which was, according to Hahn, Anaximander's basic module. The best examples Hahn mentions are drums with ratios 3.4:1 and 3.9:1. According to Hahn, Anaximander used a ratio of 9:18:27 for the heavenly bodies, which can be reduced to 1:2:3, while the architectural ratio for temple building was 1:2:4 (width:height: length).[49] Even when the ratio 1:2:3 was sometimes used in temple building, Hahn does not make it sufficiently clear why Anaximander should have converted this ratio into 9:18:27, or even into 9 (+1):18 (+1):27 (+1).[50] Hahn's suggestion that the

[49]See Hahn (2001, passim, and especially 156, 158, and 78).

[50]For an extensive criticism of Hahn's attempt, see Couprie and Pott (2002) and Couprie (2011), Chap. 12.

numbers are based on measurements used in the construction of temples shows at best that Anaximander made use of ratios, but not why precisely the ones he actually used, as I have argued elsewhere.[51]

Naddaf conjectured that "the dimensions and distances of the heavenly bodies (...) correspond in some way or other to the three social groups of which numerous πόλεις, including Miletus, were composed in Anaximander's day: the aristocracy, the (new) middle class, and the peasantry (or poor),"[52] which in Anaximander's ideal society would be considered as peers. This social relationship between peers in a political assembly may perhaps clarify the equal distances between the heavenly bodies according to Anaximander, but again it does not explain why Anaximander should have used precisely the numbers 9, 18, 27 and not any other set of numbers that could indicate equal distances.

Gregory attributes to Anaximander a "predilection for symmetry and sufficient reason,"[53] as well as a "precise arrangement of the heavenly bodies,"[54] which are also thought to give "the required symmetry for the stability of the earth."[55] Gregory also fails to explain, however, why Anaximander should have chosen precisely the numbers, 9, 18, 27 and not any other set of numbers that would result in a symmetrical universe. My impression is that Gregory does not fully accept the import of his own clever and skeptical remark: "If we accept this sequence, it has important consequences for symmetry and stability in Anaximander, *but there is need to avoid circularity here and not argue for this sequence on grounds of symmetry or stability in Anaximander.*"[56] Gregory's emphasis on symmetry is analogous to O'Brien's remark: "In Anaximander's system, as we have reconstructed it, the earth and the celestial wheels are *equidistant.* This was probably one of the primary factors influencing Anaximander's choice of measurements."[57] However, since there are many other measurements that would also result in equidistance, this statement does not answer the question of why Anaximander selected just these numbers.

Corre, convinced that the number three is omnipresent in Anaximander's conception of the celestial wheels, defends the idea that this has to do with the relation between the diameter and the circumference of the circle. According to him, Anaximander used two values of π. For practical reasons, he used the approximative value $\pi = 3$, as was usual in ancient traditions, but according to Corre he was probably also acquainted with the more precise value of 25/8 or 3 + 1/8, as in cuneiform tablets from Susa. Calculating with $\pi = 3$ gives the number 27 for the sun wheel and 18 for the moon wheel, but calculating

[51]See Couprie and Pott (2002), and Couprie (2011, 153–160).

[52]Naddaf (1998, 23).

[53]Gregory (2016, 74; see also 123): "Anaximander's system has a strong tendency to symmetry."

[54]Gregory (2016, 124).

[55]Gregory (2016, 164).

[56]Gregory (2016, 184, my italics). With "this sequence" Gregory means: "the aesthetically pleasing sequence of 9, 18, 27, which can be seen as an extension of the 3:1 ratio of the earth's width to its depth" (2016, 173).

[57]O'Brien (1967, 427).

with $\pi = 3 + 1/8$ gives 28 for the sun wheel and 19 for the moon wheel (both in rounded off figures).[58] The suggestion that precisely the best attested numbers (28 for the sun and 19 for the moon) were rounded off does not contribute to the plausibility of this interpretation.

Even if Anaximander's numbers were inspired by oriental influences, peculiarities of the Greek counting system, numismatic ratios, calendrical requirements, architectural proportions, reasons of symmetry, or by different values of π, none of these suggestions solves the fundamental problem that the numbers are fat too large for heavenly bodies that should be close to a flat earth.

An Interpretation Dating from Before Tannery

Let us then try to take a fresh look at the doxographic evidence. The most obvious interpretation of the formula "the circle of the sun (or moon) is x times the earth," as it is in most texts, is not Tannery's suggestion that this has to do with cosmic distances, measuring the diameters (or the radii) of the heavenly wheels. What these texts seem to say is that the perimeters of the wheels are measured, taking the earth as fundamental unit (module). This was how scholars before Tannery used to explain Anaximander's numbers for the sun and moon: 28 earth units in a row make the full circle of the sun wheel. O'Brien's remark, that "this in effect confuses circumference and diameter"[59] is too easy and results from his preconceived idea that we should compare like with like (diameter with diameter, radius with radius, circumference with circumference).[60] Moreover, strictly speaking, the texts do not compare the earth's diameter with the circumference of a heavenly wheel, but calculate how many earths in a row make the full circle of the wheel. In a similar way, Thales is said to have measured the sun's angular diameter by calculating how many suns go into the full the orbit of the sun.[61] The number 27 could be explained as 28 minus the aperture in the wheel, which could be identified as the sun. In the same way, the circle of the moon is 19 times the earth, meaning: 19 earth units in a row make the full circle of the moon wheel, one of which can be identified as the aperture in the moon wheel, as was already Röper's suggestion.[62] The three numbers 28, 27, and 19 are handed down in the doxography. One might argue that the

[58] See Corre (2013).

[59] O'Brien (1967, 423, n. 4).

[60] Cf. O'Brien (1967, 425).

[61] Cf. Apuleius, *Florida* 18.32 = DK 11A19 = LM THALES R13 = TP1 Th178; not in Gr and KRS. See also Diogenes Laërtius, *Vitae Philosophorum* 1.24 = DK 11A1(24) = LM THALES R14 = Gr Ths1(24) = TP1 Th237(24); not in KRS, but see p. 83.

[62] Röper (1852, 608): "(…) nachrichten, wonach Anaximander den kreis der sonne 28 mal oder, vermuthlich nach abzug des der ἐκπνοὴ zukommenden raumes, 27 mal, den kreis des mondes aber 19 mal grosser sein liess als die erde." Röper does not mention a number 18 for the moon wheel minus its ἐκπνοὴ.

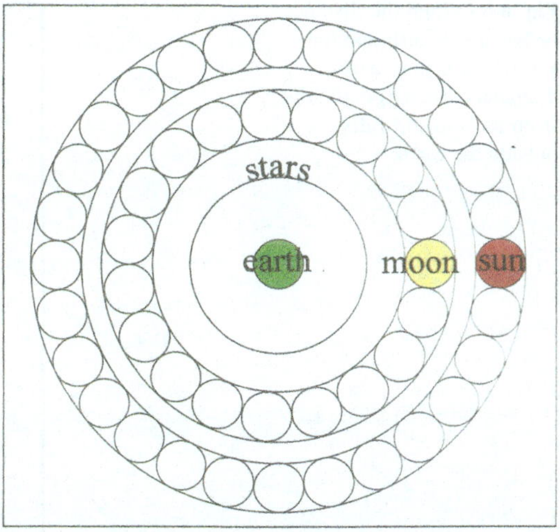

Fig. 6.5 The heavenly wheels measured in earth units

doxographers felt no need to mention the number 18 for the moon wheel minus its aperture because the method of calculating was already explained for the sun. Or perhaps the number 18 has survived in Qusṭā ibn Lūqā's above quoted report, where it is mistakenly said of the sun instead of the moon. Since there are no numbers for the stars handed down and the only thing we know is that they are under the circles of the sun and the moon, this interpretation does not add numbers for the star wheel or wheels but only assumes that the stars are the lowest luminaries.

Typical for the enticement of Tannery's suggestion for the interpretation of Anaximander's numbers is that this older interpretation is usually not even mentioned by modern scholars, as for example in Gregory's recent and extensive discussion of the numbers.[63] Conche even states, without any further argument, that "ce que l'on compare, ce sont les diamètres (*non les circonférences!)* des cercles ou anneaux (...) au diamètre de la surface circulaire de la Terre."[64] I have rendered this old interpretation in Fig. 6.5 for the circles of the sun and the moon. A dotted circle for the stars has been tentatively added somewhere between the earth and the moon circle. In the center the earth (green; fundamental unit), then the stars (dotted ring), the moon wheel of 19 earth units (18 plus one yellow aperture) and the sun wheel of 28 earth units (27 plus one red aperture).

The distances of the heavenly wheels (in other words: their radii) can easily be calculated as ratios of the earth's diameter (which we call "d"). The radius of the sun wheel is: $r = 28d \div 2\pi \approx 4.45d$. The radius of the moon wheel is $r = 19d/2\pi \approx 3d$. To give an idea of what this means: if we take it that d (the diameter of Anaximander's flat earth) is 5000 km, the distance from the center of the surface of the flat earth to the sun wheel is about 22,250 km, and the distance to the moon

[63]Cf. Gregory (2016, 169–219). Exceptions are O'Brien (1967, 423, n. 4) and Naddaf (2001, 11).
[64]Conche (1991, 209, my italics).

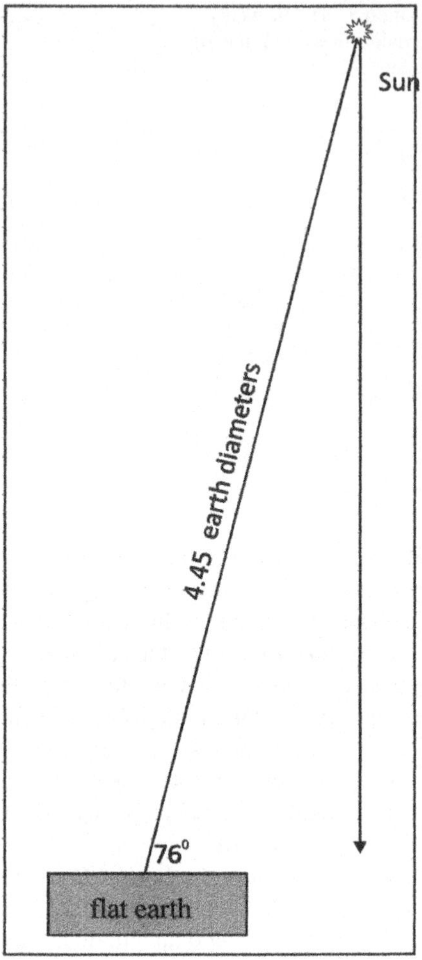

Fig. 6.6 When the distance between the earth and the sun wheel is 4.45 earth diameters, the perpendicular from the sun still falls outside the earth

wheel about 9000 km. This interpretation of Anaximander's numbers yields the best possible agreement with an important conclusion of Chap. 3: when the earth is flat the heavenly bodies are not at a great distance from the earth. Compared to the result of the calculation of the distance of the sun given there, however, even in this interpretation the sun is still too far away from the earth, as shown in Fig. 6.6. The conclusion must be that this old interpretation of Anaximander's numbers, although it performs better than that of Tannery, by bringing the heavenly bodies much closer to the earth, still proves inadequate because even here the perpendicular from the sun at noon at the summer solstice falls far outside the surface of the flat earth.

The Sun's Angular Diameter

There is yet another problem with the interpretations discussed thus far, and most obvious that of Forbiger. The usual interpretation of the text that says that, according to Anaximander, the sun is equal (in size) to the earth is that with "sun" the aperture in the sun wheel is meant. In Fig. 6.6, the sun is drawn much smaller than the earth, but Fig. 6.5 clearly shows that a number of 28 sun disks on a row would result in a huge sun with an angular diameter of $360 \div 28 \approx 12.85°$, which is more than 25 times too large. Naddaf saw this rightly: "anyone looking to the sky would see that it would take more than 27 sun disks to form a circle or ring."[65] An analogous problem exists in the interpretations that assume that the numbers express the diameters or the radii of the heavenly bodies. In Tannery's and Diels' interpretation, the perimeter of the sun wheel is $28\pi \approx 88$ sun disks and the angular size of the sun $360° \div 88$, which is more than 4°. In my former interpretation it is $28 \times 2\pi \approx 176$ sun disks and an angular size of the sun of $360° \div 176$, which is more than 2° (cf. Fig. 6.1). In reality, since the angular diameter of the sun is 0.5° and a full circle 360°, a total of 720 sun disks will make up the full circle. In other words, in all interpretations of Anaximander's numbers discussed so far, the sun (the aperture in the wheel), being as big as the earth, is far too big. On the contrary, when the earth is flat, the sun must be smaller than the earth, as we have seen in Chap. 3. Perhaps it might be argued that an angular size of the sun of about 2° would fall within the range of what was acceptable in ancient times, but an angular size of 4°, which is eight times its actual size, definitely not. In an earlier publication, I tried to provide a solution for this problem by suggesting that the clause "the sun is equal to the earth" does not refer to the aperture in the sun wheel but to the thickness of the wheel, so that the aperture can be the size required by the angular diameter of the sun (see the aperture in the sun wheel in Fig. 6.2). This interpretation has been adopted by Graham.[66]

Skeptical Conclusions and a Possible Way Out

When we list the problems of interpreting the numbers for the heavenly bodies, handed down by the doxography, we get something like this:

- Only three numbers of the doxography can be taken seriously: 28 and 27 for the sun and 19 for the moon.

[65]Naddaf (2001, 11). The total number of disks should be 28, as Forbiger's text (1877, 523, n. 57) shows. It is not clear to me, why Naddaf calls this "an obvious *petitio principii* argument."

[66]Cf. Graham (2013, 58): "the ring of the sun (...) one earth-diameter in thickness." See also his drawing at 59, Fig. 2.1. Graham's drawing is, however, strange in two other aspects: the heavenly wheels are not tilted but lie in the same plane as the earth, as if the situation before the tilt of the celestial axis is rendered, and only one wheel for the stars is drawn.

- The number 18 for the moon is not documented and is purely speculative.
- Some deviating reports (of pseudo-Galen and Qusṭā ibn Lūqā, 16 for the moon and 18 for the sun respectively) have been given hardly any attention to.
- There are no numbers at all for the stars. The numbers nine and ten for the stars are not documented but are purely speculative.
- It is uncertain whether we should visualize the stars as wheels, with or without identical diameters, or as a sphere.
- It is uncertain whether the numbers are related to circumferences, diameters or radii of the heavenly wheels.
- It is uncertain whether we are obliged to compare like with like (circumference of a wheel with circumference of the earth, etc.) or not.
- In the interpretation of the numbers as diameters of the wheels, the problem that the diameter of the wheels' rims is only half the diameter of the earth is a serious disadvantage.
- The expression "the sun equals the earth" is difficult to explain, because when the earth is flat, the sun must be much smaller than the earth.
- In all interpretations, additional provisions must be made to address the problem of the angular diameter of the sun and the moon.
- No adequate answer has yet been found to the question of why Anaximander chose precisely these numbers.

The messy doxographical tradition and the diversity of possible interpretations have regularly led to skeptical remarks as to their reliability. A critical stance seems to be expressed in Kahn's casual remarks that "it is not at all clear whether or not these similes are due to Anaximander himself," and that the whole machinery of wheels in heaven, including their distances, "reflect the style of some Hellenistic popularizer." However, Kahn immediately adds that "there is no reason to doubt the general accuracy of the images,"[67] and then he proceeds with the interpretation of Anaximander's rings, calling it "really difficult to resist (...) Tannery's reconstruction of the simple arithmetic series 9–18–27, for the circles of the stars, moon, and sun respectively."[68]

The complaint of Dicks' straightforward critical opinion is worth quoting: "The unsatisfactory nature of the evidence, which is garbled and contradictory and must be interpreted with arbitrary selectivity if a coherent account is to be obtained, makes it highly doubtful whether it has any historical worth, since there is no particular reason why one scholar's favored interpretation should be any closer to the original than another's. Therefore (...) it seems best to admit that we really know nothing about Anaximander's astronomical beliefs (...)."[69] Another example is Fehling,

[67]Kahn (1994, 87–88).

[68]Kahn (1994, 88).

[69]Dicks (1970, 46–47).

who is convinced that Anaximander's world-picture has been fundamentally misrepresented in the doxography.[70]

On the other hand, I would say, these skeptical approaches do not explain the rather strong doxographic evidence attributing to Anaximander several numbers for the sun and moon. Since none of the above interpretations of Anaximander's numbers as distances in the heavens seems to make sense, or at least can be brought into line with observations and calculations that Anaximander himself easily could have made, my conclusion is that we should look for an interpretation from a completely new perspective. I other words, perhaps we should let go of the very idea that Anaximander's numbers are meant to express something about distances in the cosmos and look for a completely different interpretation. I know of two such attempts, the first of which was put forward almost 70 years ago.

Stritzinger has brought up the idea that the records of the doxographers about the sizes of the circles of the sun and the moon are not Anaximander's at all, but were interpolated in the first century A.D. *in margine* of the texts and stem from attempts of early Christian calendar making. According to him, the word κύκλος should not be understood as a material circle but as a temporal cycle. The 19-year lunar cycle was already calculated by Meton, the 28-year solar cycle dates from after Caesar's calendar reform. The multiplication of 19 and 28 made a cycle of 532 years, which was the basis of the Christian calendar.[71] This attempt to explain Anaximander's numbers away does not sound very plausible. The other attempt has been made very recently and will be discussed in the next section.

A New Interpretation: The Numbers As a Calculator for the Lunar Cycle

The intuition of Eggermont and Stritzinger that Anaximander's numbers must have to do somehow with calendar requirements has recently been adopted in a completely different way by Thibodeau, apparently without being acquainted with their attempts. In a recent article, he suggests that the numbers for the sun and the moon (dismissing those for the stars as Tannery's invention) should be read as a kind of lunar-solar calendar, and he combines this with the old interpretation of the numbers relating to the circumferences of the heavenly wheels, as suggested by Röper (cf. note 62).

This attempt looks promising and deserves the attention of scholars, because it suggests a fresh approach to the interpretation of Anaximander's numbers. In this interpretation, only the best documented numbers 28 and 27 for the sun and 19 for the moon, as well as the additional number 18 for the moon, play a role. The author argues as follows (I only give the main lines of his argument, leaving aside the many

[70]Cf. Fehling (1985, 222).

[71]Cf. Stritzinger (1952, 65–66).

intriguing and sometimes not so easy to understand details): take the earth as module and draw 19 discs with the size of the earth in a row to make the circle of the moon wheel, and a concentric circle consisting of 28 discs of the same size to make the sun wheel. As the author puts it: "The phrase '28 times greater than the earth' conveys the size of the wheel in two ways: it represents the number of earths which the hoop contains; and if a line is drawn connecting the centres of the 28 discs, it will form a polygon with a circumference equal to 38 earth diameters. Repeating the same construction with 19 discs produces the moon wheel."[72] The resulting drawing (see Fig. 6.7) should in the first instance not be understood as a cosmological model representing the distances of the sun and the moon, but as a kind of calculator representing the relative positions of both luminaries during the lunar month. Since there are no numbers for the stars handed down, the picture does not show a circle of the stars. Moreover, a circle of stars is not needed when the picture is not conceived of as a model of the universe but rather as a kind of calendar.

The counting starts with the two white disks in the wheels of the sun and the moon, which visualize the situation of new moon. If we go in both rings from one disk to the other in the same direction in one step at a time, after about 29 and a half steps we get the situation of full moon. The length of a synodic month is approximately 29½ days, so this representation functions as a kind of lunar-solar calendar. The model visualizes the position of the moon and the sun in relation to the earth after each step. If we calculate with one step for a night-plus-day (a nukthemeron), it will take "532 steps, or 266 days, for the sun and moon to complete their respective cycles and return again simultaneously to their starting points." Dividing by nine gives the length of 1 month of 29.55555... days on the calculator, which is very close to 29.53059... days, which is the length of a synodic month.[73] Thibodeau adds several details and refinements to strengthen his interpretation, which I leave out of this short survey.

The interpretation of Anaximander's numbers as a device to represent the lunar cycle is new and inventive. If read as no more than this and not as a representation of the universe according to Anaximander, it avoids many problems inherent in the other interpretations. Therefore, it deserves the careful attention of other scholars. Some years ago, Nicholas Rescher published a note, in which he saw in Anaximander's cosmological model, as it appears in my drawings, a forerunner of the Antikythera mechanism.[74] I think he was wrong, but as an alternative I would like to suggest a comparative investigation into Anaximander's numbers 28 for the sun and 19 for the moon, as used in Thibodeau's article, and the relevant gears in the Antikythera mechanism. This would be particularly interesting for Thibodeau's claim that a model with 19 and 28 discs, based on the equation 18 months = 532 days is unique because it is the simplest possible that displays a month that is very close in length to the true lunar month, as measured by the model's steps.[75]

---

[72]Thibodeau (2017, 99).

[73]Thibodeau (2017, 102).

[74]Rescher (2014).

[75]Cf. Thibodeau (2017, 103, 106).

Fig. 6.7 Anaximander's numbers as a lunar-solar calendar (after Thibodeau)

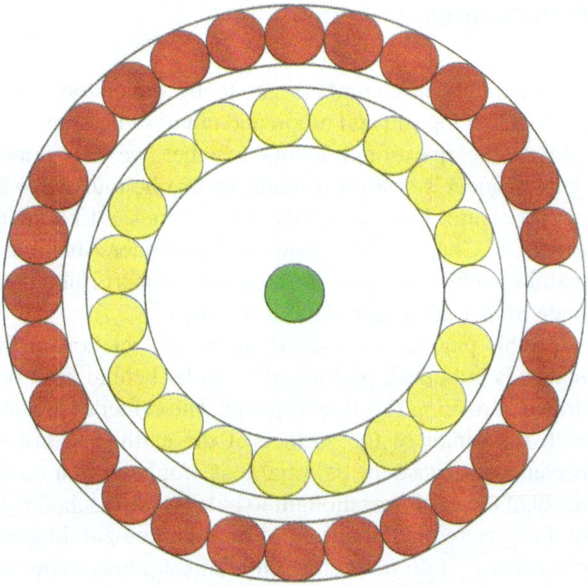

However, Thibodeau also claims that his representation as in Fig. 6.7 somehow functions as a model of Anaximander's cosmos. In this second interpretation, as he remarks, the disks must be imagined as facing the earth instead of looking "upward."[76] Here he overplays his hand, or, to state it otherwise: it is either the one or the other, but he cannot have his cake and eat it. When he presents his model as a representation of Anaximander's cosmos, Thibodeau runs into the same difficulties as in Forbiger's interpretation: the heavenly bodies, and especially the sun, are too far from the flat earth (as illustrated in Fig. 6.6), and the angular diameters of the sun and moon are far too large. Thibodeau tries to explain away the problem of the angular diameters of the sun and moon, which would be 26 and 38 times too big in his rendition. He assumes that Anaximander's sun has two parts, a small central disc which is the aperture in the wheel and corresponds to the visible disc of the sun, and a wider wheel of rays which corresponds to a circle of one earth diameter, as in the picture. In the words of the author: "the rays form part of the picture of the sun."[77] This suggestion, made earlier by Forbiger,[78] is too far-fetched to be taken seriously. Moreover, it does not explain why the disk of the moon, which has a much fainter light, should have the same diameter as the disk of the sun.

[76] Thibodeau (2017, 99, n. 26).

[77] Thibodeau (2017, 110).

[78] See Forbiger (1877, 523 n. 57): "Diese verschiedenen Angaben lassen sich wohl so erklären, dass Anaxim. den Luftkreis, der den eigentlichen Kern der Sonnen umgab, für 27 mal, also den Kern und den Lichtkreis zusammen, oder die ganze Sonne, für 28 mal grösser als die Erde hielt."

Conclusions

If Anaximander's numbers are to be understood as a calculator for the relative positions of the sun and moon and not as indications of the distances of the heavenly bodies, the question remains whether we can draw a visual representation of Anaximander's cosmos. I maintain that he must have been aware that the heavenly bodies cannot be farther than a few thousand kilometers from his flat earth, as is argued in Chap. 3. So, I now tend to render Anaximander's universe a lot smaller than I used to, somehow as in Fig. 6.8. Robert Hahn previously drew a similar small-scale model, although without the stars.[79]

In this picture, no specific numerical relation of the distances of the heavenly bodies is indicated, but only their order behind each other. The stars are tentatively drawn as making together a sphere. The center of the universe is drawn as coinciding with the center of the surface of the earth, and not with the center of the earth, because otherwise the celestial axis would end far north of Delphi (or whatever the navel of the earth was thought to be). To some scholars this must look awful, because in this picture the famous equilibrium argument does not apply. As I have explained elsewhere,[80] I do not believe that this argument should be attributed to Anaximander. Probably Simplicius' remark that, according to Anaximander, the earth rests on air is right, as Kočandrle argues in recent articles.[81] In this drawing, the sun (i.e., the aperture in the sun wheel, but the same holds for the width of the sun wheel) is much smaller than the earth, as should be the case with a flat earth. Aëtius' testimony that the sun is equal to the earth must be understood, just like Anaximander's numbers, as having nothing to do with a model of the cosmos, but everything with instructions for making a lunar-solar calculator. After more than a century of unsuccessful attempts to convey an ontological meaning to the numbers, and faced with a completely different interpretation that seems to make sense, I think it is allowed to accept the implication that the doxographers must have misunderstood the meaning of Anaximander's numbers and erroneously interpreted them as indications of the sizes of the heavenly wheels.

[79]See Hahn (2010, 162, Fig. 6.11).

[80]See Couprie (2011, 109–110).

[81]Cf. Kočandrle (2017, 2018).

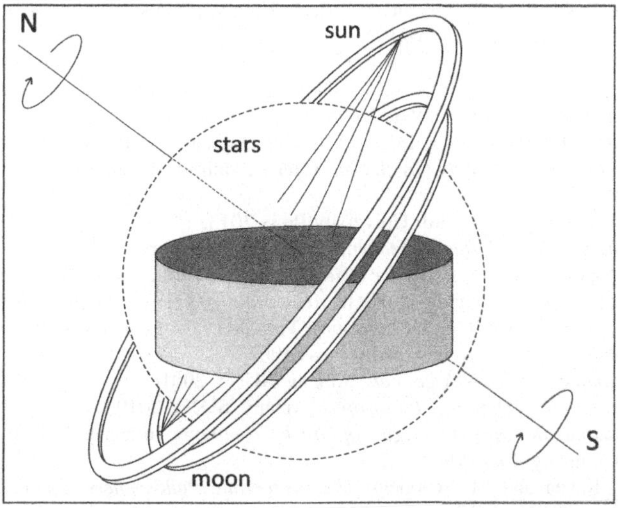

Fig. 6.8 An impression of Anaximander's universe with the heavenly bodies at shorter distances

References[82]

Bottler, H., *Pseudo-Plutarch und Stobaios: Eine synoptische Untersuchung*. Göttingen: Vandenhoeck & Ruprecht 2014.

Burkert, W., Iranisches bei Anaximander. *Rheinisches Museum für Philologie* 106 (1963), 97–134.

Conche, M., *Anaximandre. Fragments et Témoignages*. Paris: Presses Universitaires de France 1991.

Corre, J.-F., Proportions du ciel d'Anaximandre. *Phronesis* 58 (2013), 1–16.

Couprie, D.L., The Visualization of Anaximander's Astronomy. *Apeiron* 28 (1995), 159–181.

—, Anaximander's Discovery of Space. In: A. Preus, ed., *Before Plato. Essays in Ancient Greek Philosophy VI*. Albany: State University of New York Press 2001, 23–48.

—, The Discovery of Space: Anaximander's Astronomy. In: Couprie, D.L., R. Hahn and G. Naddaf, *Anaximander in Context New Studies in the Origins of Greek Philosophy*. Albany: State University of New York Press 2003, 165–254.

—, *Heaven and Earth in Ancient Greek Cosmology*. New York: Springer 2011.

Couprie, D.L. and H.J. Pott, Imagining the Universe. *Apeiron* 35 (2002), 47–59.

Dicks, D.R., *Early Greek Astronomy to Aristotle*. Ithaca N.Y.: Thames and Hudson 1970.

Diels, H., *Doxographi Graeci*. Berlin: Walter de Gruyter 1879 (=Dox).

—, Über Anaximander's Kosmos. *Archiv für Geschichte der Philosophie* 10 (1897), 228–237.

—, Anaximandros von Milet. *Neue Jahrbücher für das klassische Altertum* 51 (1923), 65–76.

Diels, H. and W. Kranz, *Die Fragmente der Vorsokratiker*. Zürich/Hildesheim: Weidmann 1951[6] (=DK).

Eggermont, P.H.L., The Proportions of Anaximander's Celestial Globe and the Gold-Silver Ratio of Croesus' Coinage. In M.A. Beek et al., eds., *Symbolae Biblicae et Mesopotamicae Francisco Mario Theodoro De Liagre Böhl Dedicatae*. Leiden: E.J. Brill 1973, 118–128.

[82]P = Aëtius in pseudo-Plutarch, *Placita* (numbering according to Dox).
S = Aëtius in Stobaeus, *Anthologium* (numbering according to Wachsmuth and Hense).

Fehling, D., Das Problem der Geschichte des griechischen Weltmodells vor Aristoteles. *Rheinisches Museum für Philologie* 128 (1985), 195–231.

Forbiger, A., *Handbuch der alten Geographie*. Hamburg: Haencke und Lehmkuhl 1842 (quoted from the second edition, 1877). Available on the Internet: https://archive.org/stream/handbuchderalten01forb#page/n0/mode/2up.

Gemelli Marciano, M.L., *Die Vorsokratiker* (3 Bände). Düsseldorf: Artemis & Winkler 2007–2013.

Graham, D.W., *The Texts of Early Greek Philosophy*. Cambridge: Cambridge University of Press 2010 (=Gr).

—, *Science Before Socrates*. Oxford: University Press 2013.

Gregory, A., *Anaximander, A Re-assessment*. London: Bloomsbury 2016.

Gruppe, O.F., *Die kosmischen Systeme der Griechen*. Berlin: Verlag von G. Reimer 1851. Available on the Internet: https://www.google.nl/?gfe_rd=cr&dcr=0&ei=o6SJWpbgENCGgAaloqigBw.

Guthrie, W.K.C., *A History of Greek Philosophy. Volume I. The Earlier Presocratics and Pythagoras*. Cambridge: Cambridge University Press 1962.

Hahn, R., *Anaximander and the Architects*. Albany: SUNY 2001.

—, *Archaeology and the Origins of Philosophy*. Albany: SUNY 2010.

Kahn, C.H., *Anaximander and the Origins of Greek Cosmology*. Indianapolis/Cambridge Hackettt Publishing Company, Inc. 1994.

Kirk, G.S., J.E. Raven and M. Schofield, *The Presocratic Philosophers*. Cambridge: Cambridge University Press 2009 (=KRS).

Kočandrle, R., The Stability of the Earth in Anaximander's Universe. *Ancient Philosophy* 37 (2017), 265–280.

—, Explaining Earth's Stability by Uniformity: Origins of the Argument. *Apeiron* 2018 (published online first: https://www.degruyter.com/printahead/j/apeiron).

Laks, A. and G.W. Most, *Early Greek Philosophy VI*, Cambridge, MA 2016.

Mansfeld, J. and D.T. Runia, Aëtiana. *The Method and Intellectual Context of a Doxographer, Volume Two, Part Two*. Leiden: Brill 2009 (=MR).

Naddaf G., On the Origin of Anaximander's Cosmological Model. *Journal of the History of Ideas* 59 (1998), 1–28.

—, Anaximander's Measurements Revisited. In: A. Preus (ed.), *Essays in Ancient Greek Philosophy VI: Before Plato*. Albany: SUNY 2001, 5–21.

O'Brien, D., Anaximander's Measurements. *The Classical Quarterly* 17 (1967), 423–432.

Rescher, N., Anaximander and the Antikythera Mechanism. http://philsci-archive.pitt.edu/3078/ (Januari 2, 2014).

Röper, W., Emendationsversuche zu Hippolyti philosophumena. *Philologus* 7 (1852), 606–637. Available on the Internet: https://archive.org/stream/philologus07deutuoft#page/606/mode/2up.

Stritzinger, H.-W., *Untersuchungen zu Anaximander*. Inaugural-Dissertation Mainz 1952.

Tannery, P., *Pour l'histoire de science Hellène: de Thalès à Empédocle*. Paris: Félix Alcan 1887.

Thibodeau, P., Anaximander's Model and the Measures of the Sun and Moon. *Journal of Hellenic Studies* 137 (2017), 92–111.

West, M.L., *Early Greek Philosophy and the Orient*. Oxford: Clarendon Press 1971.

White, S.A., Milesian Measures: Space, Time, and Matter. In: P. Curd and D. Graham, (eds.), *The Oxford Handbook of Presocratic Philosophy*. Oxford: Oxford University Press 2008, 89–133.

Wöhrle, G., *Traditio Praesocratica. Die Milesier: Thales*. Berlin: Walter de Gruyter 2009 (=TP1).

—, *Traditio Praesocratica. Die Milesier: Anaximander und Anaximenes*. Berlin: Walter de Gruyter 2012 (=TP2).

Zhmud, L., *Pythagoras and the Early Pythagoreans*. Oxford: University Press 2012.

Chapter 7
Anaximenes' Cosmology

Contents

The Cap Simile; Graham and the Top Hat

One of the strangest theories, combined with one of the most enigmatic images in Presocratic cosmology, which have puzzled many scholars, is ascribed to Anaximenes. According to him, says Hippolytus, the sun and the other celestial bodies do not go underneath the earth, but move around it like a felt hat (or a turban, or a ribbon)[1] around our head:

7.1. (Anaximenes) denies that the heavenly bodies move under (ὑπὸ) the earth, as others suppose, but he says they turn around (περὶ) the earth, like a felt cap (πιλίον) turns around our head (περὶ τὴν ἡμετέραν κεφαλὴν στρέφεται). The sun is hidden not by going under the earth, but by being covered by the higher parts of the earth (ὑπὸ τῶν τῆς γῆς ὑψηλοτέρων μερῶν) and by being a greater distance away from us (διὰ τὴν πλείονα ἡμῶν αὐτοῦ γενομένων ἀπόστασιν).[2]

[1]Several possible translations are discussed in Bicknell (1966).

[2]Hippolytus, *Refutatio Omnium Haeresium* 1.7.6 = DK 13A7(6) = LM ANAXIMEN. D3(6) = Gr Axs12(6) = TP2 As56[7.6] = KRS 156.

D. L. Couprie, *When the Earth Was Flat*, Historical & Cultural Astronomy,
https://doi.org/10.1007/978-3-319-97052-3_7

In this chapter, I will discuss several interpretations and offer some suggestions which have the intention to bring the interpretation of Anaximenes' cosmology somewhat further. The history of Anaximenes' theory of the paths of the celestial bodies, from its beginning in the doxography until the most recent interpretations, is a minefield of misunderstandings, confusions, slips of the pen, errors, and even sheer blunders, which must be dismantled to pave the way for my final suggestions for the interpretation of Anaximenes' cosmology.

To begin with, Hippolytus' text is full of strange and hard to understand details. It should be noted that neither the cap simile, nor the sun covered by the higher parts of the earth, nor the clause "by being a greater distance away from us" can be found elsewhere in the doxography on Anaximenes. Moreover, Hippolytus, in his explanation of Anaximenes' cosmology, seems to mention two causes why we do not see the sun at night, even though it is supposed to be above the earth. The first is, because the sun is covered by the higher parts of the earth, and the second, because the sun is a greater distance away from us. Panchenko reads the clause on the higher parts of the earth as referring to a "huge bulge at the centre of the earth," with which he means the subpolar region,[3] and for an explanation he points to Cosmas Indicopleustes (cf. Fig. 7.3). This connection can be questioned, but more relevant is that Panchenko explains the clause about the sun being at a greater distance from us as the expression of the idea that sunrise and sunset are optical illusions. This latter point will be addressed in Chap. 15, because Panchenko relates it to a similar issue in the Chinese conception that will be dealt with in the Chaps. 13 and 14. For now, it suffices to raise the question of how to combine the idea that sunrise and sunset are optical illusions with the idea that sunrise and sunset are caused by the sun's appearance from behind and disappearance behind a huge mountain. The problem is that the two reasons why we do not see the sun at night seem to be incompatible. One is inclined to say: either we do not see the setting sun because it is hiding behind mountains, or because the setting sun is too far away to be seen. It has been suggested that the words "and by being a greater distance away from us" could be a doxographical addition and thus have nothing to do with Anaximenes' cosmology.[4] Perhaps, Hippolytus wanted to demonstrate his knowledge of astronomy by stating that the sun is far away. He forgot, however, that this discovery follows from the conception of the earth as spherical and does not hold true for a flat earth. A more natural interpretation could be that the words "being a greater distance away from us" should be understood as an explication of "the higher parts of the earth," to emphasize that those higher parts are at the periphery of the flat earth. The conjunction "and" (καὶ) between the two clauses must in that case be understood as explicative.

According to Robinson, the only way the sun can be covered by the higher parts of a flat earth is when "it would travel from east to west close to the southern horizon,

[3]Cf. Panchenko (2015), 420.

[4]KRS, p. 156.

not overhead as it obviously does,".[5] Graham's rendition is an example of a literal interpretation, clearly showing Robinson's problem.

Graham notes: "This is bad astronomy, but it avoids the physical problem of heavenly rotations interfering with the cushion of air beneath the earth."[6] Such a cushion of air and the alleged problem of crossing it are not mentioned anywhere in the doxography. We will encounter this strange idea in a different form in McKirahan's interpretation. The problem of how the earth can be thought to rest on air does not concern us in this chapter, so I will leave it with the remark that the doxography says no more than that not only the earth, but also the sun, moon, and stars float on air, apparently without the need for cushions of air to support them. The only reason that is mentioned for the floating is their flatness:

7.2. The earth is flat riding on air, likewise the sun and moon and the other heavenly bodies, which are all fiery, float on air because of their flatness.[7]
7.3. Anaximenes (...) say[s] that flatness is the cause of [the earth's] staying in place. It does not cut, but covers like a lid (ἐπιπωμάζειν) the air below, such as flat bodies are observed to do; for they are hard to move even by wind because of their resistance.[8]
7.4. Anaximenes says (...) the earth is extremely flat (πλατεῖαν μάλα); therefore, it is understandable (κατὰ λόγον) that that the earth rides upon (ἐποχεῖσθαι) the air.[9]
7.5. Anaximenes [says] owing to its flatness it rides upon the air.[10]

If we are to believe the sources on this point, Anaximenes proposed a solution to a major problem that resulted from Anaximander's cosmology, namely why the earth does not fall, putting forward a physical argument. His solution is said to be that the earth's vastness and flatness make it float on air, but nowhere a special cushion of air is mentioned.

From the caption of Graham's drawing it is clear that the image depicts the daily orbit of the sun, circling above the earth and hiding at the night behind high mountains in the north. Graham does not draw it, but if the sun rotates as it does in Fig. 7.1, then the celestial axis must be perpendicular to the center of the earth's surface. This center cannot be Delphi, because at Delphi the celestial pole is not in the zenith. In this interpretation, Greece and the whole Mediterranean world must be thought of as being at a considerable distance from the center. A serious problem of

---

[5]Robinson (1968), 45.

[6]Graham (2013a), 64.

[7]Cf. Hippolytus, *Refutatio Omnium Haeresium* 1.7.4 = DK 13A7(4) = LM ANAXIMEN. D3 (4) = Gr Axs12(4) = TP2 As56(7.4) = KRS 151.

[8]Aristotle, *De Caelo* 294b13–21. = DK 13A20 = LM ANAXIMEN. D19 = Gr Axs13 = TP2 As3 = KRS 150.

[9]Pseudo-Plutarch, *Stromata* 3, Fr. 179 = DK 13A6 = LM ANAXIMEN. D2 = Gr Axs11 = TP2 As83 = KRS 148.

[10]P 3.15.8 (not in S) = DK 13A20 = LM ANAXIMEN. D 20b = Gr Axs15 = TP2 As46; not in KRS, but see p. 153.

Fig. 7.1 Graham's interpretation of the cap simile (Redrawn after Graham 2013a, Fig. 2.2)

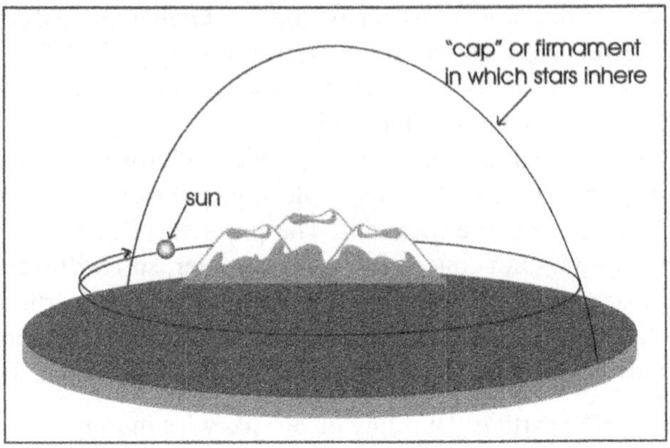

Graham's interpretation of a hemispherical cap with the sun rotating in a plane parallel to the earth's surface is that the plane of the celestial equator would coincide with the plane of the earth's surface (see Fig. 7.15a). The sun on Graham's drawing is the summer sun. Accordingly, at the equinoxes, the sun would rotate on the horizon, and during winter half a year under the horizon, contrary to what Anaximenes would have experienced in Miletus. The movements of the stars in this picture are completely incomprehensible. They are probably supposed to rotate attached to the "cap" or dome in circles parallel to the earth's surface, to avoid "the physical problem of heavenly rotations interfering with the cushion of air beneath the earth."[11] Rather than blaming Anaximenes for doing bad astronomy, we may wonder whether there is something wrong with this interpretation. Basically, what Graham describes and draws is what an observer on the north pole of a spherical earth would see or, in the conception of Presocratic flat earth cosmology, the situation before the tilt of the heavens (see Fig. 3.5a). Graham's drawing illustrates, by the way, that on a flat earth the heavenly bodies must be nearby and the sun very small indeed.

We could try to avoid the problem of the sun being half the year under the horizon, following Boll's suggestion to consider the πιλίον not as a hemispherical cap but as a top hat, as shown in Fig. 7.2.[12] The result would be that the sun can circle above the earth's surface all year round, and not only in summer.[13] The strange consequence, however, would be that the equator and the tropics become circles that are parallel to the surface of the earth and lie above each other.

[11]Cf. Graham (2013a), 64 and 65, caption of Fig. 2.2.

[12]Boll (1914), 361n. Boll's suggestion is referred to in Guthrie (1962), 138, n. 2: "Hölscher in *Hermes* (1953), 413 says on the authority of Boll, *Zeitschr. F. Assyr.* (1914), 361, n., that the idea that the sun and moon go round instead of under the earth is Babylonian."

[13]Kopf's remark that the top hat heaven of Boll's suggestion should be thought of as tilted (see Boll 1914, 361) would encounter similar problems as in McKirahan's interpretation (see below). It is not clear whether Kopf correctly reflects Boll's intentions in this respect.

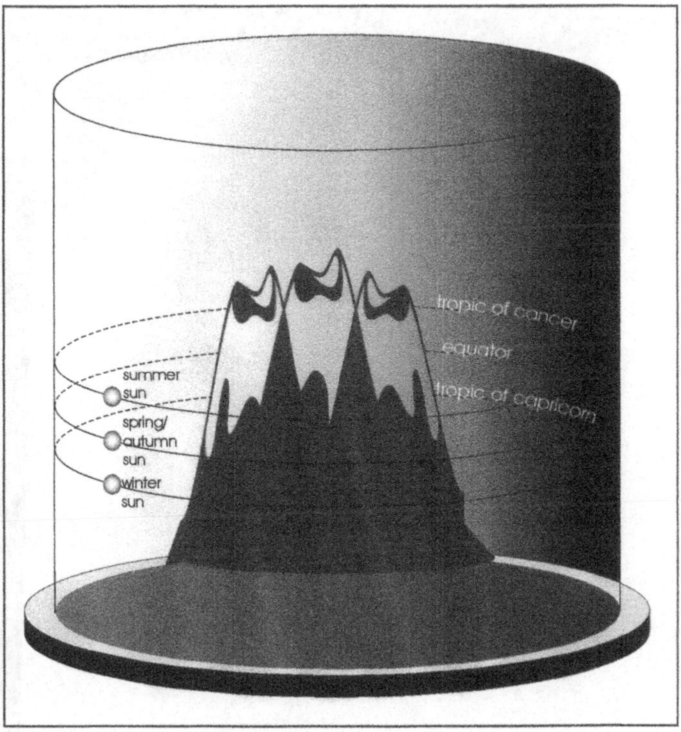

Fig. 7.2 Anaximenes' cap as a top hat

This picture clearly shows that the assumed northern mountains behind which the sun should hide at night must be enormously high, if the inhabitants of the earth's surface were to see the sun at the right altitude. The same effect is seen in Cosmas Indicopleustes' picture of his cosmology (sixth century A.D.), reproduced here as Fig. 7.3. He draws three lines representing the heights of the sun during the seasons, indicated from top to bottom by the words "short night," "mid-length night," and "long night" (μικρὰ ν[ύ]ξ), μέση νύξ, μεγάλη νύξ) respectively.[14] How the movements of the stars are supposed to be in this representation, Cosmas did not tell (and neither did Boll).

Surprisingly, although at the foot of the mountain the words γῆ πᾶσα οἰκουμένη, "all the inhabited earth" are inscribed, this picture shows the northern mountain from behind, and not from the inhabited side of the earth, as Cosmas himself indicates in his text (ἐξ ἀντιστρόφου τὸ σχῆμα τῆς γῆς, "a reversed image of the earth"; lines

[14]Anderson (2013, 43) interprets these words as "an indication that the lengths of the days of days and nights are dependent on the position of the observer (. . .) from the equator to the pole," which is certainly wrong. The picture shows different positions of the sun and not observers at different positions. The latter is also Kominko's interpretation: "The miniature showing the movement of the sun and its different path in the course of the year" (Kominko, 2013, 85).

Fig. 7.3 The movement of the sun according to Cosmas Indicopleustes (Florence, The Biblioteca Medicea Laurenziana, ms. Plut. 9.28, f. 95r. Reproduced with permission of MiBACT. Further reproduction by any means is prohibited)

3 and 4 above the picture), and as can be seen from the positions of the rising and setting sun (ἥλιος ἀνατέλλων and ἥλιος δύνων. Cf. Anderson 2013, 42–43). This does, however, not affect our discussion.

A final problem with these attempts is that they only mention or draw mountains in the far north. The sun rises and sets at the southern horizon during half of the year. This implies that, if the sun is supposed to move above the horizon and in circles parallel to the earth's surface, there must be huge mountains up to far south of the east-west diameter of the flat earth. This raises the question of why we do not see them. We will return to the issue of the mountains later, but first we will discuss some scholars who seem to have been prompted by these inconveniences to defend the idea that either Anaximenes' earth or his "cap" should be tilted.

The Tilted Earth Interpretation of the Cap Simile

Although the sources do not mention it, Kirk discusses the "attractive interpretation" that Anaximenes, just like Leucippus and Democritus,[15] held the idea of a tilted earth, and then he states: "This tilting [of the earth] would explain how the stars could set, supposing that they are somehow fixed in the heaven: they rotate on the *hemisphere* (whose pole is in the Wain) and pass below the upper, northern edge of the earth but *not below its mean horizontal axis*."[16] Anaximenes' πιλίον is, in this interpretation, a hemisphere over a tilted earth, while the top of the cap is the celestial pole and the rim the celestial equator. Kirk's "mean horizontal axis" is virtually identical with what Heidel called the "Ionian equator" of the flat earth (see Chap. 3, Sect. *Geographical Issues*), where the plane of the celestial equator intersects the surface of the allegedly tilted earth. As shown in Fig. 7.4, this interpretation would result in a big gap without stars in the southern sky. In the end, Kirk rejects this interpretation for two reasons. The first is that the "higher parts of the earth" could be better understood as the northern mountain regions, which we discussed already in the previous section. He also refers to a text of Aristotle, which will be discussed further on (see text 7.13). The second reason is that an earth with such a huge slope "would not float on air, but would slip downwards."[17]

Robinson proposes a similar explanation of the cap simile: "(. . .) the earth itself is tilted relative to the plane of rotation in which the sun is carried around, so that the sun is hidden at night (. . .) by the uptilted rim of the earth itself."[18] He immediately notices the problem that the notion of a tilted earth "carries with it so many difficulties that that it seems rash to suppose that Anaximenes could have put it forward seriously."[19] His reaction, however, "yet in his successors it is the standard answer to the problem which it was designed to solve,"[20] is not correct. Not only is the tilt of the earth not documented for Anaximenes, but in addition, as argued in Chap. 3, the standard idea was that of a tilted *heaven*, and the ascription to Leucippus

---

[15]Cf. KRS, p. 157. Kirk does not mention Democritus, but ascribes the tilt of the earth also to Anaxagoras and Diogenes. This is a mistake, because Anaxagoras and Diogenes taught the inclination of the heavens, not of the earth.

[16]KRS, p. 157, my italics.

[17]See KRS, p. 157. For the latter objection, see also Chap. 3, Sect. *The Alleged Tilt of the Earth.*

[18]Robinson (1968), 45.

[19]Ibidem.

[20]Ibidem.

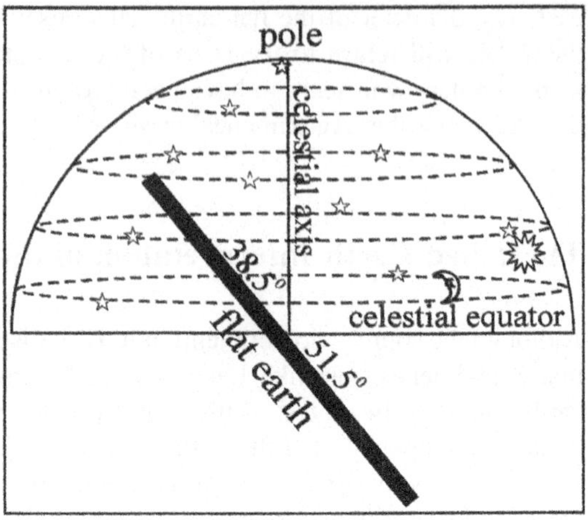

Fig. 7.4 Kirk's interpretation of the cap simile with a hemispherical heaven and the alleged dip of the earth

and Democritus of a dip of the earth must be considered a misunderstanding by the doxographers. It should be noted that neither Kirk nor Robinson have observed that this interpretation, in which not the hemispherical heavens but the earth is supposed to be tilted, would result in a big gap in the southern sky.

Bicknell's Interpretation of the Cap Simile

In an ingenious interpretation of the paths of the celestial bodies according to Anaximenes, Bicknell has tried to overcome the problem of the southern gap in the heavens.[21] He assumes, like Kirk, that Anaximenes held the idea of a tilted earth that is attributed to Leucippus and Democritus,[22] and he also claims that this explains the way Anaximenes described the paths of the celestial bodies. The difference is that Bicknell explicitly attributes to Anaximenes a spherical rather than a hemispherical heaven: "Anaximenes regarded his star-studded heaven as a sphere. My reason for rejecting [the conception of the heaven as a hemispherical dome] is that at Miletus (...) the celestial equator and the planes of diurnal rotation of all the heavenly bodies are inclined to the plane of the horizon by an angle of 53°."[23] Bicknell expresses his interpretation in rather cryptic wordings. After a description of Anaxagoras' theory of the inclination of the *heaven*, he writes: "The alternative [to Anaxagoras' theory] was to assert that in fact the heavenly bodies *did* orbit daily

[21]Bicknell (1969).

[22]Actually, Bicknell says that "Leucippus and Democritus (...) *indisputably* held that *the earth* was tilted *towards the north*" (Bicknell 1969, 78; my italics). The last words must be a slip of the pen, since the texts explicitly say that the earth was tilted towards the south (which means that the northern part was lifted). That this ascription is not indisputable is explained in Chap. 3.

[23]Bicknell (1969), 77.

in paths parallel to the equatorial plane which intersected one of the diameters of *an earth tilted upwards* in the north. The earth's obliquity to the celestial equator would correspond exactly to the observed obliquity of the paths of the luminaries to the plane of the horizon. This, I suggest, was exactly the view of Anaximenes."[24] The phrase "the equatorial plane which intersected one of the diameters of *an earth tilted upwards* in the north" must relate, just like Kirk's "mean horizontal axis," to what Heidel called the "Ionian equator" of the flat earth (see Chap. 3). Without an explanatory picture, Bicknell's interpretation remains rather enigmatic. Fortunately, Wöhrle has explained and drawn what Bicknell meant, and this is shown in Fig. 7.5: the paths of the celestial bodies go *behind* (in the picture: to the left of) the earth and thus, in a sense, not *below* the earth.[25]

In Bicknell's interpretation, not only the gap in the southern heavens has disappeared, but also the very image of a turning cap. Although he does not mention it here, we probably must consider his interpretation of the πιλίον in a previous article. There he observes that "a cap is not wound or rotated."[26] According to Bicknell, "the subject of comparison appears to be the heavenly bodies themselves, not the celestial vault," and "a spherical heaven could not be compared with a cap."[27] After having also rejected the idea of a turban,[28] he suggests "that πιλίον does not refer to any special head-dress, whether turban or cap, but merely to a strip of felt such as was used for a simple ribbon or fillet."[29] Anaximenes' image would then be comparable to Anaximander's heavenly wheels and Parmenides' στεφάναι. In this earlier article, Bicknell defends the idea that, according to Anaximenes and "in opposition to Anaximander, (...) the heavenly bodies move laterally rather than vertically beneath the earth."[30] As we have seen, in his later article he replaces that conception by the idea that the heavenly bodies go *behind* the earth's surface. Since the picture of Bicknell's interpretation (Fig. 7.5) does not look like a cap at all, we must assume that he retained the image of a ribbon or ribbons to represent the orbit or orbits of the heavenly bodies.

As noted above, the dip of the earth is not documented for Anaximenes, and its attribution to Leucippus and Democritus must be considered a misunderstanding by the doxographers. A serious objection to Bicknell's interpretation is that perhaps *on the picture* the heavenly bodies can be said to pass behind the earth, but for people living on the flat surface of an allegedly slanted earth, the setting celestial bodies can hardly be described otherwise than as passing under the earth. In a way, Bicknell reduces a phenomenological problem to a semantic issue. Moreover, Bicknell's

---

[24]Bicknell (1969), 78, second italics are mine.

[25]Cf. Wöhrle (1993), 74–75.

[26]Bicknell (1966), 17. See also Guthrie (1962), 138, n. 1: "it is the idea of movement which makes the simile so bizarre."

[27]Ibidem.

[28]Cf. Guthrie (1962), 138, n. 1; Webster (1961), 89 and 105, n. 32.

[29]Bicknell (1966), 18.

[30]Ibidem.

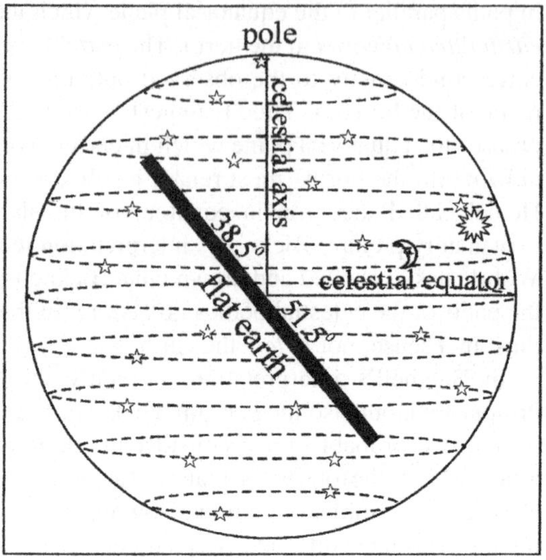

Fig. 7.5 Anaximenes' allegedly tilted earth and spherical heaven according to Bicknell (Cf. Wöhrle 1993, 74, Abb. 2. I added the daily circles of the heavenly bodies. Wöhrle drew a tilt of 53°, but this is an unimportant difference)

interpretation, which attributes this alleged dip of the earth also to Anaximenes, conflicts with several testimonies that say that according to Anaximenes the earth floats on air (see texts 7.2–7.5).

It should be noted that the none of these sources mentions any dip of the earth in connection with Anaximenes. It is hard to see how these texts can be reconciled with a 51.5° tilt of the earth (or even 53°, as in Bicknell's text). Bicknell's reply that "the atomists, whose slanted earth rested on air, explained that the air below its southern regions was rarer and therefore gave less support than the air below the upward tilted North,"[31] is weak, as I have argued in Chap. 3, Sect. *Climatological Issues.* Moreover, the first witness, Aristotle, who is not mentioned by Bicknell, clearly says that Democritus' earth covers the air like a lid. The word πῶμα is used for the lid or cover of a jar, a pot, a box, a cupboard, or even a tomb, and therefore has no connotation of being slanted or oblique. See further Chap. 3, Sect. *The Alleged Tilt of the Earth.*

Bicknell begins his discussion of the paths of the heavenly bodies according to Anaximenes with the remark: "If the early Ionian thinkers made *the obvious assumption* that the surface of the flat earth at the center of the universe coincided with the plane of the *celestial equator*, the facts of observation would be in blatant contradiction with preconceptions based on theory."[32] There is, however, no reason at all why this assumption should be obvious, nor why the early Ionians should have adhered to it. What Bicknell calls an obvious assumption is what the Presocratics, confronted with the riddle of the tilted celestial axis, offered as an explanation of its origin: *originally* the celestial axis was perpendicular to the earth (and thus the

[31] Bicknell (1969), 79.

[32] Bicknell (1969), 77–78, my italics.

surface of the flat earth coincided with the plane of the celestial equator, but later the celestial axis tilted, and the heavens with it (see Figs. 3.5a and 7.15a versus 3.5b and 7.15b).

Bicknell's strange definition of the problem also leads to an even stranger formulation of the solution, given later by Anaxagoras: "The heavenly bodies, he held, had *once circled on paths parallel to the celestial equator* (. . .). Later (. . .) Nous had given the celestial movements their presently observed obliquity."[33] This sounds as if in the present situation the heavenly bodies are no longer circling parallel but at an angle to the celestial equator, which is nonsense.[34] The consequence of what Anaxagoras (and others) said was, of course, that when the heaven tilted, the celestial equator, which originally was situated in the plane of the surface of the flat earth, went with it. The same strange idea recurs in Bicknell's rendition of Leucippus' and Democritus' alleged idea of a tilted earth: "(. . .) the heavenly bodies *did* orbit daily in paths parallel to the equatorial plane *which intersected one of the diameters* of an earth tilted upwards in the north (. . .)."[35] However, that the equatorial plane intersects one of the diameters of the flat earth is not a distinctive feature of the alleged theory of a dip of the earth, as in Fig. 7.6a, for this is also the case when the *heavens* are tilted, as held by Anaxagoras c.s. (see Fig. 7.6b). In both cases this diameter of the flat earth is what Heidel called the "Ionian equator."[36]

For his interpretation, Bicknell refers to Aristotle's *Meteorologica* 354a28–32 (see text 7.13), of which I argued in Chap. 3 (text 3.17), with Kirk, that Aristotle is not referring to a dip of the earth but to the mythical northern mountains from which the rivers flow. Bicknell's "logical supposition (. . .) that (. . .) Aristotle alludes to the slanted earth theory and that the thinkers he has in mind are Anaximenes and the two atomists who therefore held that the world's greatest rivers flow down from the north of their tilted earth,"[37] is not so logical after all. Moreover, the Presocratics knew of the existence of the great river Nile, flowing from south to north. The alleged dip of the earth would have meant that the Nile must flow uphill against a slope of 51.5°. I conclude that we should also discard Bicknell's interpretation of the path of the heavenly bodies according to Anaximenes.

[33]Bicknell (1969), 78, my italics.

[34]Bicknell is confused here with the obliquity of the ecliptic, which has nothing to do with the alleged inclination of the heaven (nor with the inclination of the earth). See Chap. 3, Sect. *The Tilt of the Celestial Axis*.

[35]Bicknell (1969), 78, second italics mine.

[36]See Chap. 3.

[37]Bicknell (1969), 78–79.

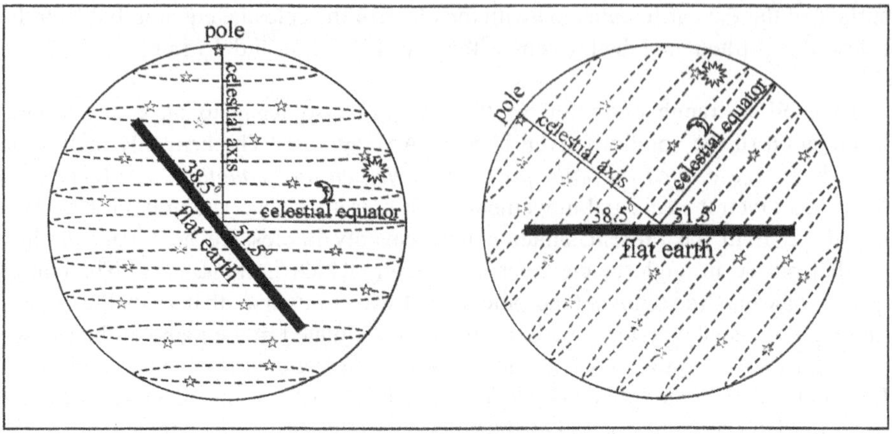

Fig. 7.6 (**a**) The tilted heaven; the equatorial plane intersects the "Ionian equator". (**b**) The tilted earth; the equatorial plane intersects the "Ionian equator"

McKirahan's Interpretation of the Cap Simile

In McKirahan's interpretation, the earth is no longer regarded as tilted, which is certainly a step forward compared to Bicknell's interpretation. He presents an illustration to clarify his interpretation, which is shown, with minor adjustments, in Fig. 7.7.[38] The main modification consists in adding two letters, A and B, which will be explained shortly. McKirahan makes the right observation that although "the cap is a handy model, because as it turns, the various points on its surface maintain constant relative positions."[39] But he also remarks: "this model cannot account for all the visible stars (. . .). Worse, it cannot account for the sun's and moon's motions."[40] One might wonder what the point of such a poorly performing image could be. The "cap" on McKirahan's drawing is bigger than Kirk's hemispherical one and extends a little further than the celestial equator, but it is not clear whether this is intentional or not.

In fact, McKirahan's interpretation is virtually identical with that of Kirk and Robinson (see Fig. 7.4), the main difference being that the heaven, not the earth, is tilted. This implies that it suffers from the same difficulties. More specifically, looking from the surface of the flat earth towards the south in the direction of B, there are no stars at all in a large part of the sky below the celestial equator. In order to save his model, McKirahan needs to assume that Anaximenes was not keen enough to realize this. Rather than blaming Anaximenes for the fact that his model

[38]See McKirahan (2010), 57.

[39]McKirahan (2010), 56.

[40]McKirahan (2010), 56, n. 15.

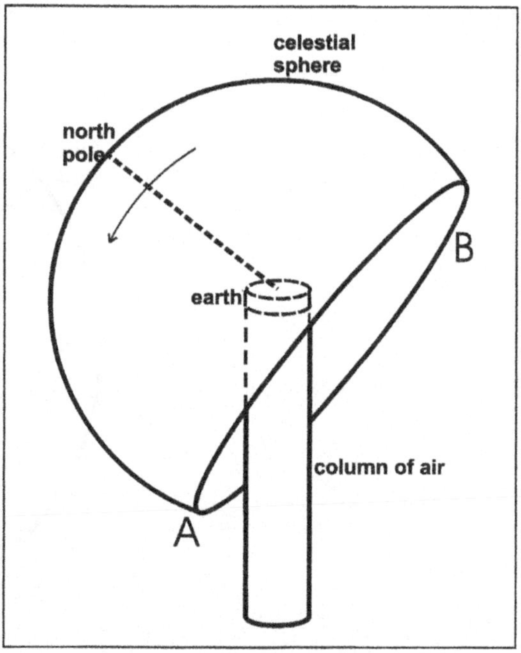

Fig. 7.7 Anaximenes' cosmos according to McKirahan

cannot account for the very things it is supposed to exhibit, one might wonder whether McKirahan's rendering of the cap simile is right.

Under the earth McKirahan draws a column of air that supports it. This rather strange feature, which Kirk, Robinson, and Bicknell did not exhibit but which is much like Graham's supposed "cushion of air," is McKirahan's interpretation of the reports that say that, according to Anaximenes, the earth because of its flatness rides on the air or covers the air underneath like a lid. This air column is not mentioned anywhere in the doxography. McKirahan's gives the following explanation for his alleged column of air: "If Anaximenes envisaged the earth as supported on a sea of air, he might have thought that the heavenly bodies, especially the sun, could not pass under the earth without disturbing its serene poise."[41] In other words, the heavenly bodies on the "cap" are supposed not to interfere with the column of air, so as not to disturb the earth's balance on top of that column. However, a celestial body that is somewhere on the celestial sphere at point A would naturally be called to be *under* the earth, although not exactly perpendicularly below it (where the alleged column of air that supports the earth is supposed to be). A body at point A is under the earth in the sense that it cannot be seen from the surface of the earth. In this sense, McKirahan's picture does not show what it intends to show, namely that the celestial bodies do not go under the earth.

[41]McKirahan (2010), 56.

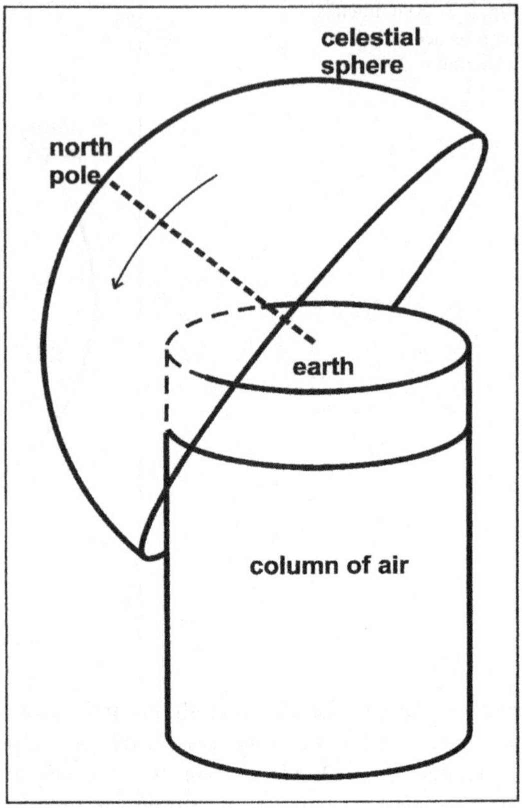

Fig. 7.8 Revised version of Anaximenes' cosmos according to McKirahan

Another problem with McKirahan's drawing is that the earth is rendered far too small or, to put it another way, that the distances to the celestial bodies are far too great. For people who think that the earth is flat, the celestial bodies are quite close and consequently quite small. The further to the north one goes, the higher the polar stars stand,[42] and the more to the south one goes, the lower they stand above the horizon. On a flat earth, the only way to explain this phenomenon is to take for granted that the stars are not far away, as shown in Fig. 3.11. Similarly, the further one goes south, the higher the sun stands at noon, until one reaches a place where the sun stands in the zenith in the summer solstice. Again, the only way to explain this phenomenon, when standing on a flat earth, is that the sun (which is lower than the stars) must be close by and therefore smaller than the earth, as can be seen in Fig. 3.10. The Milesians, who traveled from the Black Sea to Egypt, knew both phenomena for sure. How McKirahan's drawing of the cap simile would look like when the stars are nearby, is shown in Fig. 7.8.

[42]In Anaximenes' time there was not one star (almost) at the celestial pole, as is now the Polar star. People hat to orientate themselves by means of the circumpolar stars, such as the Two Bears.

A significant difference with McKirahan's drawing is that the rim of the "cap" in Fig. 7.8 now coincides with the celestial equator, in order not to let the "cap" interfere with the column of air, as McKirahan requires. The gap without stars in the southern sky remains, just as in Fig. 7.7. All in all, McKirahan's interpretation of Anaximenes' cap simile remains unsatisfactory.

Fehling and the Flat Heaven

Fehling has proposed a completely different interpretation from those discussed above. According to him, Anaximenes' world picture is still the same as what Fehling considers to be the archaic world picture, namely a flat round earth with the heavens above it not as hemisphere or a bell jar, but as a flat roof.[43] Fehling's accompanying picture is very simple (Fig. 7.9).

In Fehling's view, Anaximenes imagined that the heavenly bodies (except for the always visible circumpolar stars) would circle on the flat heaven and be covered at night by mountains on the edge of the earth.[44] Fehling mentions the northern mountains[45] that would hide the heavenly bodies, and thus does not account for those heavenly bodies that rise and set at the southern horizon. If the heavenly bodies are assumed to circle *on* the flat heaven, then these supposed mountains must be even higher than in the top hat simulation and must reach into heaven. On the other hand, if the heavenly bodies are assumed to circle *under* the flat heaven ("unter ihm ziehen Wolken und Gestirne"[46]), one wonders why a flat heaven is needed at all. Similar problems will be encountered in Panchenko's interpretation, to be discussed in Chap. 15.

According to Fehling, the conception of a flat heaven over a flat earth is the archaic model. It is not only the timeless basis of naïve thinking but also an elementary aspect of the ancient cultures and adopted from the Orient by the ancient Greeks.[47] Fehling maintains that this world picture can still be found in Democritus

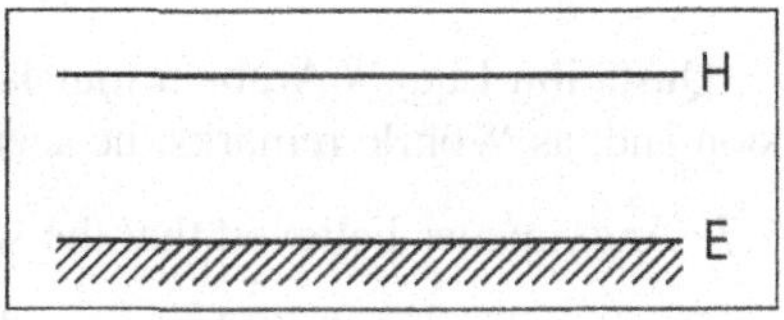

Fig. 7.9 Anaximenes' cosmos according to Fehling (Drawing after Fehling 1985, 215, Abb. 2)

[43]Fehling (1985), 206: "In diesem Weltbilde ist die Erdeoberfläche kreisrund, vom Okeanos umflosssen. Der Himmel, als etwas Festes gedacht, liegt als Flachdach (nicht Halbkugel und nicht Käseglocke) darüber."

[44]Cf. Fehling (1985), 208.

[45]Ibidem.

[46]Fehling (1985), 206.

[47]Cf. Fehling (1985), 206.

and Herodotus.[48] I think he is wrong to assume that the conception of a flat heaven parallel to the flat earth is archaic or even primitive. In the second part of this book I will show that it is, on the contrary, rather sophisticated and needs several ingenious features to make it work acceptably. The ancients of course saw the heavenly bodies set at and rise from the horizon, and the most obvious way to account for their paths during their disappearance was to assume that they traveled around the horizon of the circular flat earth, as shown in Fig. 7.13, where they were invisible because of mountains (which in that case did not need to be extremely high) at the periphery of the earth, a great distance away from us (cf. text 7.1).

The most intriguing aspect of Fehling's interpretation is that it looks like a primitive version of the ancient Chinese *gai tian* cosmology that is dealt with in the second part of this book. Panchenko goes even further and argues for a Greek influence on this Chinese cosmological conception.[49] Since a good understanding of his point presupposes knowledge of the *gai tian* system, I will postpone a discussion of it until after a thorough explanation of the *gai tian* in the second part of this book.

A Fresh Look at the Doxography

The discussion of the extant interpretations of Anaximenes' cap simile, with or without an alleged earth tilt, with a hemispherical or spherical heaven, seems to end in a deadlock. There seems to be no way left to make sense of the image of the star-studded heaven as a cap turning around the head. The interpretations of the above discussed authors are mainly dependent on the text of Hippolytus that was quoted at the start of this chapter (text 7.1). The other relevant texts about the movements of the heavenly bodies are also worth quoting. The most important is that of Aëtius, of which two versions exist. The first is that of pseudo-Plutarch:

7.6. Anaximenes [says] likewise [ὁμοίως, sc. like Anaximander] that the stars revolve under the earth and around it (ὑπὸ τὴν γῆν καὶ περὶ αὐτὴν στρεφεσθαι).[50]

Qusṭā ibn Lūqā's Arabian translation of Aëtius supports pseudo-Plutarch's version and, as Wöhrle remarks, he seems to have read ὑπέρ instead of περί:

7.7. Anaximenes believed that the stars move above and under the earth.[51]

On the other hand, Stobaeus' and Eusebius' versions, as well as another source, Diogenes Laërtius, say exactly the opposite:

---

[48]Cf. Fehling (1985), 207.

[49]See Panchenko (2015).

[50]P 2.16.6 = MR 487 (2.16.4) = TP2 As38; not in DK (but see Dox 346, n.1), LM, Gr, and KRS.

[51]Qusṭā ibn Lūqā, quoted from MR 488, after Daiber (1980),153; also in TP2 As205 and Bottler (2014), 388; not in LM, Gr, and KRS.

7.8. Anaximenes [says] the stars revolve not under (οὐχ ὑπὸ) the earth but (δὲ ἀλλὰ) around (περὶ) it.[52]
7.9. Anaximenes [says] the stars revolve not under (οὐχ ὑπὸ) the earth, but (δὲ) around (περὶ) it.[53]
7.10. (Anaximenes) says that the heavenly bodies do not travel under (οὐχ ὑπὸ) the earth, but (ἀλλὰ) around (περὶ) it.[54]

It should be noted that none of these texts mentions the cap simile, or the sun hiding behind the higher parts of the earth, or the great distance of the sun, which are three important issues in Hippolytus' text. Moreover, as far as I know, this is the only time the different versions of Aëtius are so overtly in contradiction with one another. Therefore, they deserve more attention than they have received up to now. The source of the disregard of pseudo-Plutarch's version is Diels, who substitutes Eusebius' text for it. In *Doxographi Graeci* and in *Die Fragmente der Vorsokratiker*, Diels only prints his emendations of the texts of the manuscripts of pseudo-Plutarch, which bring these manuscripts into conformity with Eusebius' version, so that the resulting text says the opposite of what is in the manuscripts.[55] Let us look at the page in question of *Doxographi Graeci* to see how Diels managed to make the text say the opposite of the reading of the manuscripts. On top of that page, Diels prints both his emendated text of pseudo-Plutarch and Stobaeus' version of a doxa from Aëtius on Anaximenes and the paths of the heavenly bodies. In Diels' eyes, they express exactly the same thing: "According to Anaximenes, the heavenly bodies do not move under the earth" (Fig. 7.10).

346 AETII PLAC. II 16 6. 7 17 1—6

PLUTARCHI EPIT. II 16. 17	STOBAEI ECL. I 24 1
Ἀναξιμένης οὐχ ὑπὸ τὴν γῆν, περὶ αὐτὴν δὲ στρέφεσθαι τοὺς ἀστέρας.	(Ἀναξιμένης) οὐχ ὑπὸ τὴν γῆν [δέ], ἀλλὰ περὶ αὐτὴν στρέφεσθαι τοὺς ἀστέρας.

Fig. 7.10 Part of the text of *Doxographi Graeci* 346

[52]S 1.24.1 = DK 13A14 = Gr Axs19 = MR 488 = TP2 As124; not in LM and KRS.

[53]Eusebius, *Preparatio Evangelica* 15.47.3 = MR 487 = TP2 As91 = Bottler (2014), 368; not in DK, LM, Gr, and KRS. Eusebius leaves out the word ἀλλὰ.

[54]Diogenes Laërtius, *Vitae Philosophorum* 2.3 = DK13A1 = Gr Axs1 = TP2 As72; not in LM and KRS.

[55]See Dox 346 and DK 12A14.

Obviously, Diels was so convinced of the correctness of his emendations of pseudo-Plutarch's version that the reader must consult the footnotes of this page in *Doxographi Graeci* to see how the not emendated text of the manuscripts sounds. When we look at the footnotes, we see a cryptic note 1 (Fig. 7.11).

The indication "οὐχ E" means that the negation οὐχ does not appear in the manuscripts of pseudo-Plutarch's text, but comes from Eusebius' version (text 7.9), from where it has been inserted by Diels in pseudo-Plutarch's text (text 7.6). The indication ": ὁμοίως (A)BC" means that in the manuscripts of pseudo-Plutarch (text 7.6) we read ὁμοίως instead of οὐχ. The word "ὁμοίως" refers to Anaximander, who is mentioned in the previous doxa. The words "περὶ αὐτὴν δὲ E" mean that περὶ αὐτὴν δὲ has been inserted from Eusebius (text 7.9). The indication ": καὶ περὶ αὐτὴν (A)BC" means that this is pseudo-Plutarch's text (text 7.6). Finally, "cf. Prol. p.136" looks as if we have to look on that page for an explanation of the insertion of Eusebius' text (text 7.9). So, let us look at p. 136 in the *Prolegomena* of the *Doxographi Graeci* (Fig. 7.12).

"Plutarchi Strom." refers to a book of another author, also known under the name of (pseudo-)Plutarch. On this page of *Doxographi Graeci*, there is no trace of Eusebius, or even of pseudo-Plutarch's version of Aëtius, but instead we read parallel texts of Hippolytus and Diogenes Laërtius, as well as Stobaeus' version of Aëtius. We should understand that Diels wanted to state that not only Eusebius and Stobaeus, but also Hippolytus and Diogenes Laërtius say that the heavenly bodies do not pass under the earth, and that therefore he had the right to ignore the manuscripts and offer his own emendated text of pseudo-Plutarch. To demonstrate what damage was done, I print the text as it is transmitted in the manuscripts:

1 οὐχ E cf. Prol. p. 136: ὁμοίως (A)BC ||
περὶ αὐτὴν δὲ E: καὶ περὶ αὐτὴν (A)BC ||

Fig. 7.11 *Doxographi Graeci* 346, n. 1

136 X THEOPHRASTEORUM CONSPECTUS 2. DE ANAXIMENE

Hippolyti	Plutarchi Strom.	Diogenis	Aëtii	Theophrasti fragm.
6 οὐ κινεῖσθαι δὲ ὑπὸ γῆν τὰ ἄστρα λέγει, καθὼς ἕτεροι ὑπειλήφασιν ἀλλὰ περὶ γῆν ὡσπερεὶ περὶ τὴν ἡμετέραν κεφαλὴν στρέφεται τὸ πιλίον.		II 3 κινεῖσθαι δὲ τὰ ἄστρα οὐχ ὑπὸ γῆν ἀλλὰ περὶ γῆν.	II 16 6 Ἀ. οὐχ ὑπὸ τὴν γῆν ἀλλὰ περὶ αὐτὴν στρέφεσθαι τοὺς ἀστέρας.	

Fig. 7.12 Part of *Doxographi Graeci* 136

7.11. Αναξιμένης ὁμοίως [sc. Αναξίμανδρῳ] ὑπὸ τὴν γῆν καὶ περὶ αὐτὴν στρέφεσθαι τοὺς ἀστέρας.[56]

Most collections of texts and handbooks either follow Diels' emendations or contain neither pseudo-Plutarch's nor Stobaeus' version. In their footsteps, studies on Anaximenes and his cosmology follow Stobaeus' or Eusebius' version (texts 7.8 and 7.9) and neglect that of pseudo-Plutarch (7.6) without even mentioning it, or they refer only to Hippolytus (text 7.1). Although they mention that Qusṭā ibn Lūqā (text 7.7) confirms the reading of the manuscripts of pseudo-Plutarch (text 7.6), Mansfeld and Runia agree with Diels that the reading of Eusebius (text 7.9) is to be preferred, mainly because this introduces a contrast with the previous doxa on Anaximander.[57] They suggest that "the alternative views found in PQ are most likely the result of textual corruption, e.g. in response to the word οὐχ falling out,"[58] which implies that they take for granted that, if they are right, δὲ has also been dropped and that ὁμοίως and καὶ have been added. In my opinion, however, the difference between Anaximander and Anaximenes in pseudo-Plutarch's version (text 7.6) is not, as Mansfeld and Runia suppose, that the one believes that the heavenly bodies go under the earth and the other does not,[59] but that Anaximander regards the heavenly bodies as fiery holes at fixed places of turning wheels, while Anaximenes has in mind fiery bodies (or leaves) floating on air. This is reflected by the verbs φέρεσθαι ("to be carried") for Anaximander in the previous doxa[60] and στρέφεσθαι ("revolve") for Anaximenes.[61]

Unfortunately, Wöhrle and Bottler do not offer pseudo-Plutarch's text as it appears in the manuscripts, but in the following somewhat cryptic version:

7.12. Αναξιμένης <οὐχ> [ὁμοίως] ὑπὸ τὴν γῆν [καὶ] περὶ αὐτὴν <δὲ> στρέφεσθαι τοὺς ἀστέρας.[62]

Above this text, Wöhrle refers to Eusebius: "= As 91" (text 7.9). We are supposed to understand that the text in the angled brackets is identical with the version of Eusebius, and that the text in the square brackets corresponds to the manuscripts of pseudo-Plutarch. In Wöhrle's translation of pseudo-Plutarch, these brackets no longer appear. There he follows Eusebius and thus accepts Diels' emendations.[63] However, above his translation of Qusṭā ibn Lūqā's version (text 7.7), Wöhrle notes: "Die arabische Fassung bestätigt im vorliegenden Fall die

[56]Cf. MR 487.

[57]Cf. MR 491 and n. 321. See also their reconstructed text and translation in MR493 and 494.

[58]MR 488. This argument is repeated in Bottler (2014), 389: "Die korrupte Textversion von Ps. Plutarch und Qosṭā ibn Lūqā sei auf das ausgefallene οὐχ zurückzuführen."

[59]See MR 491.

[60]Cf. P 2.16.5 ≠ S 1.24.2 = DK 12A18 = Gr Axr28 = TP2 Ar56 = MR 487 and 488 = KRS 128; not in LM.

[61]This opposition only holds for the sun and moon (and the planets, and perhaps the invisible bodies), but not for the stars, who are like nails fixed into the firmament and thus carried along.

[62]TP2 As38.

[63]See TP2 As38.

Überlieferung der griechischen Handschriften der *Placita*, nicht die Emendationen von H. Diels (...) Vgl. As 38"[64] (= text 7.6). The reader wonders why Wöhrle, if the Arabian text confirms pseudo-Plutarch's text, nonetheless translates pseudo-Plutarch in Diels' emendated version. Generally speaking, I am an admirer of Wöhrle's volumes, but in this case, he only adds to the confusion. Bottler is not very helpful either. Under her quotation of pseudo-Plutarch's and Stobaeus' texts (texts 7.6 and 7.8), she adds: "omisso lemmate δὲ plus] P = E = Q(= DG) ↔ S(= M&R)."[65] The text to the left of "↔" means that she accepts Diels' emendations in *Doxographi Graeci*, while the sign "↔" indicates that the readings of the texts to the left are different from those to the right of this sign. This is misleading, because P = E = Q (meaning: the texts of pseudo-Plutarch, Eusebius, and Qusṭā ibn Lūqā are identical) is only true if one accepts Diels' emendations of pseudo-Plutarch and admits that Qusṭā ibn Lūqā translated a corrupted text. In my opinion, these scholars attach too much importance to Diels' emendations and are too hasty in rejecting the reading of the manuscripts.

I suggest that not the versions of pseudo-Plutarch and Qusṭā ibn Lūqā (texts 7.6 and 7.7) are corrupt, but that, on the contrary, the texts in which is said that according to Anaximenes the heavenly bodies do not go under the earth (texts 7.1, 7.8, 7.9, and 7.10) are corrupt. There are several reasons to consider this possibility. The first is that the texts that state that the heavenly bodies do not go under the earth encounter severe interpretive problems, as has been demonstrated extensively in the previous sections. Given this dead lock, it seems appropriate to investigate other interpretive options. The text of Pseudo-Plutarch (text 7.6), on the other hand, is perfectly understandable: according to Anaximenes, the paths of the heavenly bodies are going around (περί) the earth and therefore continue under (ὑπό) the earth. Qusṭā ibn Lūqā's text (text 7.7) confirms this.

A second, as far as I know never noticed, textual argument is the word περί in the texts of Hippolytus, Stobaeus, Eusebius, and Diogenes Laërtius (texts 7.1, 7.8, 7.9, and 7.10), in which the movement under the earth is denied. The word περί does not, however, express the spatial meaning of "above;" περί can only mean "above" when it is used in a figurative sense, indicating superiority.[66] Without further explanation, the word περί is not very suitable as a description of the movements of the heavenly bodies *above* the flat surface of the earth. Pseudo-Plutarch (text 7.6), on the contrary, uses περί in the literal, spatial meaning of "around" to reinforce the idea that the heavenly bodies in their daily orbits go under the earth as well. If he had wanted to emphasize that the heavenly bodies do not go under the earth, Aëtius would have used ὑπέρ ("above") to get a real contrast with οὐχ ὑπὸ ("not under"); ὑπέρ is the usual word for "above," "over," "across," and is also used of motions. The same argument is used by Mourelatos when he states that according to Parmenides the moon goes around and thus also under the earth: "Parmenides does not say *hyper gaian*, 'above the earth,' but rather *peri gaian*, 'around the earth.' This indicates that

[64]Caption of TP2 As205.

[65]Bottler (2014), 388.

[66]See LSJ, περί III.

he has grasped that the moon's diurnal course, and, by implication, the sun's diurnal course as well, are not arcs but complete loops, part of each loop lying above and part lying below the earth."[67] The appearance of the word περί in all these texts is, in my opinion, a strong argument in favor of the idea that pseudo-Plutarch's version of Aëtius is to be preferred.

In the third place, it is hard to imagine that pseudo-Plutarch would have accidentally omitted the words οὐχ and δέ, and then, as a kind of compensation, added ὁμοίως and καί. On the other hand, as will be shown below, I think it can be argued how the texts of Hippolytus et al. could have originated from a text in Aristotle's *Meteorologica*.

Anonymous Texts and Kirk's Interpretation

The problem with anonymous texts is, of course, that one must be cautious in attributing their contents to a specific thinker, while the author of the text, for whatever reason, has refrained from doing so. Some texts that do not mention Anaximenes are occasionally brought forward as to support the idea that the heavenly bodies do not travel under the earth. The first, already mentioned in the Sect. *Bicknell's Interpretation of the Cap Simile* is, since Diels included it into the doxography on Anaximenes, usually considered to describe his cosmology:

7.13. Many of the ancient meteorologists are convinced that the sun does not travel under (ὑπὸ) the earth, but rather around (περὶ) the earth and that (northern) region, and it disappears and causes night because the earth is high toward the north.[68]

As already stated in Chap. 3, Sect. *The Alleged Tilt of the Earth*, this text has nothing to do with an alleged dip of the earth, as many authors believe. Mansfeld and Runia use Aristotle's text (7.13) as an argument for the correctness of Stobaeus' and Eusebius' reading (texts 7.8 and 7.9).[69] In my opinion, it is a strange way of arguing to regard a text of Aristotle, in which he attributes an opinion to many anonymous meteorologists, as proof that Anaximenes must have said the same thing. When we read Aristotle's text (7.13) with an unbiased eye it is clear that Aristotle does not refer here to Anaximenes, but to ancient mythological stories that "told how the sun, when he set in the west, was carried round the encircling stream of Ocean in a golden boat to rise in the east again."[70] This image, rendered in Fig. 7.13, which entails a

---

[67]Mourelatos (2013), 99.

[68]Aristotle, *Meteorologica* 354a28–32 = DK 13A14 = LM ANAXIMEN. D16 = Gr Axs18 = TP2 As4. = KRS 157. Graham translates "this region," but meant is the northerly region mentioned just before.

[69]Cf. MR 488.

[70]Guthrie (1962), 138, where he also notes: "Hölscher (. . .) says, on the authority of Boll (. . .), that the idea that the sun and moon go round instead of under the earth is Babylonian." See also Hölscher (1970), 218, and Boll (1914), 361, n.

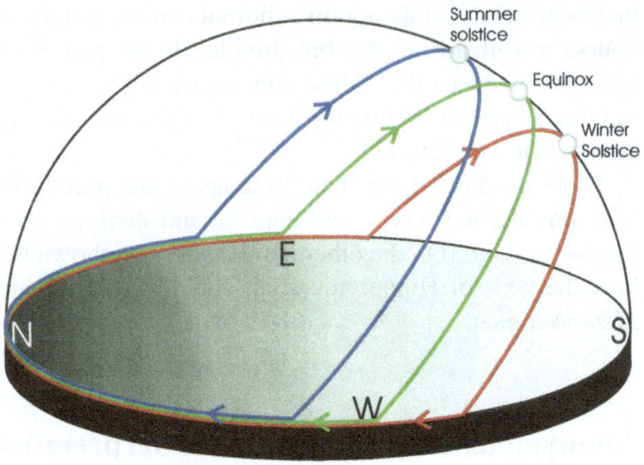

Fig. 7.13 The archaic conception of the paths of the sun

double bend in the paths of the heavenly bodies, during their rising and setting, is of a certain naivety, as Wöhrle dryly remarks.[71] That may be true, but this conception at least does not need such huge mountains as in Graham's and Boll's interpretations of Anaximenes' cap simile, but mountains of an average size. If we consider that not only the sun but all the heavenly bodies (except the circumpolar stars) set regularly at the horizon, it will be clear that, in this archaic conception, mountains are not only needed in the north, but on the entire periphery of the earth to explain their paths. In the same sense, Panchenko remarks as to the conception of the northern mountains, "none of the Presocratics (. . .) could have overlooked [the] fact", that "the idea is not workable for at least half a year,"[72] namely, when the sun does rises and sets on the southern half of the horizon. It must also be noted that Hippolytus (in text 7.1) does not speak of "northern mountains" but of "the higher parts of the earth," which could be interpreted as the mountains all around the periphery of the earth. These mountains could be said to be invisible being at a great distance from us (see text 7.1). Moreover, they can serve as a barrier to keep the water of the Ocean from flooding the earth.[73]

In the context of an archaic conception, the word περί is best suited to describe the paths of the heavenly bodies along the periphery of the earth between their settings and risings. This archaic conception is Kirk's final interpretation of Anaximenes cap simile: "The heavenly bodies do not pass under the earth, but (as in the pre-philosophical world-picture, where the sun, at least, floats round river Okeanos to the north) they move round it, like a cap revolving round our head."[74]And then: "Thus Anaximenes seems to have accepted the broad structure of the naïve world-picture."[75] Strangely enough, Kirk sees no problem in combining

[71]Wöhrle (1993), 74.

[72]Panchenko (2015), 420.

[73]In Chap. 3, Sect. *The Shape of the Earth*, we saw that in this archaic conception there must also be mountains at the outer edge of the Ocean to prevent the water from flowing off the earth.

[74]KRS, p. 156.

[75]KRS, p. 157.

the archaic world-picture with the cap simile, of which he asserts: "This image is scarcely likely to have been invented by anyone except Anaximenes."[76] It is not clear to me how an image that implies a double bend in the paths of the heavenly bodies can be compared with a cap revolving around our head. As far as I can see, the only way to preserve the archaic image of Fig. 7.13 as an explanation of how Anaximenes understood the movements heavenly bodies is to assume, as Fehling does,[77] that Hippolytus invented the simile of the cap turning around a head to make his account more vivid.[78] After all, it is remarkable that this striking image does not appear in other accounts.

The abovementioned mountains at the periphery of the earth are mentioned in an anonymous text about the paths of the heavenly bodies that has only recently been added to the doxography on Anaximenes:

7.14. [A polemic against earlier theories:] They construct walls in a circle [around the earth] so that they may screen us against the vortex, as it whirls around (περιφερο[μ]ένης) outside the earth, for all those who drive the heavenly bodies around (περιάγου[σι]ν) overhead ([ὑ]π[ὲρ κε]φά[λ]ης).[79]

The task of the peripheral mountains, in this text imagined as a wall, is described as to screen us from the heavenly vortex. Perhaps this can be understood a somewhat strange expression of the archaic belief, in which the setting sun and other heavenly bodies disappear behind mountains. The second half of the text specifies who are supposed to have introduced this ring of mountains, namely "all those who drive the heavenly bodies around overhead." It should be noted that both prepositions περί and ὑπέρ are present here, which could be interpreted as indicating both parts of the paths of the heavenly bodies, one above (ὑπέρ) the earth and the other around (περί) it (cf. again Fig. 7.13). Another possible interpretation, however, is that this text is concerned with the situation before the tilt of the celestial axis (see Fig. 3.5a), in which the heavenly bodies with the lower orbits can be said to turn around the earth, behind the peripheral mountains, and the heavenly bodies with the higher orbits can be said to turn overhead. In this interpretation, text 7.14 should be understood in relation with text 3.6, in which it is said that according to Archelaus, before the tilting of the heavens, the sun circled around the horizon and did not shine upon the earth, because it was invisible behind the raised edges of the concave earth.

---

[76]KRS, p. 156.

[77]Fehling (1985), 208 n. 33: "Solche hausbackenen Vergleiche gehören typisch zum etwas gönnerhaften Umgang des 4. Jahrhunderts mit den Vorsokratikern. Es gibt keinen Grund, den Vergleich Anaximenes zu geben."

[78]This would, of course, not implie that Fehling's interpretation of a flat heaven (cf. Fig. 5.9) should be followed. Moreover, and very inconsistently, Fehling (1985, 208 n. 33) sees the image of a tilted cap as an excellent illustration of Anaximenes' intentions.

[79]Epicurus, *On Nature* ĪĀ [33] Arrighetti, from Herculaneum Papyri 1042.8.vi = Gr Axs 20, not in DK, LM, TP2 (but see p. 243, n. 2), and KRS. Graham's translation inserts the word "and" (which is not in the Greek text), between the two halves of this text. This suggests a juxtaposition of two functions of the ring of mountains, whereas the last clause is meant to indicate who the people are that introduced the idea of a ring of mountains.

Perhaps a confirmation of the interpretation of the circular mountains behind which the setting heavenly bodies hide could also be read in yet another difficult text from the doxography on Anaximenes:

7.15. Anaximenes declares that the outermost periphery [of the heaven] is earthy (γηίνην).[80]

This is Mansfeld and Runia's controversial translation of Aëtius' text in the version of pseudo-Plutarch, in which they make "periphery" the subject of the sentence. It is also possible to make "heaven" the subject and read, "Anaximenes declares that heaven is the outermost earthy periphery,"[81] but this seems to make less sense. However, maybe Aëtius meant something else, for Stobaeus' version, which DK prefers, reads as follows:

7.16. Anaximenes and Parmenides say that the outer periphery of the earth (τῆς γῆς) is the heaven.[82]

Laks and Most's translation, "Anaximenes [...] the revolution farthest from the earth is the heavens," makes this text better understandable, but at the same time makes it look a truism.[83] We may conclude that Stobaeus' version would not support the interpretation of an earthy periphery of mountains.

Fehling also made another suggestion, which undermines the interpretation of Anaximenes as a representant of the archaic world-picture. He suggests that Hippolytus' text (text 7.1) is a rephrasing, adapted to Anaximenes, of Aristotle's more general statement about the paths of the heavenly bodies according to "the archaic meteorologists" (text 7.13).[84] I would add that possibly the same holds true for the other items in the doxography in which it is denied that, according to Anaximenes, the heavenly bodies go under the earth. In that case, not only Hippolytus, but also Eusebius, Stobaeus, and Diogenes Laërtius (or a source they used) thought that Aristotle was talking about Anaximenes and therefore they inserted the words οὐχ and δὲ ἀλλὰ in the text of Aëtius (cf. texts 7.1 and 7.8–7.10). This would also explain the unexpected appearance of the word περὶ instead of ὑπέρ in these texts. On the other hand, the word περί is used correctly both in Aristotle's text (7.13) and in that of pseudo-Plutarch (7.7). In Aristotle's text, it indicates the movement of the sun and the other heavenly bodies around the horizon. In pseudo-Plutarch, περί is used as a further qualification of the heavenly bodies, which are not only seen above the earth,

[80]P 2.11.1 = MR 434; not in Gr and KRS.

[81]Dox 339 and TP2 read τὴν περιφορὰν τὴν ἐξωτάτω γῆς εἶναι; Bottler (2014, 350) translates: "Anaximenes (die οὐσία des Himmels sei) der äußerste irdische Umkreis."

[82]S 1.23.1 = DK13A13 = LM ANAXIMEN. D12; not in Gr and KRS.

[83]Similar translations can be found in Dumont (1988), 54 and Reale (2017), 297.

[84]Fehling (1985), 208: "Seine Formulierung ist sonst der allgemeinen des Aristoteles so ähnlich daß man dahinstehen lassen muß, wieweit sie als individuelle Aussage über Anaximenes zu werten ist."

Fehling (1994), 145: "Er hat deshalb mangels eines anderen Kandidaten die oben zitierte Bemerkung des Aristoteles über die 'alten Meteorologen' auf Anaximenes bezogen und sich damit auch in der modernen Forschung durchgesetzt."

but also travel under (ὑπό) the earth and thus orbit around (περί) the earth. In the versions of Hippolytus et al., however, (texts 7.1, 7.8, 7.9, and 7.10), this περί is confusingly used as a denial of ὑπό, where one would have expected ὑπέρ. The ultimate confusion is found in Hippolytus' text, where he tries to explain the word περί by means of the cap simile.

In another anonymous text, the idea that the celestial bodies do not travel under the earth seems to be suggested by the image of a turning millstone:

7.17. [on the cosmos] some held that it turns around (περιδινεῖσθαι) like a millstone (μυλοειδῶς), others like a wheel (τροχός).[85]

Diels considered this text as derived from Aëtius,[86] and in this he is followed by Mansfeld and Runia, who hypothesize that it, together with a text on the Stoics and the extending and contracting cosmos, belongs to a separate chapter Περὶ κινήσεως κόσμου, which they number 2.2A.[87] Diels also proposes, and again followed by other scholars, that Anaximenes (millstone) and Anaximander (wheel) are meant.[88] Mansfeld and Runia, however, remark that "it is unlikely that A[ëtius] originally gave name-labels, for then S[tobaeus] would not have left it out."[89] The text is confusing, in the first place because, for Anaximander it is not the cosmos that turns like a wheel, but the heavenly bodies, and especially the sun and moon, are imagined as turning wheels. Secondly, it is perhaps too easily assumed that a millstone turned horizontally and could thus be used as an image for Anaximenes' alleged conception of the cosmos. A millstone could also be placed vertically, for instance in an olive press or, in general, in an animal-driven mill. This vertical stone could even have been made somewhat conical, so that it rolled around at a slight angle. I would like to add: if Hippolytus' cap simile dates back to Anaximenes, as some scholars assume, Aëtius would not also have used the image of the millstone to indicate the same motion. All in all, there are too many problems with this text to make use of it for interpreting Anaximenes.

[85]Theodoretus, *Graecarum affectionum curatio* 4.15–16 = DK 13A12 = TP2 As112 = Ar135 = MR 331, not in Gr, LM, and KRS. Wöhrle's translation, which says that the text is about the earth ("Erde") must be a slip of the pen, because the word περιδινεῖσθαι clearly has to do with the cosmos, and τροχός is Anaximander's word for the heavenly bodies.

[86]In DK 13A12, it is simply announced as a text from Aëtius and not from Theodoretus; this information is given only in Dox 329, at the bottom of the page).

[87]Cf. MR 333.

[88]See Dox 329 and, e.g., Conche (1991), 212, n. 40.

[89]MR 335.

Towards an Interpretation of Anaximenes' Cosmology

As far as I am concerned, the discussion in this chapter allows for three interpretive possibilities that would make Anaximenes' cosmology understandable within the context of contemporary Greek cosmology. In all three options, the word περί (around), which is used in several reports on Anaximenes to describe the motion of the heavenly bodies finds a natural explication. All three options would also imply that the cap simile cannot be ascribed to Anaximenes but was Hippolytus' invention, in a failed attempt to elucidate Anaximenes' intentions.

The first option would be to accept Kirk's interpretation, which makes Anaximenes a representative of the archaic world-view as discussed above and shown in Fig. 7.13. This interpretation would imply that Stobaeus and Eusebius had the right text of Aëtius (texts 7.8 and 7.9) and thus that the heavenly bodies do not go under the earth (in which they are supported by Diogenes Laërtius and Hippolytus, texts 7.10 and 7.1). It would also imply that the word περί which they use must be understood as "going around the earth's periphery," where the setting sun, moon and stars hide behind mountains that are as far away as possible on a flat earth. This interpretation has the charm of making sense of most of the relevant texts.

On the other hand, we must consider that Anaximenes was a younger fellow citizen of Anaximander, who was the first, as far as we know, to teach that the celestial bodies make full circles and thus also go under the earth. It is hard to believe that Anaximenes would have fallen back into the archaic world picture as rendered in Fig. 7.13. This consideration makes us look with some suspicion at reports which say that Anaximenes held that the celestial bodies do not go under the earth. Even more decisive is what we read in another doxographical report:

7.18. Anaximenes [says] the stars are fixed like nails to the crystalline [heaven] (τῳ κρυσταλλοειδεῖ), so as to form designs (ὥσπερ ζωγραφήματα). [According to some, however, (ἔνιοι δὲ) Anaximenes says they are fiery leaves].[90]

It is commonly agreed that the text of this report is distorted. I agree with Schwabl, who argues that the text between square brackets in the last line has been wrongly inserted in the manuscripts between τῳ κρυσταλλοειδεῖ and ὥσπερ ζωγραφήματα, where it disrupt the coherence of the text.[91] That is why, like Mansfeld[92] and Graham, I placed them at the end of the text. Moreover, I agree with Schwabl and Graham, that the word ἔνιοι refers "to sources that present Anaximenes' doctrine in a different way, rather than to other philosophers."[93] More specifically, with Schwabl,[94] I think that the "fiery leaves" do not refer to

[90] P 2.14.3 (= S 1.24.1) = DK 13A14 = LM ANAXIMEN. D14[3] = Gr Axs17 = MR 470 = TP2 As37 = KRS 154.

[91] See Schwabl (1966), 37.

[92] Mansfeld (1983), 97.

[93] LM in a note to ANAXIMEN. D14[3].

[94] Cf. Schwabl (1966), 36.

the stars, but to the sun (which is said to float like a leaf in text 4.3. quoted in Chap. 4, section *The Cosmic Tree*), to the moon, and also probably to some planets, all of which move (like floating leaves) between the fixed stars. The fixed stars, on the other hand, are grouped in pictures, with which of course the constellations are intended.

The stars that set at the western horizon and rise again in the same constellations at the eastern horizon must have passed under the earth, since they are fixed to the crystalline heaven. The implication of this is that Anaximenes cannot have meant that the sun does not go under the earth, for the orbit of the sun in its daily movement is parallel and similar to the daily movement of the stars. As for those scholars who believe that Anaximenes taught a hemispherical firmament in which the heavenly bodies do not go under the earth, either they simply do not notice the discrepancy with the report that says that the stars are like nails in the crystalline vault, or they should consider this report as erroneous. The latter option is advocated by Longrigg and Kirk, who suggest a backward projection from Empedocles.[95] Against this, Bicknell rightly argues that a confusion with Empedocles, a thinker from another tradition, is improbable.[96] Longrigg and Kirk also argue that a crystalline vault "is foreign to the little that is known of Anaximenes' cosmogony and to the other details of his cosmology."[97] Schwabl has exactly the opposite opinion: "Jedenfalls paßt der bildhafte Vergleich ganz zu dem, was sich über Anaximenes sonst noch ausmachen läßt."[98] It might be argued, however, that the idea of a crystalline heaven conflicts with the fiery nature of the heavenly bodies:

7.19. Anaximenes says the nature of the heavenly bodies (ἄστρα) is fiery (...).

This argument would make sense, if κρυσταλλοειδής were to be translated as "like ice," emphasizing coldness, as it is in KRS, and Bicknell,[99] but it can also be translated as "like crystal" or "crystalline," emphasizing transparency.[100] In my view, however, the real problem with the word κρυσταλλοειδής is not whether it conflicts with the fiery nature of the stars, but how to understand that the black heaven we see between the stars at night can be called "crystalline" at all. Ice is usually associated with white rather than black, and transparency also seems to be incompatible with the black color of the night sky. In any case, the crucial point for my argument is not the meaning of the word κρυσταλλοειδής, but the description of

---

[95] See KRS, p. 155 and Longrigg (1965), 249.

[96] See Bicknell (1969), 55.

[97] KRS, p. 155.

[98] Schwabl (1966), 34.

[99] See KRS, translation of 154 and argument on p. 155; Bicknell (1969), 54. Wöhrle (1993), 44 has "eisartig," but in TP2 As37 "kristallartig."

[100] Longrigg (1965), 249 states that "the word κρύσταλλος and its derivatives *never* in the Classical period refer to anything but *ice*." This does not, however, preclude the text from emphasizing the transparency of ice rather than its coldness.

the stars as fixed nails grouped in images, which precludes an archaic interpretation of their paths at night (as in Fig. 7.13).

In view of these considerations, I would like to suggest a second and a third possible interpretation of Anaximenes' cosmology. The second interpretive possibility, which I now prefer, is to simply accept pseudo-Plutarch's version as the best rendering of Aëtius' original text (7.6). In that case, Anaximenes taught, just like Anaximander, that the heavenly bodies go under the earth. Pseudo-Plutarch's text depicts the present situation with a tilted heaven and the celestial bodies passing under the earth while they are turning around (περί) the earth. In the context of ancient Greek cosmology, the implication of the idea of the tilted heaven is that in the past there has been a situation in which the heaven was not tilted and the celestial axis stood perpendicularly to the earth's surface. I agree with McKirahan that, according to Anaximenes, it was not the earth that was tilted, but the heaven (or the cosmos, or the celestial axis). In the Sect. *The Tilted Tree* of Chap. 4, I argued that the idea of a tilt of the celestial axis goes back to Anaximander. I think that this idea can also be linked to Anaximenes. Anaximenes' conception of the cosmos could, in this option, be rendered as in Fig. 7.14. We have no information of Anaximenes' ideas about the distances of the heavenly bodies, but he probably imagined a spherical heaven, onto which the stars were attached, and thus the sun and the moon (and the planets) turned around the earth either on the screen of the heaven or not very much below it.

This interpretive option implies that the reports which say that the heavenly bodies do not pass under the earth (texts 7.1 and 7.8–7.10) are corrupt, as argued above in the Sect. *A Fresh Look at the Doxography*. I also suggested, at the end of the previous section, that the origin of their texts could have been a misinterpretation of Aristotle's text (7.13) on the ancient meteorologists, which is actually about an archaic conception of the paths of the heavenly bodies, and not about Anaximenes.

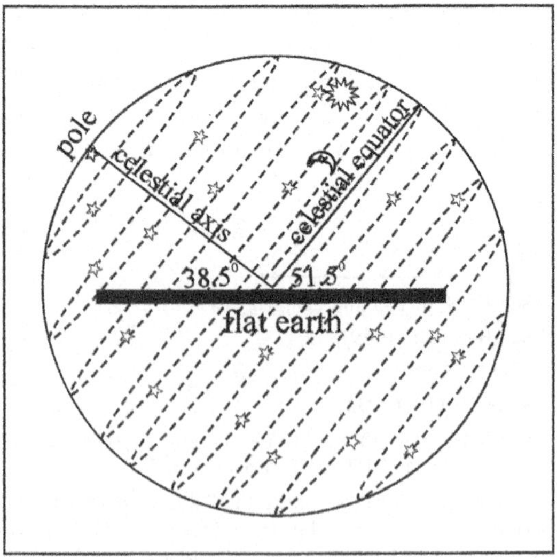

Fig. 7.14 The cosmos according to Anaximenes

In the third interpretive option, which I proposed in an earlier publication,[101] Anaximenes is supposed to have used the cap simile to describe what the heaven looked like in the original situation, before the tilt of the celestial axis. Nevertheless, these interpretations are not mutually exclusive. Rather, the third can be seen as supplementary to the second, a kind of extended version. In this option, we must accept that there are two different versions of the paths of the heavenly bodies according to Anaximenes. The reports which say that the heavenly bodies pass under the earth refer to the present situation (texts 7.6 and 7.7; see Fig. 3.5b). On the other hand, the reports on the image of the felt cap and the motions of the heavenly bodies above the earth would not refer to the *actual* situation, but to the *original* state before the inclination of the celestial axis (texts 7.1 and 7.8–7.10; see Fig. 3.5a). In Fig. 3.5a and b, not the whole cosmos is pictured, including the heavenly bodies under the earth, but only the celestial vault above the surface of the flat earth, as seen by an observer at the center of the earth's surface. The picture of the original state, Fig. 3. 5a, can be said to make perfect sense as an illustration of the image of the heavenly bodies that revolve like a felt cap on our head. However, when a person on the pole of the spherical earth tells us that the stars do not set but rotate around his head, this does not mean that he believes that there are no stars under the horizon, but only that he cannot see those stars. And when he tells us that during the summer the sun does not set but turns around his head, he does not mean that in winter there is no sun under the horizon, but only that then the sun is invisible. Similarly, in this interpretation, the implication of Fig. 3.5a is not that in the situation before the tilt of the heavens there were no celestial bodies under the flat earth, but only that they were not visible. The original situation, before the tilt of the heavens, depicted in Fig. 3.5a, is the same as what a person sees standing on one of the poles of a spherical earth: a hemispherical vault, turning horizontally around him. Wöhrle seems to have already acknowledged this. Having compared the *original* situation on a flat earth with the situation on the north pole of a spherical earth, he concludes "(damit) dreht sich natürlich das ganze Himmelsgewölbe—wie eine Kappe um den Kopf."[102] What an observer on the surface of a flat earth sees in the situation *after* the tilt of the heavens, are the slanted orbits of the celestial bodies as shown in Fig. 3.5b, of course not with the gap without stars above the southern horizon as in Figs. 7.4 and 7.7.

If we draw, in the original situation before the tilt of the heavens, the paths of the heavenly bodies under the earth, which are invisible to someone who lives on the surface of the flat earth, the picture looks like Fig. 7.15a. And when we make the celestial axis tilt in order to get the present situation, the resulting picture is Fig. 7.15b. These pictures exemplify Anaximenes' conception of the cosmos in the third interpretive option.

---

[101]See Couprie (2015a).

[102]Wöhrle (1993), 73.

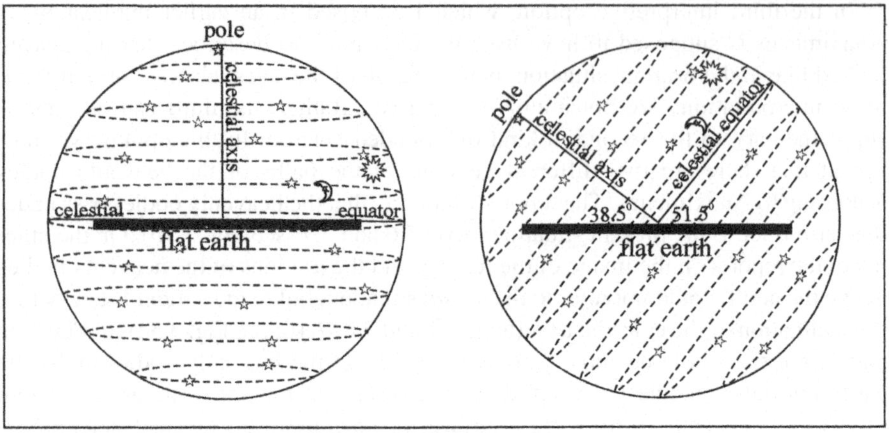

Fig. 7.15 (**a**) The original situation of the heavens. (**b**) The present situation after the tilt of the celestial axis

Concluding Remarks

As far as I can see, the possible interpretations proposed in the previous section, although they may not be as spectacular as attributing to Anaximenes all kinds of strange and observationally impossible cosmological ideas, are the only ones that make sense within the context of Presocratic flat earth cosmology. All other interpretations, as we have seen, are confronted with interpretive absurdities and oddities. Of the three options proposed in the previous section, the first one presupposes that Anaximenes had regressed into an archaic conception of the movements of the heavenly bodies and a hemispherical heaven. This is why I prefer the second option, which is in fact an improved version of Anaximander's world-picture, with a spherical heaven, a tilted celestial axis, and celestial bodies that pass under the earth. Curiously enough, I fully agree with Bicknell, who maintains "that Anaximenes regarded his star-studded heaven as a sphere" and rejects the view "that it was a hemispherical dome." It is worthwhile to quote him at length, because I agree with every word of it: "My reason for rejecting this view is that at Miletus, which lies roughly 37° north, the celestial equator and the planes of diurnal rotation of all the heavenly bodies are inclined to the plane of the horizon by an angle of 53°. To an observer at Miletus or anywhere near it, it would immediately be evident that the apparent movements of the fixed stars could not be explained on the supposition that they were attached to a hemispherical dome. Such an account would fit the facts of observation only at the terrestrial poles where the celestial pole corresponds to the zenith and the planes of the horizon and the celestial equator coincide. (...) the diurnal paths of the fixed stars are parallel to those of the rest of the luminaries, and therefore whatever Anaximenes said of the latter must have applied to the former too. (…) At Miletus, the sun, moon, and planets and the majority of the fixed stars appear to pass beneath the earth."[103] It is a pity that Bicknell spoiled these right observations and considerations by his strange interpretation of the cap simile which was discussed earlier in this chapter.

As already announced, I have postponed until Chap. 15 a discussion of Panchenko's interpretation, which would, in my opinion, make Anaximenes a complete outsider in Presocratic cosmology. Since his book has been published in Russian only,[104] my discussion of Panchenko's views are based on what is available in English. The reason for the postponement is that his interpretation presupposes knowledge of the ancient Chinese *gai tian* system, which will be explained in Chaps. 13 and 14. Panchenko's interpretation is, in a sense, akin to that of Fehling. The few lines in which this latter interpretation is proposed have already been discussed above. Moreover, Fehling does not make the Chinese connection. Kočandrle's article, which advocates an interpretation similar to that of Panchenko has so far only appeared in Czech.[105]

References[106]

Anderson, J.C., *The Christian Topography of Cosmas Indikopleustes*. Roma: Edizioni di Storia e Letteratura 2013.

Bicknell, P. J., Anaximenes' ΠΙΛΙΟΝ Simile, *Apeiron* 1 (1966), 17–18.

—, Anaximenes' Astronomy. *Acta Classica* 12 (1969) 53–85.

Boll, F., Footnote in: Aus einem Briefe des Herrn Proff. August Kopf an C. Bezold. *Zeitschrift für Assyriologie* 28.1914, 352–361, 361, n. 1.

Bottler, H., Pseudo-Plutarch und Stobaios: Eine synoptische Untersuchung. Göttingen: Vandenhoeck & Ruprecht 2014.

Conche, M., *Anaximandre. Fragments et Témoignages*. Paris: Presses Universitaires de France 1991.

Couprie, D.L., The Paths of the Celestial Bodies According to Anaximenes. *Hyperboreus* 21 (2015), 5–32.

Daiber, H., *Aetius Arabus: die Vorsokratiker in arabischer Überlieferung*, Wiesbaden: Steiner 1980.

Diels, H., *Doxographi Graeci*. Berlin: Walter de Gruyter 1879 (= Dox).

Diels, H. and W. Kranz, *Die Fragmente der Vorsokratiker*. Zürich/Hildesheim: Weidmann 1951[6] (= DK).

Dumont, J.-P., *Les Présocratiques*. Paris: Gallimard 1988.

Fehling, D., Das Problem der Geschichte des griechischen Weltmodells vor Aristoteles. *Rheinisches Museum für Philologie* 128 (1985), 195–231.

—, *Materie und Weltbau in der Zeit der frühen Vorsokratiker: Wirklichkeit und Tradition*. Innsbruck Verlag des Instituts für Sprachwissenschaft der Universität Innsbruck 1994.

---

[103] Bicknell (1969), 77.

[104] See Panchenko (2016).

[105] See Kočandrle (2018).

[106] P = Aëtius in pseudo-Plutarch, *Placita* (numbering according to Dox).
S = Aëtius in Stobaeus, *Anthologium* (numbering according to Wachsmuth and Hense).

Graham, D.W., *The Texts of Early Greek Philosophy*. Cambridge: Cambridge University Press 2010 (= Gr).
—, *Science Before Socrates*. Oxford: University Press 2013.
Guthrie, W.K.C., *A History of Greek Philosophy. Volume I. The Earlier Presocratics and Pythagoras*. Cambridge: Cambridge University Press 1962.
Hölscher, U., Anaximander und die Anfänge der europäischen Philosophie. *Hermes* 81 (1953), 257–277 and 385–418.
Hölscher, U., Anaximander and the Beginnings of Greek Philosophy. In: D.J. Furley & R.E. Allen (eds.), *Studies in Presocratic Philosophy. Volume I. the Beginnings of Philosophy*. London: Routledge & Kegan Paul 1970, 281–322.
Kirk, G.S., J.E. Raven and M. Schofield, *The Presocratic Philosophers*. Cambridge: Cambridge University Press 2009 (= KRS).
Kočandrle, R., Charakteristické aspekty Anaximenovy kosmologie [Characteristic Aspects of Anaximenes' Cosmology]. *Reflexe* 54 (2018), 19–42.
Kominko, M., *The World of Kosmas. Illustrated Byzantine Codices of the Christian Topography*. Cambridge: Cambridge University Press 2013.
Laks, A. and G.W. Most *Early Greek Philosophy VI*, Cambridge, MA. 2016 (= LM).
Liddell, H.G., R. Scott H.S. Jones and R. McKenzie, *Greek-English Dictionary*. Oxford: Clarendon Press 1996. Also available on the internet, e.g.: http://stephanus.tlg.uci.edu/lsj/#eid=1714&context=search (= LSJ).
Longrigg, J., Κρυσταλλοειδῶς. *The Classical Quarterly* 15 (1965), 249–251.
Mansfeld, J., *Die Vorsokratiker I*. Stuttgart: Reclam 1983.
Mansfeld, J. and D.T. Runia, *Aëtiana. The Method and Intellectual Context of a Doxographer, Volume Two, Part Two*. Leiden: Brill 2009 (= MR).
McKirahan, R.D., *Philosophy Before Socrates*. Indianapolis/Cambridge: Hackett Publishing Company, Inc. 2010.
Mourelatos, A.P.D., Parmenides, Early Greek Astronomy, and Modern Scientific Realism. In: J. McCoy, ed., *Early Greek Philosophy*, Washington, D.C.: The Catholic University of America Press 2013, 91–112.
Panchenko, D., Anaximenean Astronomy in the Light of Chinese Parallels.*Tsinghua Studies in Western Philosophy* 1(2) (2015), 412–426. Text of a paper read at the *Fourth* (not 14th, as mistakenly written in a note on p. 412) *Biennial Conference of the International Association for Presocratic Studies* at the Aristotelian University of Thessaloniki, June 30–July 4, 2014.
—, *On the Eastern Slope of Olympus: The Impact of Greek Ideas on Chinese Cosmology*. St. Petersburg: Academy of Sciences [Nauka], 2016 (in Russian).
Reale, G. (ed.), *I Presocratici. Prima traduzione integrale con testi originali a fronte delle testimonianze e dei frammenti nella raccolta di Hermann Diels e Walther Kranz*. Milano: Bompiani, Il pensiero occidentale 2017.
Robinson, J.M., *Introduction to Early Greek Philosophy*. Boston: Houghton Mifflin Company 1968.
Schwabl, H., Anaximenes und die Gestirne. *Wiener Studien* 79 (1966), 33–38.
Webster, T.B.L., *Greek Art and Literature 700–530 BC* New York: Praeger 1961.
Wöhrle, G., *Anaximenes aus Milet. Die Fragmente zu seiner Lehre*. Stuttgart: Franz Steiner Verlag 1993.
—, *Traditio Praesocratica. Die Milesier: Anaximander und Anaximenes*. Berlin: Walter de Gruyter 2012 (= TP2).

Chapter 8
Xenophanes' Cosmology

Contents

A Cosmological Quotation from Xenophanes' Poem

The flat earth cosmology of the Presocratics is not always easy to understand. The interpretation of Xenophanes' ideas on cosmology in particular is notoriously difficult. This has led to the suggestion that Xenophanes allowed himself a certain degree of fantasy and possibly humor and irony. Moreover, his bizarre original

D. L. Couprie, *When the Earth Was Flat*, Historical & Cultural Astronomy,
https://doi.org/10.1007/978-3-319-97052-3_8

statements would have led to misunderstandings on the part of the doxographers.[1] One might add: as well as confusion among modern scholars. Most of Xenophanes' cosmological teachings have come to us through the reports of later authors. Only one of them is documented by a quotation from his poem:

8.1 Γαίης μὲν τόδε πεῖρας ἄνω παρὰ ποσσὶν ὁρᾶται
ἠέρι προσπλάζον, τὸ κάτω δ' ἐς ἄπειρον ἱκνεῖται.[2]

This can be translated as:

8.2 The upper boundary of the earth is here at our feet, touching the air; the lower (boundary) reaches down without limit (or: underneath, the earth reaches down without limit; or: that what is under the earth reaches down without limit).[3]

At the end of this chapter, I will suggest another translation. The complex difficulties involved in interpreting of Xenophanes' cosmology can be readily illustrated by the way the just quoted text was understood by ancient authors.

Xenophanes' Text in the Interpretation of Aristotle, Achilles Tatius, Empedocles, Pseudo-Aristotle, and Simplicius

Aristotle, assuming that Xenophanes' text is to be understood as a contribution to the discussion on the stability of the earth, refers to these lines (text 8.1) when he discusses the question of whether the earth is at rest or in motion:

8.3 Some for this reason say the lower part is infinite (ἄπειρον τὸ κάτω) [saying that its roots go down infinitely (ἐπ' ἄπειρον ἐρριζῶσθαι)], as does Xenophanes of Colophon, in order that they will not have problems finding the explanation [for the earth's stability].[4]

As we will see, the metaphorical words "root" or "rooted," which are not in Xenophanes' text, were also used by several doxographers. Untersteiner has pointed out that the critical apparatus of Diels-Kranz is insufficient and that the metaphor of the roots of the earth, which I have put between square brackets, is missing in the best manuscripts of *De Caelo*.[5] However, when he states that Xenophanes' own

---

[1]Cf. KRS, p. 175.

[2]Achilles Tatius, *Introductio in Aratum* 4 = DK 21B28 = LM XEN. D41 = Gr Xns52 = KRS 180.

[3]In this text, Graham translates "earth" and "air," without articles. However, Xenophanes' omission of the articles has to do with the requirements of meter.

[4]Aristotle, *De Caelo* 294a22–25 = DK 21A47 = LM XEN. R13a = Gr Xns53; not in KRS, but see p. 175.

[5]Cf. Untersteiner (2008, CLIV, n. 97). See also Mourelatos (2002, 333), and Mueller (2005, 115, n. 295). LM notes that these words are lacking but nevertheless translates them; Gr has the whole text including these words.

words guarantee that the text between square brackets is authentically Aristotelean, I do not agree. Mueller argues that Simplicius, in the text also discussed below, must have had Aristotle's text without the metaphor, because "he makes no mention of the unusual word *errizôsthai*, which implies that the earth stretches down infinitely."[6] Achilles Tatius is the only ancient author who follows Aristotle's interpretation in its two main points: that the problem at issue was the stability of the earth, and that, therefore, the earth extends infinitely downwards, but he does not use the metaphor of the earth's roots. In the introductory lines to the direct quotation of Xenophanes, he writes:

8.4 Xenophanes does not think that the earth is suspended, but rather that it extends infinitely downwards (κάτω εἰς ἄπειρον καθήκειν), for he says: (...).[7]

In my opinion, it is decisive that in Xenophanes' own text not the verb ἐρριζῶσθαι but a form of ἱκνέομαι is used. This strongly suggests that the metaphor of the earth's roots was not used by Xenophanes himself in this context. It remains uncertain whether some doxographers borrow this metaphor from a version of Aristotle's text in which it was included or from some other source. My guess is that the metaphor of "rooting" goes back to Theophrastus and was later inserted into Aristotle's text.

Aristotle quotes Empedocles as a witness to his interpretation that Xenophanes was dealing with the problem of the stability of the earth:

8.5 (...) that is why Empedocles is so critical of him, saying, "if the depths of the earth (γῆς βάθη) and the wide aether are boundless (ἀπείρονα), as the words coming vainly through the tongue of many mouths are poured out, of those who have little seen of the totality."[8]

Empedocles, however, does not mention the problem of the stability of the earth, but seems to criticize Xenophanes for calling the depths of the earth (whatever he might have meant by this expression) and the aether "boundless." Empedocles' verses are not only interesting because he adds the boundlessness of the aether, while Xenophanes mentions the air without calling it "boundless," but also because he speaks of the depths of the earth, which appears to be his interpretation of Xenophanes' τὸ κάτω. The combination of the boundless depth of the earth and that of the air returns in a text of pseudo-Aristotle:

---

[6] Cf. Mueller (2005, 115, n. 295).

[7] Achilles Tatius, *Introductio in Aratum* 4, not in DK, LM, Gr, and KRS, but see Heitsch (1983, 62) (Greek text without translation). Cf. Mourelatos (2002, 332).

[8] Aristotle, *De Caelo* 294a26–28 = DK 31B39 = LM EMP. D113 = Gr Xns53; not in KRS, but see p. 175. Gr translates "plentiful heaven," while LM has "vast aether," and Untersteiner "l'immenso etere," which seems the better translation. Cf. LSJ, δαψιλός, referring to δαψιλής 2, "of space, ample, wide," and not to δαψιλής 1, "abundant, plentiful."

8.6 As also Xenophanes says that the depth of the earth (τό τε βάθος τῆς γῆς) and that of the air are boundless.[9]

In Aristotle's text (8.3) it is not immediately clear what should be inserted as the subject of ἄπειρον τὸ κάτω, which words obviously are linked to τὸ κάτω δ' ἐς ἄπειρον in Xenophanes' text (8.1), of which it is also not immediately clear what τὸ κάτω is supposed to refer to. Simplicius, in his Commentary on Aristotle's *De Caelo*, apologizing for not having been able to find Xenophanes' text to which they refer, wonders what Aristotle could have meant by τὸ κάτω and Empedocles by γῆς βάθη:

8.7 (...) Xenophanes of Kolophon declared that what is beneath the earth is infinite and as a result it remains at rest. Since I have not encountered words of Xenophanes dealing with this subject, I do not know whether he said that the underneath part of the earth (τὸν ὑποκάτω τῆς γῆς) is infinite and as a result remains at rest or that the region under the earth and aether is infinite so that[10] the earth falls ad infinitum (ἐπ' ἄπειρον καταφερομένην) and seems to be stationary. For Aristotle does not make this clear and the words of Empedocles do not determine this clearly. Since "depth of the earth" (γῆς βάθη) might also mean that into which it descends.[11]

We can readily assume that Simplicius, whose commentary is always clever, would have interpreted the last words of Xenophanes above quoted text 8.1 (if he could have found it) as follows:

8.8 (...) that which is *under the earth* (τὸ κάτω) goes on without limit.

Since he could not find the quotation, Simplicius relies on the text of Aristotle (8.3) and accepts that Xenophanes' intention (in 8.1) was to prevent the earth from falling, which was Aristotle's main concern. We might agree with Simplicius that Aristotle's words τὸ κάτω and Empedocles' words γῆς βάθη may indicate the depths under the earth, without sharing Aristotle's supposition that Xenophanes was speaking here about the problem of the stability of the earth and thus without accepting Simplicius' explanation that the earth must be thought of appearing stable just because it is falling *ad infinitum* (see text 8.7). If we look at Xenophanes' words unbiased by Aristotle's interpretation, it is certainly not impossible to understand them as expressing that either the underside of the earth or whatever there is under the earth is unfathomable. It also seems plausible that Empedocles was right in

---

[9]Pseudo-Aristotle, Pseudo-Aristotle, *De Melisso Xenophane Gorgia* 976a32 = DK 30A5(22); not in LM, Gr, and KRS, but see Dumont (1988, 303).

[10]Lesher (1992, 219) abusively inserts here the words, "the earth extends to infinity and" into his translation.

[11]Simplicius, *In Aristotelis De Caelo Commentaria* 522.7 (my italics) = DK 21A47 (missing the first sentence) = LM XEN. R13b (missing the first sentence and the last line); not in Gr and KRS. Mueller (2005, 63) reads the last words as: "since 'depths of earth' might also mean the things into which it descends."

stating that Xenophanes wanted to express that the "depths of the earth" mirror the "wide aether" or in more mundane words, that the one is as unfathomable as the other. The reason why I use the word "unfathomable" instead of "boundless" here will become clear at the end of this chapter.

Lesher takes a somewhat strange stand on this issue. On the one hand, he translates in Aristotle (text 8.5): "the (region) below the earth," but on the other hand, commenting on Simplicius' text (8.7), he maintains that: "Fragment 28 [= text 8.1] makes it virtually clear that the former [sc. "that the underneath part of the earth is infinite" or, in Lesher's own rendition, "the part of the earth that lies below is infinite"] (in some sense of ἄπειρον) was really Xenophanes' view," thereby excluding that Xenophanes meant the region below the earth.[12]

Xenophanes' Text in the Interpretation of Aëtius, Strabo, and Cicero

Other ancient authors do not mention the problem of the earth's stability that was so important to Aristotle. They seem to understand that it is not the earth itself that is extended infinitely downwards, but that the earth is rooted in the infinite:

8.9 Xenophanes says that the earth from its lower part down (ἐκ τοῦ κατωτέρου μέρους) is rooted in an unlimited depth (εἰς ἄπειρον βάθος ἐρριζῶσθαι).[13]

8.10 (According to) Xenophanes, (the earth is) first—for it is rooted in the infinite (εἰς ἄπειρον γὰρ ἐρριζῶσθαι).[14]

8.11 (. . .) no such revolution (viz. of the celestial bodies) could take place if the earth were rooted to an infinite depth (ἐρριζωμένης ἐπ' ἄπειρον).[15]

8.12 (. . .) in order to see whether the earth is fixed deep down (*penitusne defixa sit*) and is fastened as if by its roots (*radicibus suis*) (. . .).[16]

Whereas Strabo says (in 8.11) that according to Xenophanes the earth "is rooted" ἐπ' ἄπειρον, Aëtius uses (in 8.9 and 8.10) εἰς ἄπειρον, the same words as Xenophanes (in 8.1). Immediately after the words εἰς ἄπειρον in text 8.9, the manuscripts hesitate between βάθος (depth) and μέρος (part). Laks/Most choose for βάθος. Diels has μέρος, but puts it between square brackets, which means that he proposes to read only εἰς ἄπειρον. For this, he refers to Aëtius in pseudo-Plutarch (text 8.10) and to

---

[12]Lesher (1992, 219 and n. 63).

[13]P 3.9.4 (not in S) = DK 21A47 = LM XEN. D42; not in Gr and KRS. Lesher (1992, 219) translates, "is rooted in an infinite portion."

[14]P 3.11.2 (not in S) = DK 21A47 (abusively numbered II 11.2) = LM XEN. D43; not in Gr and KRS.

[15]Strabo, *Geographica* 1.1.20, not in DK, LM, Gr, and KRS.

[16]Cicero, *Academica* 2.39.122 = DK 21A47, not in LM, Gr and KRS.

Aristotle (text 8.3).[17] Lesher, obviously preferring the reading with μέρος, translates in text 8.9, somewhat confusingly: "[the earth] from its lower portion is rooted *in an infinite portion*."[18]

Xenophanes' Text in the Interpretation of Diogenes of Oinoanda, Hippolytus, and Pseudo-Plutarch

Some ancient authors give it yet another twist, assuming that the infinity of the earth has not so much to do with the earth's stability, nor with the idea that the earth is rooted in an infinite depth, but with the idea that the earth is not surrounded by air, which implies that the heavenly bodies do not rise and set and therefore do not go under the earth. Xenophanes certainly taught that the celestial bodies do not go under the earth, as we learn elsewhere in the doxography, but in his quoted verses (text 8.1) he does not link it to the air not surrounding the earth.

8.13 What do you mean, gentlemen, when you think fit to explain the earth in this way as boundless? Do you limit the earth throughout its length from above, circumscribing it with a vault of sky, and from that starting-point do you extend it indefinitely into the region below (εἰς τὴν κάτω ζώνην ἐπ' ἄπειρον ἐκτείνετε), dismissing the unanimous opinion of all men, both laymen and philosophers, that the heavenly bodies pursue their courses round the earth both above and below, and withdrawing the sun sideways outside the cosmos and reintroducing it sideways?[19]

8.14 The earth is boundless (ἄπειρον εἶναι) and surrounded (περιέχεσθαι) neither by air nor by heaven.[20]

8.15 He declares also the earth is boundless (ἄπειρον εἶναι) and not surrounded everywhere (κατὰ πᾶν μέρος μὴ περιέχεσθαι) by air.[21]

Although Xenophanes is not mentioned in the text of Diogenes of Oinoanda (8.13), the first lines probably refer directly to him. At the end of text 8.12, Diels refers to Diogenes' text but does not quote it. As for the text of pseudo-Plutarch (8.15), Mourelatos notes that Diels' reading is not that of the manuscripts, which have κατὰ πᾶν μέρος μὴ περιέχεσθαι and not μὴ κατὰ πᾶν μέρος περιέχεσθαι.[22] It

---

[17]Cf. Dox 376, n. at line 15.

[18]Lesher (1992, 219, my italics).

[19]Diogenes of Oinoanda, fr. 66, not in DK (but see the reference in the last line of 21A47), LM, Gr, and KRS.

[20]Hippolytus, *Refutatio Omnium Haeresium* 1.14.3 = DK 21A33(3) = LM XEN. D22[3] = Gr Xns59(3) = KRS 175.

[21]Pseudo-Plutarch, *Stromata* 4 = DK 21A32 = LM XEN. D23 = Gr Xns58, not in KRS. The negation μὴ must be read before περιέχεσθαι and not before κατὰ, as in DK. See Mourelatos (2002, 333).

[22]Mourelatos (2002, 333).

not clear to me why Mourelatos twice, less precisely, translates περιέχεσθαι as "contenue" instead of "environnée" or "entourée". He states that the doxographers tried to understand Xenophanes as teaching that the earth not only has infinite roots, but also that the surface of the earth is an infinite plane.[23] As such this is a bold conclusion from these texts alone, but Mourelatos intends to show that Xenophanes' original words (in 8.1) say the same thing. I will discuss his claim further on, in the section *Xenophanes' Text in the Interpretation of Mourelatos*.

Xenophanes' Text in the Interpretation of Some Recent Authors

The result of the investigation in the previous section is that there are at least three possible ways to understand Xenophanes' words τὸ κάτω: (1) "the lower part of the earth,' reading τὸ κάτω [sc. γαίης; *genitivus subiectivus*];" (2) "that what is beneath the earth," also reading τὸ κάτω [sc. γαίης; *genitivus loci*]; (3) "the lower limit of the earth," reading τὸ κάτω [sc. πεῖρας]. Schäfer translates τὸ κάτω [sc. πεῖρας] as "das untere Ende [eigtl. 'Grenze'']," but takes a somewhat strange position when he writes: "Xenophanes (…) sieht die Erde als Mittelpunkt zweier unendlicher Räume; der erste erstreckt sich von ihr aus nach oben und ist Luft (…). Nach unten erstrecken sich die Enden der Erde ins Grenzenlose" (and similar words on the next page).[24] It is not clear to me how the earth could be both the center and infinitely extended downwards. Barnes envisaged Xenophanes' earth as an infinite column,[25] as I did myself, comparing our position on top of the earth with Simeon the Stylite sitting on his pillar.[26] Drozdek wonders whether Xenophanes' earth should be thought of as an infinite column and he cares about winds that could possibly make it topple. That is why he accepts Mourelatos' interpretation, of which I will come to speak immediately, of an earth that is also extended infinitely sideways.[27]

Several authors who interpret Xenophanes' text as the opposition of two limits of the earth, the upper one in the first line and the lower one in the second line, are troubled by the strange combination of πεῖρας and ἄπειρος. For example, Lesher, who on the one hand translates: "the upper limit of the earth (…) that below,"[28] remarks on the other hand: "Since there is no lower limit to Xenophanes' earth it would be odd for τὸ κάτω to mean 'the lower limit' and for it (i.e. the limit) to 'reach without limit,' although 'this limit' in line one and the μὲν-δὲ structure seems to call

[23]Mourelatos (2002, 334).
[24]Schäfer (1996, 140).
[25]Barnes (1982, 27).
[26]Couprie (2011, 166).
[27]Drozdek (2008, 30).
[28]Lesher (2001, 128).

for just a contrast of two limits."[29] Instead of seeing a logical problem or suggesting a physical solution, Heitsch defends a more poetic position. For him it is certain that the antithesis πεῖρας—ἄ-πειρον is intentional, which means that this *contradictio in terminis* ("unbounded boundary") is used to express an unimaginable depth. In the same sense, Homer and Hesiod spoke of the boundaries of the earth, both at the end of the Ocean and in the depths of Tartaros.[30] This interpretation can be seen as a further explication of Diels' suggestion: "ἄπειρον *indefinitum*, nicht *infinitum*."[31] In the final analysis, and referring to Diels' note, Lesher seems to incline towards this interpretation. Initially, he is somewhat hesitant and writes, rather confusingly: "ἐς ἄπειρον (...): minimally translated as '*indefinitely*' or 'without limit' (...) and interpreted either as truly infinite extent or *merely indefinitely large*."[32] This would be more consistent if the words "or 'without limit' ... as truly infinite extent or" had been left out and replaced by a comma. On the next pages, however, Lesher adds: "Other aspects of Xenophanes' teaching point in the general direction of an earth of '*indefinite*' or '*indeterminate*' depths. (...), what Xenophanes could affirm with considerable basis in observed fact was that—for all locations—the earth below our feet stretches down *indefinitely* far."[33]

Xenophanes' Text in the Interpretation of Mourelatos

Mourelatos has made the above quoted lines from Xenophanes' poem (text 8.1) the subject of close reading. The first line, and especially the word τόδε, Mourelatos says, indicates the one and only ("unique") cosmic boundary: the earth at one side and the air at the other.[34] In other words, the surface of the earth is the boundary that separates the two cosmic realms of the earth and the air. This interpretation has already been suggested by Dicks[35] and West,[36] and earlier by Tannery,[37] although all three without further arguments. More recently, similar interpretations are brought forward by

---

[29]Lesher (2001, 128, n. 4).

[30]Heitsch (1983, 162). Somewhat illogically, Heitsch adds "Mit anderen Worten: Nach unten gibt es keine Grenze." What Homer and Hesiod (and Xenophanes) wanted to express is only that we do not know where the boundaries of sea and earth are.

[31]DK, note at 21B28.

[32]Lesher (2001, 128, n. 4 and 129, n. 5, my italics).

[33]Lesher (2001, 130, 131); '*finite*' Lesher's italics, the others mine.

[34]Mourelatos (2002, 334): "Ainsi, le premier verse insiste clairement sur l'existence d'une frontière unique (τόδε): la Terre d'un côté, et l'air de l'autre."

[35]Dicks (1970, 48): "(...) the earth, which he regarded as boundless in extent."

[36]West (1971, 229): "(...) this boundless earth to the north and south of us." It is strange that West only mentions the latitudes of the earth.

[37]Tannery (1882, 630): "La terre, plate, n'a point de limites ni de côté ni en dessous."

Schäfer, Drozdek, and Graham. Tannery,[38] Schäfer[39] and West[40] drew the consequence already articulated by Empedocles (text 8.5), that in this interpretation the heaven above the earth must also be considered infinite. Similarly, Mourelatos speaks of "le double apeiron de la Terre en-dessous et de l'air au-dessus."[41] On the other hand, however, he speaks of "the inverted bowl of the heavens," compares it with the Dome of the Capitol in Washington calling it a "tholos," and states that Xenophanes' cosmology can be understood as an adaptation of that of Anaximenes and his cap simile, all of which strongly suggest a finite heaven and, accordingly, an earth that is not infinitely extended sideways.[42]

After his extensive analysis of Xenophanes' text (8.1), Mourelatos proposes a somewhat free rendering that would better express the meaning of Xenophanes' words:

8.16 The boundary of the earth is this: up here, visible at our feet,
touching the air; below, the earth reaches down without limit.[43]

According to Mourelatos, the main contrast between the two lines is not between πεῖρας and ἐς ἄπειρον. The word ἄνω in Xenophanes' lines stands metrically apart from τόδε πεῖρας and is contrasted with τὸ κάτω, so we must read: "the boundary of the earth is this: *up here* (...); *below* etc."[44] Graham, who says to follow Mourelatos' interpretation, ignores the separation between τόδε πεῖρας and ἄνω and translates: "The upper boundary of earth is visible here at our feet."[45] In Mourelatos' view, the text expresses that there is only one boundary, namely the earth's surface that we see under our feet. This implies not only that the earth extends downwards without a limit, but also that the earth has no limits at the periphery but stretches out infinitely in the horizontal plane. If Xenophanes would have thought that the earth had a finite surface, he would have expressed this in so many words, Mourelatos argues.[46] This argument, however, could also easily be raised the other way around: if Xenophanes would have held that the earth has an infinite surface, he would have expressed it in so many words, all the more so as it conflicts with the ideas of Anaximander and Anaximenes. Mourelatos sees confirmation of his interpretation in two ancient

---

[38]Tannery (1882, 630): "(...) au-dessus, l'air est également infini."

[39]Schäfer (1996, 140): "Xenophanes sieht also offenbar die Erde als Mittelpunkt zweier unendlicher Räume: der erste erstreckt sich von ihr aus nach oben und ist die Luft (...)."

[40]West (1971, 228): "Seeing no evidence for a solid heaven, he substitutes infinite air from the earth upwards." On which text the first part of this sentence is based I do not know.

[41]Mourelatos (2002, 348).

[42]Mourelatos (2002, 348; 2008, 146).

[43]My translation of Mourelatos (2002, 334): "La limite de la Terre est ceci: ici au-dessus, visible à nos pieds, poussant contre l'air. En dessous, la Terre ne cesse d'avancer (plus profondément) sans limite." See also Graham (2013, 70).

[44]Cf. Mourelatos (2002, 334), and Xenophanes' two hexaneters divided into into four cola each at 332.

[45]Graham (2013, 70).

[46]Cf. Mourelatos (2002, 336).

commentaries, cited above (8.14 and 8.15), in which it is said that the earth is boundless and not surrounded by air: "Très probablement, la doxographie, derrière les résumées cherchait à comprendre que la Terre de Xénophane avait une étendue infinie à la fois en surface et en profondeur."[47] Mourelatos concludes that in this interpretation, all movements of celestial bodies that would make them go wholly or partly under the earth are excluded.[48]

Mourelatos' conviction that both Xenophanes' text (8.1) and those of Hippolytus and pseudo-Plutarch (8.14 and 8.15) imply the idea of a boundless earth in the horizontal plane is articulated by the pregnant translation of τόδε (in text 8.1) as "unique." This is, however, too heavy a burden for such a simple word. The consequence of this interpretation is that, in a sense, the boundless extent of the earth's surface is expressed by pointing to the earth's boundary. In my opinion, a more cautious observation would be that the limits of the earth in the horizontal plane are simply not mentioned in Xenophanes' text. Perhaps we could even venture a wild guess, reject Diels' generally accepted emendation ἠέρι and instead retain the words καὶ ῥεῖ of the manuscripts. One could take this to mean that the earth at its periphery touches (προσπλάζον) the circumfluent Oceanus.[49] This would imply that the earth is not extended infinitely sideways. It might be argued that Oceanus is meant in a genuine fragment of Xenophanes in which it is said that the great Sea is the originator of clouds. This could be relevant, because, as we shall see, according to Xenophanes, the heavenly bodies are a special kind of clouds that originate at the periphery of the earth:

8.17 The Sea is the source of water, the source of wind;
for neither <would there be wind> without the great Sea,
nor currents of rivers nor rainwater from the sky,
but the great Sea is the begetter of clouds, winds,
and rivers.[50]

One could even think of a kind of cybernetic circular process if a text of Diogenes Laërtius is also considered:

8.18 The clouds are formed by the sun's vapor raising and lifting them up to the surrounding (air) (εἰς τὸ περιέχον).[51]

However, as far as I know, no one has read text 8.1 in this way. A safer way to investigate the sustainability of Mourelatos' interpretation is to look at the

[47]Mourelatos (2002, 333–334).

[48]Mourelatos (2002, 336).

[49]Homer uses προσπλάζω for a wave that approaches the coast (*Ilias* 12, 285), or for the water that reaches to Tantalus' chin (*Odyssea* 11.583), but I cannot see why, in the meaning of "reaching to," it could not be used the other way around as well.

[50]Geneva scholium on *Ilias* 21.196 = DK 21B30 = LM XEN. D46 = Gr Xns55 = KRS 183.

[51]Diogenes Laërtius, *Vitae Philosophorum* 9.19 = DK 2A1(19) = LM XEN. D24 = Gr Xns1(19); not in KRS. The words "up to the surrounding (air)" are strange, because elsewhere we read that the air does not surround the earth (cf. texts 6.14 and 6.15).

consequences of the alleged boundlessness of the surface of the earth. This requires a preliminary investigation of the heavenly bodies and their movements according to Xenophanes.

The Nature and Movements of the Celestial Bodies

For the understanding of the movements of the sun the following texts are crucial, although their meaning is sometimes difficult to grasp:

8.19 Xenophanes [says the heavenly bodies (ἄστρα) come to be] from incandescent clouds (ἐκ νεφῶν πεπυρωμένων); being quenched every day they flare up at night, like coals; their risings and settings are kindlings and quenchings (ἐξάψεις καὶ σβέσεις).[52]

The heading of the chapter of Aëtius from which this text is taken reads: "What is the substance of the heavenly bodies, both planets and fixed stars" (Τίς ἡ οὐσία τῶν ἄστρων, πλανήτων καὶ ἀπλανῶν).[53] Here, the word ἄστρα clearly means "heavenly bodies," subdivided into planets and fixed stars. The ancient Greeks counted the sun and moon among the planets, and these two are the only ones mentioned in the doxography on Xenophanes. The second clause of 8.19, however, mainly has to do with the stars (but also with the not mentioned planets, Mercury, Venus, Mars, Jupiter, and Saturn), which become visible at night. According to this clause and the third, the becoming invisible of the stars and the setting of the heavenly bodies in general is understood as "quenching," and the rising of all heavenly bodies in general as "kindling." Mourelatos argues that the becoming visible of the stars by night is intentionally not called "kindling" but "flaring up."[54] We may wonder, however, whether this is not one subtlety too many. As already said, Aëtius text (8.19) is from his chapter about "the nature of the heavenly bodies, both planets and fixed stars." In this text, both the disappearances of the stars by day and when they set are called "quenchings," and the reason why the word "kindling" is replaced by "flaring up" in the case of the disappearances of the stars by day, is that the latter word fits better with the comparison of stars with coles. The last clause is missing in Theodoretus' version.[55] Achilles Tatius uses slightly different wordings that are worth quoting:

8.20 Xenophanes says the stars come to be from incandescent clouds and (he says that) they are quenching and glowing up as if they are coals. When they are

[52]P 2.13.14 (=S 1.24.1) = DK 21A38 = LM XEN. D36 = Gr Xns60 (referring abusively to (DK) 8 instead of 38) = MR 454; not in KRS, but see p. 173.

[53]Cf. MR 452 and 468.

[54]Cf. Mourelatos (2002, 338–342).

[55]Cf. Theodoretus, *Graecarum Affectionum Curatio* 4.19 = LM XEN. D36a = MR 456; not in DK, Gr, and KRS.

kindled, in our imagination it looks like their rising and when they are quenched their setting.[56]

Several texts give us more information about the moon and sun:

8.21 Xenophanes says [the moon] is an (incandescent) condensed cloud.[57]

In this text, the reading of the adjective belonging to "cloud" is controversial. DK reads πεπιλημένον ("felted," "condensed"), as in one of the manuscripts of pseudo-Plutarch and in Stobaeus. Graham adds "incandescent" (πεπυρωμένον), so that we should have to read "the moon is an incandescent condensed cloud." This emendation with two qualifications of the noun "cloud" results from Runia's meticulous examination of this text.[58] His conclusion is that "from the viewpoint of the systematic structure the majority reading νέφος πεπυρωμένον must be preferred."[59] He also suggests, however, that "it is plausible that the other participle was added by way of further differentiation."[60] Theodoretus uses the single qualification πεπυρωμένος in one sentence for both the sun and the moon:

8.22 To be sure, Xenophanes says that both the sun and the moon are incandescent (πεπυρωμένα) clouds.[61]

The header of the chapter of Aëtius, into which the next doxa (6.23) belongs, reads "On the moon's eclipse" (Περὶ ἐκλείψεως σελήνης), but this text is clearly about the phases of the moon.

8.23 Xenophanes [says] its monthly disappearance too comes about as a result of quenching (κατὰ σβέσιν).[62]

Mansfeld and Runia remark: "note again the confusion between eclipses and phases."[63] They probably refer to a text (8.27) on an eclipse of the sun, in which, obviously, the setting of the sun is meant. This text will be discussed below, where I will argue that what sounds like a confusion to us does not necessarily have to be so in the context of Xenophanes' thinking.

In texts 8.24–8.26, the mechanism of how each day a new sun originates and becomes an incandescent cloud is elaborated further than the simple "kindling" of text 8.19, which is used there for the heavenly bodies in general.

---

[56] Achilles Tatius, *Introductio in Aratum* 11 = LM XEN. D36b = MR 458; not in DK, Gr, and KRS.

[57] P 2.25.4 = S 1.26.1 = DK 21A43 = LM XEN. D29a = Gr Xns67 = MR 572; not in KRS.

[58] See Runia (1989, 245–269) and MR 580 and 586.

[59] Runia (1989, 265, cf. 267). On the same page, the strange word πεπυρωλημένον in one of the manuscripts is explained as a form of parablepsis of the two qualifications, πεπυρω<μένον πεπι>λημένον.

[60] Runia (1989, 267).

[61] Theodoretus, *Graecarum Affectionum Curatio* 4.21; cf. MR 577 and Bottler (2014, 447).

[62] S 1.26.3 (not in P) = DK 21A43 = LM XEN. D29c = GR Xns69 = MR 614; not in KRS.

[63] MR 618.

8.24 He says that the sun is produced by the collecting together of many small fires (ἐκ μικρῶν καὶ πλειόνων πυριδίων) (...) and that the sun and the heavenly bodies come to be from clouds.[64]

8.25 The sun comes to be every day from tiny flares (ἐκ μικρῶν πυριδίων), gathered together, (...) and there are numberless (ἀπείρους) suns and moons.[65]

8.26 Xenophanes [says] the sun comes from incandescent clouds (ἐκ νεφῶν πεπυρωμένων). Theophrastus in the *Physics* has written that the sun is composed of tiny flares (ἐκ πυριδίων) being gathered together from the moist evaporation (ἐκ τῆς ὑγρᾶς ἀναθυμιάσεως), and gathering together the sun.[66]

I read the second sentence of 8.26 as meaning: "Theophrastus has written that, *according to Xenophanes*, etc."[67] Mourelatos has argued that Xenophanes cannot have used the word πυριδίον because in none of its grammatical tenses it would fit his meter.[68] We have to assume that the doxographers had read something they thought could be rendered as πυριδία.

The header of the chapter of Aëtius, into which the next doxa (8.27) belongs, reads "On the sun's eclipse" (Περὶ ἐκλείψεως ἡλίου), but Diels already notes that in the first line of 8.27, instead of a solar eclipse the setting of the sun must be meant, because the rest of the sentence is about the rising sun.

8.27 Xenophanes [says an eclipse (ἔκλειψις) of the sun (or rather: the setting of the sun)] is a quenching (κατὰ σβέσιν), and another (new, different: ἕτερον) [sun] in turn comes to be in the east. He also reported in passing an eclipse of the sun that lasted a whole month, and indeed a total eclipse, in which day appeared as night.[69]

Diels' interpretation of the first line of this text has been followed by most commentators.[70] The issue has a more general bearing, as Laks & Most rightly point out: "This chapter of Aëtius is about eclipses, but this explanation bears rather on sunset. *The important point for Xenophanes seems to have been disappearance in general*."[71] We already saw that in text 8.23 the phases of the moon were called "quenchings" and that they appeared in Aëtius' chapter on eclipses. Nowadays, we look at eclipses, moon phases, the settings of the sun, moon, planets, and stars, and the disappearance of stars in the daylight as completely different phenomena, but according to Xenophanes they are all characterized as quenchings. More specifically, lunar and solar eclipses and moon phases are of the same kind as their daily

---

[64]Pseudo-Plutarch, *Stromata* 4 = DK 21A32 = LM XEN. D23 = Gr Xns58 = KRS 176 (only last clause).

[65]Hippolytus, *Refutatio Omnium Haeresium* 1.14.3 = DK 21A33(3) = LM XEN. D22(3) = Gr Xns59(3) = KRS 175.

[66]S 1.25.1 (≈P2.20.3) = DK 21A40 = LM XEN. D28a and b = Gr Xns60 = MR 516 = KRS177.

[67]Cf. Dox 348, note at line 11, and MR 523.

[68]Cf. Mourelatos (2008, 144).

[69]P 2.24.4 = S 1.25.1 = DK 21A41 = LM XEN. D34 = Gr Xns65 = MR 563; not in KRS.

[70]DK, note at 21A41; cf. Graham (2010), in Xns 65: "rather its setting."

[71]LM, note 1 at XEN.D34, my italics.

disappearances, due to the total or partial quenching of their lights. In an analogous way, Anaximander explained both the phases of the moon and lunar eclipses as the closing of the hole in the moon wheel. We may assume that Xenophanes, according to whom all heavenly bodies were essentially the same, namely incandescent clouds, called all quenchings of heavenly bodies "eclipses," and that this was the source of Aëtius' confusion, because he was familiar with the differences between sunset and a solar eclipse. This is also Lesher's guess: "(...) ἔκλειψις (throughout) may be merely a 'cessation' or 'departure,' (ἐκλείπειν) of the sun (...)."[72] For these reasons, I disagree with Mourelatos, who calls Diels' gloss "wholly unwarranted."[73]

West explains the solar eclipse that lasted longer than a month as the result of the eruption of a volcano,[74] which was seen as yet another quenching of the sunlight. In my opinion, it is preferable, following Untersteiner's suggestion,[75] to read it as an attempt by Xenophanes to understand tales about the long disappearance of the sun in northern regions. This would mean that almost the entire text 8.27 deals with the setting of the sun and not with eclipses, and that only in the last words a (total) eclipse of the sun os indicated. These last words were presumably the reason why Aëtius, who had no separate chapter on risings and settings, decided to put the whole text under the heading of "Eclipses of the sun." Similarly, he put another mention of the risings and settings of the celestial bodies as ignitions and quenchings in his chapter "On the nature of the heavenly bodies, both planets and fixed stars," because the text of his source also mentioned that, according to Xenophanes, the heavenly bodies consist of ignited clouds (see text 8.19). In an analogous way, as we saw, he placed a text on Xenophanes' conception of the phases of the moon as quenchings in the chapter "On the moon's eclipse" (see text 8.23).

The overall picture that emerges from these texts is that Xenophanes is said to have explained the celestial bodies as incandescent clouds (which, incidentally, means that, being a particular kind of clouds, they cannot be too far away), their risings and nightly appearances as kindlings, and their settings and daily disappearances, phases, and eclipses as quenchings. This implies that the heavenly bodies do not go under the earth, but are kindled every time they seem to rise at the eastern horizon and quenched when they seem to set at the western horizon. There is a new sun every morning, and the same holds true, *mutatis mutandis*, for the other celestial bodies. There are some subtle differences between the behavior of the individual heavenly bodies. The stars display a twofold appearance and disappearance: they may rise and set, and they glow at night and vanish by day. The moon undergoes a special kind of kindling and quenching, because it not only rises and sets but also decreases and grows during the month. The kindling of the sun is more especially explained as a gathering of tiny flares that arise from the moist evaporation of incandescent clouds. During a partial or total eclipse, both the sun and moon show

[72]Lesher (2001, 217 n. 56).

[73]Cf. Mourelatos (2008, 161, n. 26).

[74]See West (1971, 229, n. 2).

[75]Cf. Untersteiner (2008, 83, n. at A41). See also Mourelatos (2008, 165, n. 86).

an extraordinary kind of quenching. As already suggested above, Xenophanes did not yet distinguish between different kinds of disappearances of heavenly bodies, or stated positively, he explained them all by one general theory of disappearance, which he named "quenching", and he could call all of them "eclipses." This overall picture seems to be disturbed in another quotation from Aëtius:

8.28 Xenophanes says there are many suns according to the climes and regions and zones of the earth (κατὰ κλίματα τῆς γῆς καὶ ἀποτομὰς καὶ ζώνας), and at a certain time the disk falls (ἐμπίπτειν) into an area (εἴς τινα ἀποτομὴν) of the earth not inhabited by us, and just as if it were stepping into a hole (ὥσπερ κενεμβατοῦντα) it produces an eclipse. The same said the sun goes on beyond our experience (εἰς ἄπειρον μὲν προιέναι) but seems to move in a circle because of its distance (δοκεῖν δὲ κυκλεῖσθαι διὰ τὴν ἀπόστασιν).[76]

This text is by far the most difficult of Xenophanes' accounts of the heavenly bodies.[77] The next five sections of this chapter will be devoted to it in a step-by-step analysis, with particular regard to Mourelatos' influential interpretation. It will also be explained why I translate εἰς ἄπειρον as "beyond our experience" instead of the usual "without limit." The chapter will end with a suggestion for a new interpretation of Xenophanes' conception of the cosmos.

The Interpretation of an Enigmatic Text: Drozdek and Mourelatos

Text 8.28 begins with the enigmatic words "There are many suns." As such, these words could be understood as a poetic expression of the idea that every day a new sun appears, composed of tiny flares, but the combination with the climes, regions, and zones makes this interpretation problematic. The accumulation of the words "climes and regions and zones of the earth" is strange. Untersteiner notes that the word "zones," which belongs to Parmenides' conception of a spherical earth, is anachronistic.[78] Drozdek refers, rather ingeniously, to a text in which the belief in an infinity of worlds is attributed to Xenophanes:

[76]P 2.24.9 ≈ S 1.25.3 = DK 21A41a = LM XEN. D35 = Gr Xrs66 = MR 563 = KRS 179.

[77]Cf. MR 567: "The final lemma returns to Xenophanes, this report being even odder than the earlier one."

[78]Untersteiner (2008, 83, note at A41a): "di ciò non deve aver parlato Senofane; si tratta di espressioni dossografiche tardive."

8.29 He says that there are (...) numberless worlds, but not overlapping one another (οὐ παραλλακτοὺς).[79]

Drozdek states that "overlapping" can be meant spatially, for "if the earth is assumed to extend sideways, there is enough room for multiple worlds on its surface."[80] Usually, however, οὐ παραλλακτοὺς is understood as "not overlapping in time,"[81] although Lesher points to παράλλαξις, used for the overlapping of broken bones in Hippocrates' *Concerning Fractures* 15.[82] Drozdek also points to another testimony that "ascribes to Xenophanes a belief in infinity of worlds 'in every direction,' i.e., concurrently." This is questionable, because only the version of pseudo-Plutarch, in which no mention is made of Xenophanes, has the spatial word περίστασις ("surrounding space"), while Stobaeus' version, which mentions Xenophanes, has the temporal word περιαγωγή ("turning around, revolution"):

8.30 Anaximander, Anaximenes, Archelaus, Xenophanes, Diogenes, Leucippus, Democritus, and Epicurus held that there are infinite worlds in the infinite during each cycle (κατὰ πᾶσαν περιαγωγήν).[83]

I translate "during each cycle," in conformity with Bottler: "gemäß jeder 'Periode der Weltbildung'."[84] For Mourelatos, to whom Drozdek refers, Aëtius' text (8.28) expresses that there is not only an infinite number of suns (and moons) succeeding one another, so that a new sun comes every morning, but also that in the regions to the north and south of us there is an infinite number of suns whose paths run parallel to that of our sun. As for what is happening in the regions to the east and west of our region, he offers two possibilities because, as he says, the texts are unclear on this point. I quote: "Il y a une incertitude dans les témoignages quant à savoir s'il y a quotidiennement un 'allumage' de l'exemplaire de soleil (...) à l'Est, et une 'extinction' correspondente à l'Ouest (...), ou si chaque exemplaire de soleil (...) évolue indéfiniment sur un ligne Est-Ouest (...)."[85] As far as I can see, the interpretive problems result mainly from Mourelatos' idea that Xenophanes' earth is horizontally extended infinitely to all sides, so that every section of this infinite

79 Diogenes Laertius, *Vitae Philosophorum* 9.19 = DK21A1(19) = Gr Xns1(19); not in LM and KRS.

80 Drozdek (2008, 30).

81 Cf. Gr Xns1(19): "not overlapping in time;" Untersteiner (2008, 7): "i mondi succedentisi non diversi l'uno dall'altro."

82 Cf. Lesher (2001, 197, n. 4).

83 S 1.22.3 (P 2.1.3 does not mention Xenophanes) = DK 12A17 = LM ANAXIMAND. D13 = MR 308 = TP2 Ar145; not in Gr and KRS.

84 Bottler (2014, 284). TP2 Ar145 translates: "bei jeden Umlauf." DK adds after κατὰ πᾶσαν περιαγωγήν, "sc. γίνεσθαι καὶ φθείρεσθαι" (becoming and decay). LM ANAXIMAND. D13, on the other hand, inserts pseudo-Plutarch's περίστασις instead of Stobaeus' περιαγωγή; similarly, MR 319 in their "Reconstruction" of this testimony. In MR 309, it is argued, rather cryptically, that this reading is "clearly preferable because there is no need to assume that the entire universe would have a single unified revolution."

85 Mourelatos (2002, 337).

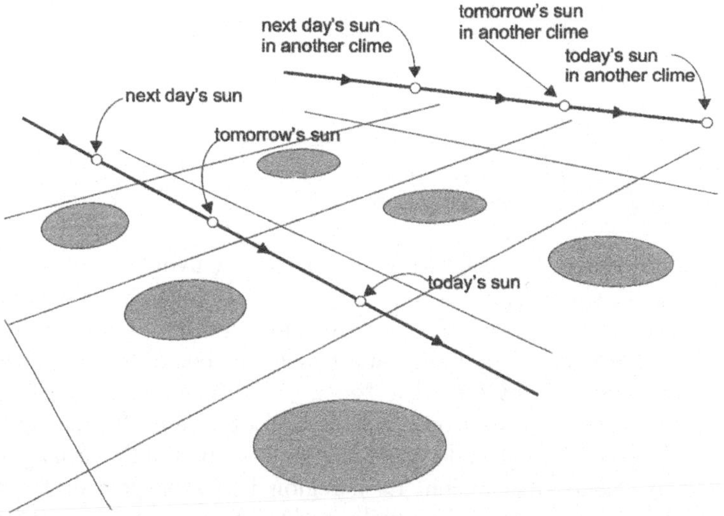

Fig. 8.1 Six regions of Xenophanes' infinite earth with their suns (after Graham, with some additions). Cf. Graham (2013), 71, Fig. 2.3

surface of the earth must be provided with sunlight in one way or another. Moreover, Mourelatos forgets to mention that, to account for the seasons, all these suns should be considered to be moving in the course of the year, from the summer solstice to the winter solstice, from north to south and back again.

Mourelatos' Interpretation Illustrated by Graham

It is always instructive to visualize the interpretations of ancient cosmologies, to see what consequences they lead to, and fortunately Graham has delivered one that is essentially dependent on Mourelatos' ideas. It looks like Fig. 8.1.

Graham's text does not offer much explanation of this picture, but in email correspondence he wrote to me: "The lines from lower left to upper right are meant to provide something like imaginary meridians of longitude (on a flat surface of course) to indicate the progress of the suns. The circles I meant as bodies of water, like the Mediterranean Sea, to provide a bit of perspective." The arrowed lines from top left to bottom right indicate the paths of the sun and are thus above the earth's surface. Since the surface of the earth is supposed to be infinitely extended in all directions, we must imagine an infinite number of suns both from east to west and from north to south. To provide a better impression of the infinite plane of areas with their respective suns, I have added in Fig. 8.1 two more suns and seas than in Graham's original drawing, and also some "parallels of latitude." This picture, resulting in rectangular areas, each with its own sun "seems to capture the picture

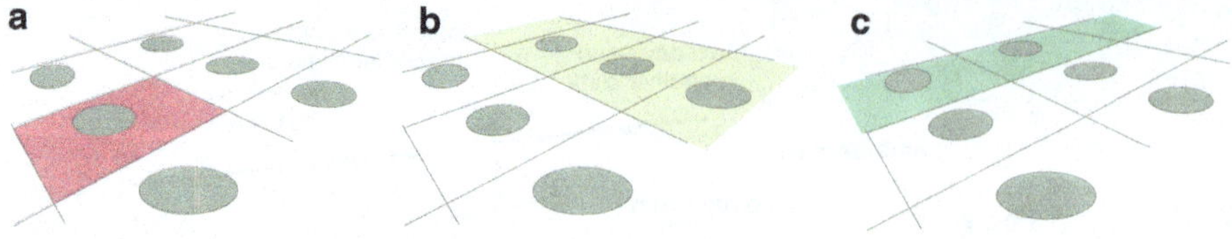

Fig. 8.2 (**a**) A region, (**b**) a clime, and (**c**) a zone

I was trying to convey," Graham wrote me. In order to avoid conceptual misunderstandings, I call each rectangle or square a "region" (ἀποτομή, see the pink area in Fig. 8.2a), each east-west row of regions a "clime" (κλίμα, see the yellow row of areas in Fig. 8.2b), and each north-south row of regions a "zone" (ζώνη, see the green row of areas in Fig. 8.2c).

Unfortunately, the word ἀποτομή is also used for the uninhabited part of the earth into which the sun falls. I consider this to be a sub-section of the rectangular region and call it an "area" of a region. Each region is a separate world and can be considered as one of Drozdek's "multiple worlds." In a sense, this interpretation of Xenophanes is a translation into a flat plane of Plato's spherical or dodecahedron-shaped earth, on which other people live "in many regions similar to those who live in our part of the earth" (ἐν πολλοῖσι τοιούτοις τόποις).[86] Plato's regions (τόποι) correspond to the regions (ἀποτομάι) of this interpretation of Xenophanes' earth. For Plato's spherical earth, however, only one sun was needed.

Graham's interpretation of the second of Mourelatos' above-quoted two possibilities for the east-west path of the sun can be called an ingenious solution of the combination of three items in the doxography: the sun going on infinitely in a straight line, the sun being ignited in the morning and quenched at the end of the day, and the existence of many suns according to the climes, regions, and zones of the earth. In his own words: "Apparently, there is some nucleus that continues 'on without end,' but which can be extinguished or ignited according to the atmospheric conditions; this nucleus attracts fire particles and unifies them into a diurnal conflagration."[87] However, such a nucleus of the sun traveling from one region to another and so on without end is not mentioned *expressis verbis* in the doxography. According to Graham's interpretation, there must be an infinite number of such nuclei, because in each of the parallel regions a sun ignites and is extinguished every day, as can be seen in Fig. 8.1.

The interpretations of Graham and Mourelatos do not fully coincide, for the latter wrote me that according to him every part of the earth that is called here "region" has its own "*klimata* and *apotomai* and *zônai*, each of which experiences (has) its proper sun and moon." I must confess that here the waves of interpretation are going too high for my comprehension, so I will leave this out of consideration.

---

[86]Plato, *Phaedo* 109b4.

[87]Graham (2013, 71–72).

A Cosmic Railway System and a Cosmic Ballet

The ultimate consequence of Mourelatos' and Graham's interpretation is that there exists an infinite number of regions, in all directions, comparable to the one we live in. Since, in this interpretation, the surface of the earth is supposed to be infinitely extended, Xenophanes' "many suns" have become an infinite number of parallel rows of moving suns, which are lit, glow for several hours, and fade, traveling in parallel lines from east to west like a cosmic railway system. Neither Mourelatos nor Graham mentions it, but to account for the changing path of the sun during the seasons, all these infinite rows of suns will have to move northward and southward during the seasons in a kind of infinite cosmic light show ballet. This is not the only strange consequence of this interpretation. According to Mourelatos, there is also an infinite number of moons, which behave like the sun.[88] Graham, too, writes "there are multiple moons as there are suns,"[89] which words suggest that these moons must be thought to behave in a way like that of the suns, in straight lines from east to west. It appears that we have to imagine a kind of duplication of the infinite cosmic railway system and cosmic light show ballet.

We may wonder why we can see all the lanterns in a street throwing their light circles, but not the suns in the other regions. Maybe Mourelatos and Graham would answer: because in Xenophanes' imagination they are too far away. However, for a person living near the eastern border of his region it is hard to believe that he could not see the quenching evening sun of the neighboring area, and the same goes for someone living on the western border and the igniting morning sun of the neighboring region. And what about the space between two regions, one of which is to the south of the other? Either someone living there would be able to see one sun in the north and another sun in the south, or, if both suns are considered too far away to be seen, he would have to live in an area where it is always night. Perhaps, an answer to these problems is hidden behind the conception of the stars in Mourelatos' interpretation.

Strangely enough, the movements of the stars, although they are also supposed to move above the earth's surface and not to go under the earth, must be imagined completely different from those of the sun and moon: "The stars presumably circle rather than going on without end."[90] Or, in the words of Mourelatos: "(...) ces cercles des étoules doivent-ils tous se situer toujours au-dessus de la Terre. (...) toutes les étoiles devraient être circumpolaires."[91] The orbits of the stars must be imagined in parallel circles on the dome of the heaven.[92] Mourelatos and Graham do not explain how we should combine this conception with that of an infinitely extended earth. According to Mourelatos, the air above the earth is as infinite as

[88]Cf. Mourelatos (2002, 337, quoted above).

[89]Graham (2013, 72).

[90]Ibidem. In the doxography on Xenophanes, the planets are not mentioned.

[91]Mourelatos (2002, 347).

[92]Cf. Mourelatos (2002, 348).

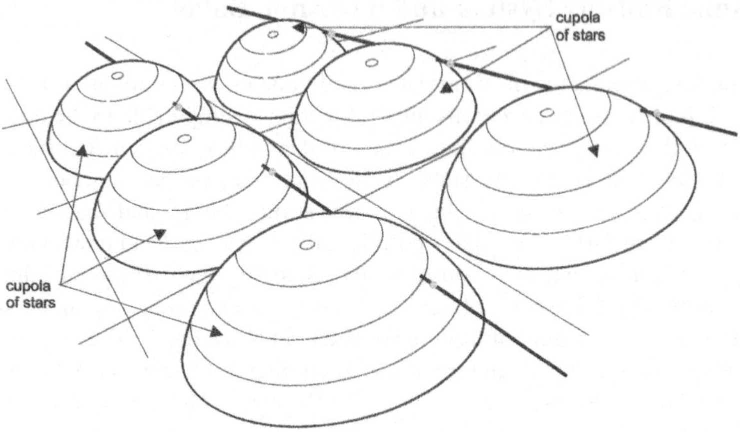

Fig. 8.3 Six of the infinite number of cupolas of stars, tentatively after Mourelatos

the earth below our feet: "(. . .) le double apeiron de la Terre en-dessous et de l'air au-dessus (. . .)."[93] This suggests that we should imagine one infinite cupola of stars all over the infinite earth. How to imagine an infinite dome, and how to reconcile this with the conception of the heavenly bodies as clouds, which suggests that they are nearby, is difficult to understand. Another possibility would perhaps suit this interpretation better, namely that every square region is adorned with its own cupola of stars with its own constellations. In Fig. 8.3, I have tried to give an impression of this idea. Within each cupola on the drawing it is night, because each quenched sun (the small grey circles) is outside the cupolas on its own rail from east to west.

Each region with its own hemisphere of stars is a "world" in Drozdek's sense. Presumably, this drawing also comes close to what Mourelatos means, for he wrote me "Let us agree that by 'region' we mean *kosmos*, i.e., a huge earth segment that is circumscribed by a cupola of stars." This version of Mourelatos' interpretation of the movements of the stars might explain why we do not see the sun in an adjacent region. If we imagine these cupolas as made of some condensation of air that makes them impenetrable to eyesight, then we can only see the sun inside our cupola. On the other hand, the conception of an infinite number of finite cupolas is difficult to combine with the idea of an air that is infinitely extended upwards.

In Mourelatos' interpretation of Xenophanes, the real problem is how to explain why we do not *see* the paths of the stars as circles parallel to the earth's flat surface, while they are all *supposed* to orbit above the earth, parallel to its surface. When we try to visualize this, the problem is how to explain that the movements of the stars are *perceived* as in Fig. 8.4a, while they are supposed to move "*in reality*" as in Fig. 8.4b. In my opinion, when the heaven is conceived of as a (hemispherical)

[93]Mourelatos (2002, 348).

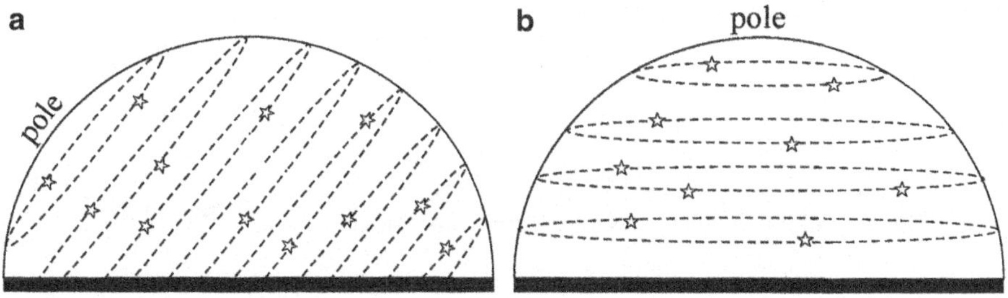

Fig. 8.4 (**a**) The movements of the stars as they are seen on a flat earth. (**b**) The movements of the stars as they are supposed to be in Mourelatos' and Graham's interpretation

cupola, there is no optical illusion conceivable that would be able to distort circles as in Fig. 8.4b into curves as in Fig. 8.4a.

The Different Paths of the Heavenly Bodies According to Mourelatos and Graham

According to Xenophanes, what happens with the heavenly bodies at the horizon is not what we think we see, namely that they go under the earth to rise again at the opposite end of the horizon. Instead of rising, they are kindled and instead of setting they are quenched. In the interpretation of Mourelatos and Graham, the sun and moon move in a straight line above the earth while the stars rotate in concentric circles parallel to the plane of the earth. The problem is that what we observe as similar movements—and this applies in particular to the stars of the Zodiac that follow the same path as the sun—should be explained completely differently. Mourelatos adduces as a kind of support that there is no testimony of the stars going forward in straight lines, just like the sun,[94] but this *argumentum ex nihilo* is hardly convincing. Mourelatos' second suggestion is even stranger: due to a poorly understood rule of perspective ("en désaccord flagrant avec l'apparence du ciel"), Xenophanes should have concluded that all stars should be understood as to make full circles, as Anaximander already taught it, but this time not under the earth as well, but entirely above the earth.[95] In other words, when we see the sun or moon moving from one side of the horizon to the other in an arc, this should be considered an optical illusion, the actual movement being a straight line; and when we see a star moving in an arc from one side of the horizon to the other, this should be considered another optical illusion, the actual movement being a circle parallel to the earth's surface. We might also think of the rainbow, which, according to Xenophanes, is a

[94]Cf. Mourelatos (2002, 347).

[95]Ibidem.

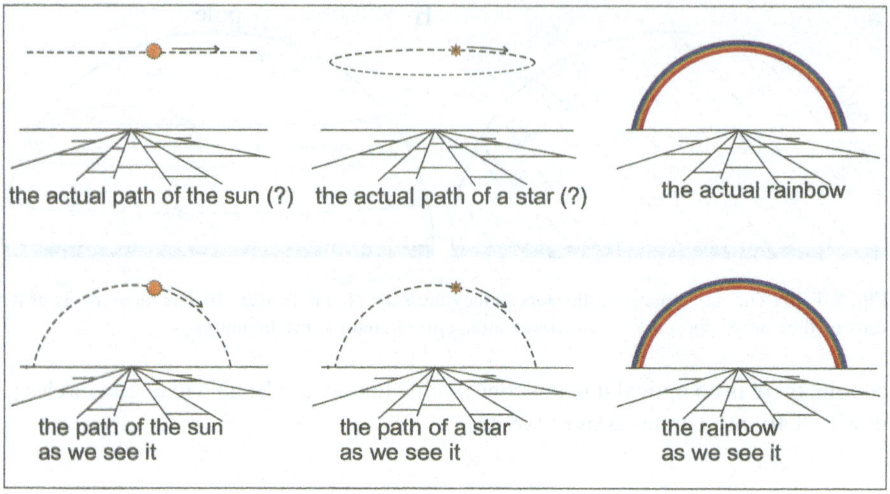

Fig. 8.5 The paradox of Mourelatos' explanation of the movements of the heavenly bodies

kind of cloud like the sun,[96] but this rain-bow is not said to be the optical illusion of a rain-line or a rain-circle. How can it be that the curved paths of the sun and moon are to be interpreted as an optical illusion, the paths of the stars as yet another optical illusion, while the rainbow, which is an optical illusion, looks like the paths of the heavenly bodies? These paradoxes are shown in Fig. 8.5.

I think this will suffice to show that Mourelatos' and Graham's interpretations lead to insurmountable difficulties, which, to say the least, would make Xenophanes a complete outsider, not fitting in the row of the other Presocratics who tried to articulate the cosmology of a flat earth. As we will see in Chaps. 13 and 14, in China there has existed a conception of a flat earth cosmology in which the celestial bodies circle around the pole and above the surface of the earth, but I will argue in Chap. 15 that this has nothing to do with Xenophanes' ideas, nor with those of Anaximenes, for that matter. Some essential differences between the ingenious Chinese system and the interpretation of Xenophanes' cosmology discussed above are that the flat surface of Chinese earth was finite, that there was only one sun, one moon, and only the stars we perceive, that the heaven was not a cupola, but a flat plane parallel to the earth's surface, and that not only the stars but all heavenly bodies orbited around the celestial pole.

[96]See *Scholium in Iliadem* 11.27 = DK 21B32 = LM XEN. D39 = Gr Xns72 = KRS 178.

Some More Textual and Conceptual Problems

In text 8.28, the word ἀποτομή is also used for the uninhabited part into which the sun falls. I consider this to be a sub-section of a rectangular region and call it an "area" of a region. The idea of an *eclipsing* sun falling on earth (in text 8.28) is admittedly very strange. Bicknell's explanation, repeated by Mourelatos,[97] is that the sun ceases to burn because it enters an area of the earth devoid of water. In a confusing passage, Mourelatos refers to DK 21A41 (which is text 8.31 below) and writes: "*we are told* that eclipses occur when the sun passes over arid areas of earth that do not provide sufficient evaporation to sustain the luminary."[98] Elsewhere, he mentions "an episodic 'quenching' of the sun because a failure in the supply of vapor" and refers to DK 21A41a, which is our much-discussed text 8.28. Something must have gone wrong here, for these texts read:

8.31 [the relevant part of 8.27] Xenophanes [says an eclipse (ἔκλειψις) of the sun (or rather: the setting of the sun)] is a quenching (κατὰ σβέσιν), and another (new, different: ἕτερον) [sun] in turn comes to be in the east.[99]

8.32 [the here relevant part of 8.28] and at a certain time the disk falls (ἐμπίπτειν) into a section (εἴς τινα ἀποτομὴν) of the earth not inhabited by us, and just as if it were stepping into a hole (ὥσπερ κενεμβατοῦντα) it produces an eclipse.[100]

Both texts have nothing to do with a lack of vapor or evaporation and also nothing to do with arid areas over which the sun passes. As already argued at text 8.28 (and its excerpt in 8.31), this text is not about eclipses but about sunset. Nevertheless, as argued above in the section *The Nature and Movements of the Celestial Bodies*, I agree with Mourelatos, that in Xenophanes' "cloud astrophysics" an eclipse is a special kind of, namely episodic,[101] quenching. The "not inhabited areas" of text 8.32 are not necessarily dry, as Mourelatos' interpretation seems to be;[102] the sea is also an uninhabited area. To translate the word ἐμπίπτειν as "pass over", as Mourelatos does,[103] instead of simply "fall into," is rather artificial. I assume that Mourelatos' interpretation is based on an already quoted testimony of Aëtius (text 8.26), in which it is said that the rising sun at the horizon is gathered together from the moist evaporation. I repeat the sentence in question:

---

[97] Cf. Bicknell (1967).

[98] Mourelatos [2008, 138 (my italics)].

[99] P 2.24.4 = S 1.25.1 =DK 21A41 = LM XEN. D34 = Gr Xns65 = MR 563; not in KRS.

[100] P 2.24.9 ≈ S 1.25.3 = DK 21A41a = LM XEN. D35 = Gr Xrs66 = MR 563 = KRS 179.

[101] Cf. Mourelatos (2008, 153).

[102] Cf. Mourelatos (2008, 138): "arid areas" (abusively referring to DK21A41 instead of 21A41a).

[103] Cf. Mourelatos (2008, 138). It is strange that Mourelatos quotes all the relevant texts in full, except for the crucial text of DK21A41a.

8.33 Theophrastus in the *Physics* has written that the sun is composed of tiny flares (ἐκ πυριδίων) being gathered together from the moist evaporation (ἐκ τῆς ὑγρᾶς ἀνθυμιάσεως), and gathering together the sun.[104]

Perhaps it could be inferred that the sun, being an (incandescent) cloud, constantly needs fuel from moist evaporation during its daily journey, but this is not implied in these texts. If the lack of support of moist evaporation, which supposedly nourishes the sun, were to be an explanation for the quenching of the eclipsed sun, this would not explain its falling (ἐμπίπτειν). A cloud that lacks support of water will shrink, but will not fall. At the end of this chapter, I will suggest that a much simpler interpretation of text 8.31 is possible, namely that Xenophanes just offers another description of what happens when the sun sets, which is here called an "eclipse." As explained in the discussion of text 8.27, Xenophanes saw all disappearances of heavenly bodies as "quenchings" and he could call all of them "eclipses."

In the last sentence of text 8.28, we meet again the expression εἰς ἄπειρον, which Graham[105] translates as "without end." But, as Guthrie rightly remarks, here the meaning cannot be "without end" or "infinitely," because the sun quenches at the end of the day.[106] Nevertheless, most commentators explain Aëtius words "the sun goes on without end" as meaning that the sun goes in a straight line, apparently as a contrast (μὲν-δὲ) with "to circle around," and claim that this is the result of an optical illusion, indicated by the word δοκεῖν, and that this is "because of the distance" (διὰ τὴν ἀπόστασιν). I could, however, not find one commentator explaining how to combine the idea of the sun's path as a straight line with the falling sun, mentioned in the same report. The term προιέναι, here used for the sun, does not necessarily mean going forward in a straight line, and the expression εἰς ἄπειρον does not necessarily refer to a rectilinear movement or a straight line.

The Earth Not Infinitely Extended, Neither in Surface Nor in Depth

In the last sections of this chapter, I will explore the possibility of proposing another interpretation that is more in line with what we know of those ancient Greek cosmologists who thought that the earth is flat and finite. To achieve this goal, several textual and conceptual problems, which have occupied us already so many pages, still need to be solved. I will successively expose them and suggest ways to understand them. For a start, let us return to the idea of a flat earth, infinitely extended in all directions in the horizontal plane. It follows from the above that

---

[104] S 1.25.1.b (≈P2.20.3) = DK 21A40 = LM XEN. D28b = MR 516 = KRS177.

[105] Graham (2010, 125, no. 66).

[106] Guthrie (1962, 318, n. 1): "(…) εἰς ἄπειρον, 'indefinitely,' not 'to infinity'; it burns out in a short time."

the evidence for this interpretation is rather thin. It is not in Xenophanes' text (text 8.1), which only says that the upper limit of the earth can be seen under our feet, but not that it is infinitely extended in all directions. If Xenophanes would have claimed that the flat surface of the earth stretches infinitely he would have expressed it in so many words, because it conflicts with the conception of the earth of his predecessors. The doxography that is adduced to support the interpretation of an infinitely vast surface of the earth can better be understood as referring to something else, namely the depths of the earth. When pseudo-Plutarch and Hippolytus (texts 8.15 and 8.14) say that according to Xenophanes the earth is boundless (ἄπειρον εἶναι), they clearly refer to what Aristotle (in text 8.3) calls "its lower part." The idea they want to express is not that the earth's surface is infinitely extended but that when you think of the earth as having infinite roots, the celestial bodies cannot be supposed to go under the earth.

This leads us to the question of what can be meant with the depths of the earth. Both in Mourelatos' interpretation and in those of the doxographers, the idea of the earth having infinite roots remains strange. Its origin is Aristotle's preoccupation with the problem of why the earth does not fall and his assumption that Xenophanes' solution to this problem was an earth that was infinitely extended downwards. It is my conviction that this is not what Xenophanes wanted to express. There is no indication that he has worried about a falling earth. Above, in the section *Xenophanes' Text in the Interpretation of Some Recent Authors*, I distinguished three possible ways to understand Xenophanes' words τὸ κάτω in text 8.1, namely (1) "the lower part of the earth,' reading τὸ κάτω [sc. γαίης];" (2) "that what is beneath the earth," also reading τὸ κάτω [sc. γαίης]; (3) "the lower limit of the earth," reading τὸ κάτω [sc. πεῖρας]. I think the third option is intended, but in my interpretation, the differences between the three options are less relevant because, as I will argue, Xenophanes' cosmos is not infinite but finite, although it is beyond human capacities to know where its limits are.

The two lines of text 8.1 deliberately contrast with each other, as indicated by the words μέν and δέ. The intended contrast is between the clearly visible limit of the earth under our feet and our complete ignorance of where the other end or limit of the earth is. I think the most natural way to understand this contrast is the way it is usually read, namely that it has to do with two limits, the upper limit as opposed to the lower limit: μὲν τόδε πεῖρας ἄνω (. . .) τὸ [πεῖρας] κάτω δ'. Anaximander had said that the depth of the earth is one third of its diameter. In the next section, I will argue that Xenophanes wants to emphasize that it is fair to admit that we simply do not know what the depth of the earth is.

The Two Meanings of ἄπειρος

The next issue to be clarified is, therefore, why the lower limit of the earth is said to reach downwards ἐς ἄπειρον. Several commentators, following in Diels' footsteps, have remarked that when the earth is said to have a lower limit, the meaning of the expression ἐς ἄπειρον cannot be "infinitely," because an infinite limit would be a *contradictio in terminis*. So, they translate it as "indefinitely" (Diels: *indefinitum*, nicht *infinitum*).[107] In order to underpin this idea, I would like to propose another, provocative, option for the translation of ἐς ἄπειρον in Xenophanes' original fragment (text 8.1), following a suggestion made by Tannery more than a century ago in his interpretation of Anaximander, later by Tumarkin in her interpretation of both Anaximander and Anaximenes, and also by Lumpe in his Inaugural-Dissertation on Xenophanes.[108] There is also another word ἄπειρος, derived from πεῖρα and meaning "inexperienced," "not acquainted with." Tannery and Tumarkin put forward the idea that Anaximander used it in its passive form (which is not documented elsewhere in the literature but is grammatically correct), meaning "that which is not experienced," "not sensible," which we could render by the unusual term "unexperienced," or more generally "out of our sight," "beyond our experience."[109] Tumarkin also suggests this meaning for Anaximenes' air, as a "nähere Bestimmung des 'Unerfahrbaren' (ἄπειρον) als das nicht Wahrnehmbare."[110] Lumpe, following a hint by Egermann, was the first to propose this translation in connection with Xenophanes[111]

This meaning of ἄπειρος is not too far away from Diels' "indefinitely."[112] Homer can call the earth, the sea, and even the Hellespont ἄπειρος, and Herodotus can speak of an ἄπειρος plain, as far as the eye can see.[113] In this use of the word ἄπειρος, the idea of an unseen boundary is implicit; neither the sea, nor the Hellespont, nor the plain, nor the earth are considered infinite in the literal sense. Kahn mentions a text in Aristotle's *Physica*, in which he discusses the meaning of ἄπειρος and explicitly says: "But you might also mean that, though it is of such nature that you can traverse it, it does not admit (whether you are speaking absolutely *or practically*) of your getting 'through' it so as to come out beyond it."[114] The depth of the earth could also be taken to be traversable in principle, by digging a hole or by descending into a cave, but practically speaking we will never reach the end: the depth of the earth is unfathomable, out of our reach and beyond our experience. When Kahn renders "boundless" as "what cannot be passed through to the end," while mentioning in

---

[107]DK, note at 21B28. See also, e.g., Lesher (1992, 131).

[108]Tannery (1904). Tumarkin, 56–58. Lumpe (1952, 38–39). Both Tumarkin and Lumpe are obviously independent from Tannery, and Lumpe also from Tumarkin.

[109]Tumarkin (1943, 56), also suggests that in this meaning the word is used in the neuter, which is a needless limitation.

[110]Tumarkin (1943, 58).

[111]Lumpe (1952, 38).

[112]Tannery (1904, 704) even suggests that the two words might have the same root, although he calls this less probable.

[113]Homer, *Ilias* 7.446, *Odyssea* 4.510, *Ilias* 15.545; Herodotus, *Historiae* 1.204. Cf. Lesher (1992, 130).

[114]Aristotle, *Physica* 204a2–7, my italics.

passing Xenophanes' lines of text 8.1, he means that it is beyond the capacities of mortals, but not beyond those of gods like Hera.[115] This, too, is not too far away from "beyond our experience," and here, too, the idea of an unseen, remote and unreachable boundary is implicit. Even now, we can speak of "an endless prairie" or an "infinite road," meaning that we do not know or cannot see where the prairie or the road ends, just as Homer spoke of the boundless see and Herodotus of an endless plain.

One might even say that, after all, the meanings of the two words ἄπειρος do not differ that much from each other. What is boundless, or infinite is, as such, beyond our experience. As Untersteiner says: "'all' infinito' viene a significare presso a poco lo stesso che '*nello sconosciuto*,'" even though he calls Lumpe's translation "tesi errata," without any explanation.[116] Lesher also refers to Lumpe, but argues that the words προσπλάζον and ἱκνεῖται "suggest strongly that we are dealing throughout with a question of the earth's *extension*."[117] Against this, I suggest that Xenophanes is playing with the two meanings of ἄπειρος, "boundless" and "not experienced." His wordplay also entails that the words εἰς ἄπειρον ("beyond our experience") in his fragment contrast (μὲν-δὲ) with ὁρᾶται ("be seen") in the first line.[118] The upper limit of the earth is seen at our feet, the lower limit of the earth cannot be seen, is beyond our experience. A free rendering of Xenophanes' lines in text 8.1 would read:

8.34 The upper boundary of the earth is visible here at our feet, touching the air; the lower boundary of the earth reaches down beyond our experience.

To put it bluntly: an ἄπειρον πεῖρας is a limit that is beyond our experience, further than we can see, out of our reach, a limit we do not know where it is, "*nello sconosciuto*." This meaning of ἄπειρος has nothing to do with philosophical skepticism or pessimism, as Mourelatos suggests,[119] but rather with intellectual honesty which does not allow us to claim more than we can verify with our own eyes (cf. ὁρᾶται in the first line of text 8.1).

The Presocratic cosmologists, who did not yet use the word ἄπειρος in the technical sense of "infinite," which became its exclusive meaning in the thinking of the Pythagoreans, Plato, and Aristotle, still felt, I would suggest, both meanings when they used the word ἄπειρος. A cosmological example is the use of the word ἄπειρος in the κόσμοι ἄπειροι of the atomists. These infinite or innumerable worlds are not the stars, as one might think from our modern cosmological conception. In ancient Greek cosmology, the stars belong to our world or cosmos. As Furley rightly says: "Both the Atomists, who believed in the infinite universe, and the Aristotelians,

[115] Kahn (1994, 232–233).

[116] Untersteiner (2008, CLVI, my italics, and n. 108).

[117] Lesher (1992, 129), writing προσπλάζων instead of προσπλάζον.

[118] Cf. Lumpe (1952, 38): "das ἄπειρον in frg. 28 steht im Gegensatz zum ὁρᾶται (v. 1); es heißt hier nicht 'grenzenlos' (ohne πέρας), sondern 'unerfahrbar" (ohne πεῖρα)."

[119] Cf. Mourelatos (2002, 335).

who did not, agreed that our world is itself a finite system, bounded by the sphere of stars."[120] This means that the κόσμοι ἄπειροι are not only infinite in number, but also that they lie outside of our world, beyond our horizon of experience. The latter meaning is incorporated, as it were, in the former.

Without explicitly noticing it, for several scholars the notion of ἄπειρος as "unexperienced," "beyond our experience," is part of their interpretation of Anaximander's ἄπειρον as "the boundless." Accordingly, Kahn calls it "some more permanent source that is *partially or wholly unknown to us*" and "this *unknown* world source."[121] What the insertion of "partially" before "or wholly unknown" means is not clear to me, for this alleged body or mass that surrounds the world is as much beyond our experience as are the infinite worlds of the atomists. Another example is West, who summarizes: what Anaximander meant to say is that "earth, sea, sky, and Tartarus (. . .) have their sources in a boundless *Beyond*."[122] And yet another is Freeman, who writes: "for though the Non-Limited was material and therefore perceptible, *it was removed from our perceptions by being out of reach*."[123] Recently Graham, who endorses Kahn's characterization of Anaximander's boundless as "a spatially unlimited stuff," and "a kind of reservoir which assures that the particular kinds of matter will never run out," wrote explicitly: "the boundless remains outside the cosmos, surrounding and controlling it in some fashion, but it is not, so far as Anaximander tells us, in our world. It is forever *inaccessible and mysterious, beyond empirical scrutiny*."[124] These authors do not even mention that there are two different words ἄπειρος, and yet they interpret "boundless" as "beyond our experience." These examples could easily be multiplied by quoting other authors on Anaximander's ἄπειρον.

Perhaps the following observation is also indicative: Xenophanes (in text 8.1) uses ἐς ἄπειρον, while Aristotle has ἐπ' ἄπειρον. This difference may look trivial, but ἐπ' ἄπειρον seems to be the more technical term, undoubtedly meaning "infinitely." Euclid, speaking about infinite lines in the famous 5th postulate of his *Elements*, uses ἐπ' ἄπειρον. The words εἰς ἄπειρον, it seems to me, are not used in this technical, mathematical sense but in a metaphorical or figurative sense (Cf., e.g., Plato, *Laws* 910b3: people multiplying their crimes infinitely; Aristotle, *Physics* 209a25: an argument *ad infinitum*; *Nicomachean Ethics* 1113a2: deliberation

[120] Furley (1987, 136), see also Furley (1989, 2): "(. . .) no one in classical antiquity believed that the *world* is infinite."

[121] Kahn (1994, 237, my italics).

[122] West (1971, 78 and 79).

[123] Freeman (1966, 56, my italics).

[124] Graham (2006, 31 and 34, my italics).

ongoing infinitely).[125] Untersteiner ignores this difference when he reads Xenophanes as if he wrote ἐπ' ἄπειρον instead of ἐς ἄπειρον.[126]

Be that as it may, I think that Xenophanes' lines have been more or less misunderstood since Aristotle (text 8.3), supposing that they were about the question of why the earth does not fall, read ἐπ' ἄπειρον instead of ἐς ἄπειρον, and since the introduction of the image of the earth's roots. Strabo, pseudo-Plutarch, Hippolytus, Aëtius, Cicero, and Diogenes (all quoted above in texts 8.9–8.15) clearly depend on Aristotle. They try to imagine the consequences of his words ἐπ' ἄπειρον: if the earth has infinite roots, the heavenly bodies cannot pass under the earth and the earth cannot be completely surrounded by air or heaven. When they talk about the earth as boundless, they do not refer to its periphery, but to its "roots." However, not only the earth's periphery, but also these "roots" are not mentioned in Xenophanes' text 8.1, and we will never know if he spoke of them in other lines that have not been preserved. The last example in this long line of interpretation that started with Aristotle is that of Mourelatos and Graham, despite all the differences.

A Spherical Cosmos and a Hemispherical Heaven

When Heraclitus and pseudo-Aristotle call Xenophanes' aether or air ἄπειρος (texts 8.5. and 8.6), this does not mean "infinite" but "unfathomable," as explained above. Both pseudo-Plutarch and Hippolytus inform us that Xenophanes' earth "is surrounded neither by air nor by heaven" or "is not surrounded from all sides by air" (texts 8.14 and 8.15). According to Mourelatos, both texts convey the idea that the earth is infinitely extended both in extent and in depth.[127] I do not see, however, that these texts imply that the earth has no limits at its periphery. The fact that the earth is said not to be surrounded by air or heaven does not necessarily involve that the earth's surface is infinite. We can easily imagine a finite flat earth of unknown size with the hemispherical dome above it resting on the rim of the earth, as a kind of extrapolation of what we observe on the horizon, where heaven and earth meet. An indication that something like this must have been Xenophanes' conception is given in the accounts that say that stars, moon, and sun, are incandescent clouds. This is to say that he considered the sun and the other celestial bodies as a kind of meteorological phenomena. When the celestial bodies are conceived of as clouds, they cannot be far away. Apparently, Xenophanes realized that on a flat earth the celestial bodies had

[125]The remark in Lesher (2001, 130), that "the phrase *ep' apeiron* occurs in Homer" is incorrect. He mentions *Odyssea* 7.286 and 4.510, and *Ilias* 7.445 (must be 7.446) and 14.545 (must be 24.545). The first two have κατά instead of ἐπί. The first of these is a case of *tmesis*: in the phrase κατ' ἀπείρονα χεῦεν, κατ' belongs to the verb χεῦεν. In the second case, κατὰ belongs to πόντον and means "down into." In the fourth case, there is no preposition at all before ἀπείρων. Only in the third case the preposition ἐπ' is used, but here it belongs to γαῖαν: "on the boundless earth."

[126]Untersteiner (2008, CLIV and note 96).

[127]Cf. Mourelatos (2002, 334).

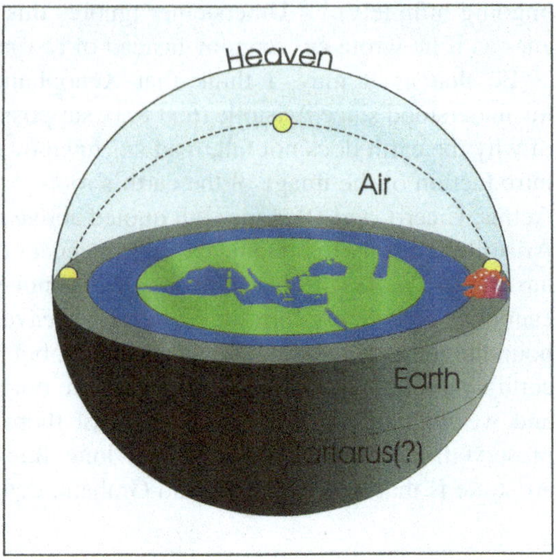

Fig. 8.6 An alternative representation of Xenophanes' cosmological ideas

to be relatively close by. It does not seem obvious, therefore, that Xenophanes imagined the heaven as reaching out infinitely above the earth, but rather that he understood it as a dome resting on the rim of the earth. If the radius of the dome of the heaven is equal to or less than the radius of the earth's surface, which could easily be the case if the celestial bodies are considered meteorological phenomena such as clouds, then neither the air nor the celestial bodies can go under the earth. Just as he expressed his ignorance about the depth of the earth and about the size of the earth's surface, Xenophanes expressed his ignorance about the height of the heaven. All these distances, he must honestly admit, were "unfathomable." There are indications of this view in some texts in the doxography saying that Xenophanes regarded the cosmos as limited and spherical. Theodoretus reports:

8.35 Xenophanes (...) said the totality was one, spherical and limited.[128]

Although Theodoretus might have based this on Xenophanes' text, of which he quotes another line, he probably also refers to Aristotle, who wrote:

8.36 Parmenides (...) says the one is limited (...). Xenophanes (...), of whom Parmenides is supposed to have been a student (...), with a view to the whole heaven he says the one is god.[129]

We can imagine such a spherical cosmos with the dome of the heaven as the upper hemisphere and the earth as the lower hemisphere, as I have done in Fig. 8.6. Another possibility would be to make the earth a slice of the sphere (like Anaximander's

[128]Theodoretus, *Graecarum Affectionum Curatio* 4.5 = DK 21A36 = LM XEN. R11 = Gr Xns46; not in KRS.

[129]Aristotle, *Metaphysica* 986b18–27 = DK 21A30 = LM XEN. R12 = Gr Xns40 = KRS 164 and 174.

column drum) and the Tartarus situate underneath it as the lowest part of the spherical cosmos. Xenophanes would have said: "this must necessarily remain speculative. We do not know where the lower limit of the earth is because it is beyond our experience." Summarizing: In my version of Xenophanes' cosmos the earth is not surrounded by air or heaven, not infinite in the horizontal plane, and not infinitely extended downwards. When asked how great these distances really are, Xenophanes must confess his ignorance, only knowing that they are greater than humans can measure.

The "Many Suns"

If the earth' surface is not infinitely extended, the question arises as to the meaning of the phrase about the many suns according to the climes, regions, and zones of the earth (in text 8.28). In my opinion, these words can be read as a poetic expression of the discovery that in places in the north the days in summer are (much) longer than in Greece and in winter (much) shorter, and that in places east and west of Greece the sun rises and sets earlier or later. These were important discoveries because, if properly understood, they potentially challenged the very conception of a flat earth. Xenophanes, however, could not have grasped their full impact, for otherwise he would not have thought that the earth was flat. Instead of ignoring these tales, which his intellectual honesty forbade him, he expressed his wonder. To blow these words up to an infinite number of parallel suns and moons hovering over an infinity of regions as our own is to entrust Xenophanes with a weird imagination that is far beyond what the words of text 8.28 can bear. In my opinion, Xenophanes was not that kind of speculative thinker, but rather one who was convinced of the limitations of our knowledge, who tried to find explanations in accordance with what he saw with his own eyes, and who expressed his ignorance of things that were beyond human experience.

The Curved Paths of the Celestial Bodies

Let us look again at the paths of the celestial bodies. The usual interpretation of the sun's daily path according to Xenophanes is that by an optical illusion it seems to be curved, but that in reality it is straight. Above I already wondered how the path of the sun, supposedly a straight line, can appear curved from our perspective, as an optical illusion, I also wondered how to explain the combination of the images of the straight line of the sun's path on the one hand and that of the sun falling into an uninhabited area of the earth on the other. As to Mourelatos' interpretation, I wondered how the path of the sun can be an optical illusion and the similar path of rising and setting stars a completely different optical illusion. I argued that προιέναι does not necessarily mean going in a straight line. I already quoted with consent that the words εἰς ἄπειρον in the last line of text 8.28 cannot mean "*ad infinitum*," "without end" or "infinitely."

Several commentators have tried to solve this problem by using the translation "indefinitely," which, like a Houdini trick, makes εἰς ἄπειρον almost mean "finite." I think that here, too, these words can be better understood as conveying the meaning "out of our sight." The sun, having ended its daily journey along the firmament as a part of a circle, moves out of our sight (εἰς ἄπειρον προϊέναι). That we lose sight of the sun is "because of the distance" (διὰ τὴν ἀπόστασιν). When the sun sets at the horizon, it is at the greatest distance possible on a flat earth, namely at its periphery.

Taking all this together, I think that the meaning of the last line of text 8.28 is not that, *mirabele dictu*, the sun's path is straight although its seems curved, but that what Xenophanes wants to say is that we get the impression (δοκεῖν) that the sun makes a full circle, because we see the sun describe a part of a circle during the day. The word κυκλεῖσθαι, used in the last sentence of text 8.28 means "to make a full circle," like a ring, and not "to make a curve," like a part of a ring. I agree with Lesher's suggestion that Xenophanes' intention is to oppose to what Lesher calls "the psychological phenomenon" that makes us believe "that the sun that sets in the west is identical with the one that rises in the east in the next morning."[130] In Xenophanes' opinion, this impression is deceptive, especially since the sun and the other heavenly bodies are the most distant objects, describing huge curves. This is what the words "because of the distance" (διὰ τὴν ἀπόστασιν) at the end of text 8.28 express. All we can honestly say, based upon observation, is that every day a sun appears on the eastern horizon and disappears on the western horizon, at the farthest possible distance on earth. And since Xenophanes considers the sun as a kind of meteorological phenomenon, a glowing cloud, he cannot believe that the sun goes under the horizon and even less that the sun goes under the earth, describing a full circle. Consequently, Xenophanes says that there are countless suns and moons, namely every day a new sun and another moon.

The paths of the stars are a major problem in a cosmology that does not allow the celestial bodies to make full orbits to go under the earth. Mourelatos notes that there must be a twofold kindling and quenching, one when they all appear at sunset and disappear at sunrise, and another one for those that seem to rise and set one after another during the night, except for the circumpolar stars that never set.[131] If I understand him well, Mourelatos chooses to treat the fixed stars completely different from the sun and the moon, because it is counterintuitive to assume that new stars will be ignited every evening.[132] In his opinion, Xenophanes distinguished between two types of disappearance and taught that the disappearance of the stars during the day was due to their being outshone by the sun. As already argued above, Mourelatos' explanation of the paths of the stars as optical illusions is paradoxical (see Fig. 8.5). I do not see any reason why Xenophanes should not conceive of the paths of the stars (except those of the circumpolar stars that do not set) as arcs from

---

[130]Lesher (1992, 218, n. 59). A do not agree, however, when Lesher thinks that Xenophanes' sun goes indefinitely downwards, just like the earth. See also Mourelatos (2008, 161, n. 29).

[131]Mourelatos (2002, 339–340).

[132]Cf. Mourelatos (2002, 343).

one side of the horizon to the other. Consequently, in Xenophanes' conception of flat earth cosmology, where the heavenly bodies do not pass under the earth, we must take it for granted that new stars somehow manage to regroup into the same constellations every evening and to rise and set exactly where we would expect them. Similar phenomena happen every day with the new sun and the moon, which also manage to rise and set exactly where we would expect them and, in the case of the moon, to show exactly the right phase every day, being full when it is opposite the sun and new when it is near the sun.

All Disappearances of Heavenly Bodies Are Quenchings

In an earlier section, I already pointed out the manifold applicability of the word "quenching", applicable to the setting of heavenly bodies, to the disappearing of the stars by day, to eclipses, and to the phases of the moon (see 8.23). In text 8.31 (quoted earlier in text 8.27), the word "eclipse" (ἔκλειψις) is used for sunset, as explained above. The image of the *setting* sun disk as falling down as if it were stepping into a hole is a vivid poetic depiction of the disappearance of the setting sun in a remote and uninhabited area. For the translation of ὥσπερ κενεμβατοῦντα as "stepping into a hole," see LSJ, s.v. κενεμβατέω, "step on emptiness, step into a hole." The quenching sun falls "into a section of the earth not inhabited by us," which obviously is far away, at a great distance. The words "falling down," which are difficult to understand of a sun moving straight, go very well with a sun moving along the bow of the cupola of the firmament. We may assume that the setting sun falls into a hole in an uninhabited section of the earth, where it quenches and disappears into nothingness. Mansfeld and Runia's "stepping into the void," Graham's "treading on nothing" and Bottler's "über das Leere" express this association less adequately.

Final Remarks

In the context of his time, Xenophanes' ideas about the cosmos were not so strange after all. Xenophanes was not a bold and innovating cosmologist like Anaximander. Rather, he was a more mundane agnostic, who did not so easily believe in things he could not see, such as the heavenly bodies going under the earth, and who did not so easily make guesses about dimensions he was not able to measure. On the other hand, believing like Anaximander that the earth is flat, he realized that the celestial bodies could not be as far away as Anaximander imagined, and that they must therefore be relatively small. This enabled him to dispose of Anaximander's conception of the celestial bodies as wheels and to explain them as meteorological phenomena like clouds. This may look strange to us, but in the context of a flat earth cosmology it is a plausible idea.

My interpretation of Xenophanes' cosmology is illustrated in Fig. 8.6, where the rising sun is seen as originating from incandescent clouds and tiny flares, and the setting sun, after it has finished its curved path along the dome of the heaven, as falling as if stepping into a hole, in opposite uninhabited areas on the edge of the earth, beyond the Ocean, the great Sea, which is the begetter of clouds. We can observe these effects in the phenomena of dawn and dusk. Sometimes it is thought that the two images of the rising sun as originating from incandescent clouds and as originating from tiny flares are irreconcilable. I do not think the question is very important, but we could imagine either a cloud becoming incandescent because of tiny flares, or tiny flares making a cloud incandescent. In Fig. 8.6, I drew both effects. The phenomenon of the sun rising from scattered fires can be seen when we climb a high mountain like Mount Ida (in the Troad, 5820 feet high), as reported in the doxography:

8.37 (. . .) from the peaks of Mount Ida
Scattered fires may be descried at daybreak
From which a mass, as it were, comes together in one and makes a sphere.[133]

The moon, which has a much fainter light, obviously does not produce such visible effects when it rises and sets. In order not to complicate Fig. 8.6, I left out the moon, planets and stars, all of which can be imagined as moving in arcs or circles along the dome of the heaven. Only the curved path of the sun during the equinoxes is shown. This noon sun is not in the zenith of the heavenly dome, because it is seen from Greece, which is supposed to be in the center of the flat earth.

Perhaps one could ask what is outside the cosmos presented like this. I think the answer must be: simply nothing, just like outside Aristotle's spherical cosmos there is nothing, or nothing of any interest or value. Or perhaps an archaic-sounding answer is more appropriate, as suggested by Schäfer: Xenophanes could have conceived the cosmos as a kind of breach or air-bubble in the surrounding boundless (ἄπειρον) Chaos: "Die (. . .) Anschauung, unser geordneter Kosmos sei nur eine Art 'Bresche' oder 'Luftblase' innerhalb eines ihm umgebenden undifferenzierten Chaos."[134] My personal assumption is that Xenophanes would have replied that an answer to this question would go far beyond the intellectual capacities of human beings.

[133]Lucretius 5.663–665 = Gr, Xns 62, not in DK, LM, and KRS. (see also Gr. Xns 63 and 64).
[134]Schäfer (1996, 141).

References[135]

Barnes, J., *The Presocratic Philosophers*. London: Routledge & Kegan Paul 1982.
Bicknell, P. J., Xenophanes' Account of Solar Eclipses, *Eranos* 65 (1967), 73–77.
Bottler, H., *Pseudo-Plutarch und Stobaios: Eine synoptische Untersuchung*. Göttingen: Vandenhoeck & Ruprecht 2014.
Couprie, D.L., *Heaven and Earth in Ancient Greek Cosmology*. New York: Springer 2011.
Dicks, D.R., *Early Greek Astronomy to Aristotle*. Ithaca N.Y.: Thames and Hudson 1970.
Diels, H., *Doxographi Graeci*. Berlin: Walter de Gruyter 1879 (=Dox).
Diels, H. and W. Kranz, *Die Fragmente der Vorsokratiker*. Zürich/Hildesheim: Weidmann 1951[6] (=DK).
Drozdek, A., *In the Beginning Was the* apeiron. Stuttgart: Franz Steiner Verlag 2008.
Dumont, J.-P., *Les Présocratiques*. Paris: Gallimard 1988.
Freeman, K., *The Pre-Socratic Philosophers. A Companion to Diels, Fragmente der Vorsokratiker*. Cambridge MA: Harvard University Press 1966.
Furley, D., *The Greek Cosmologists. Volume I: The Formation of the Atomic Theory and Its Earliest Critics*. Cambridge: Cambridge University Press 1987.
—, *Cosmic Problems. Essays on Greek and Roman Philosophy of Nature*. Cambridge: Cambridge University Press 1989.
Graham, D.W., *Explaining the Cosmos: The Ionian Tragition of Scientific Philosophy*. Princeton: Princeton University Press 2006.
—, *The Texts of Early Greek Philosophy*. Cambridge: Cambridge University Press 2010 (=Gr).
—, *Science Before Socrates*. Oxford: University Press 2013.
Guthrie, W.K.C., *A History of Greek Philosophy. Volume I. The Earlier Presocratics and Pythagoras*. Cambridge: Cambridge University Press 1962.
Heitsch, E., *Xenophanes. Die Fragmente*. München: Artemis Verlag 1983.
Kahn, C.H., *Anaximander and the Origins of Greek Cosmology*. Indianapolis/Cambridge Hackettt Publishing Company, Inc. 1994.
Kirk, G.S., J.E. Raven and M. Schofield, *The Presocratic Philosophers*. Cambridge: Cambridge University Press 2009 (=KRS).
Laks, A. and G.W. Most, *Early Greek Philosophy VI*, Cambridge MA 2016.
Lesher, J.H., *Xenophanes of Colophon*. Toronto: University of Toronto Press 1992 (repr. 2001).
Mansfeld, J. and D.T. Runia, Aëtiana. *The Method and Intellectual Context of a Doxographer, Volume Two, Part Two*. Leiden: Brill 2009 (=MR).
Lumpe, A., *Die Philosophie des Xenophanes von Kolophon*. München: W. Foth 1952.
Mourelatos, A.P.D., La Terre et les étoiles dans la cosmologie de Xénophane. In: A. Laks and C. Louguet (eds.), *Que'st-ce que la philosophie Présocratique?* Villeneuve d'Ascq: Presses Universitaires du Septentrion 2002, 331–350.
—, The Cloud-Astrophysics of Xenophanes and Ionian Material Monism. In: P. Curd and D.W. Graham, *The Oxford Handbook of Presocratic Philosophy*. Oxford: University Press 2008, 134–168.
Mueller, I., *Simplicius, On Aristotle's De Caelo 2.10-14*. Ithaca NY: Cornell University Press 2005.
Runia, D., Xenophanes on the Moon; a *doxographicum* in Aëtius. *Phronesis* 34 (1989), 245–269.
Schäfer, C., *Xenophanes von Kolophon. Ein Vorsokratiker zwischen Mythos und Philosophie*. Stuttgart und Leipzig: Teubner 1996.
Tannery, P., Pour l'histoire du mot 'apeiron'. *Revue de Philosophie* 5 (1904), 703–707.
Tumarkin, A., Der Begriff des ἄπειρον in der griechischen Philosophie. *Jahrbuch der Schweizerischen Philosophischen Gesellschaft* 3 (1943), 55–71.

[135] P = Aëtius in pseudo-Plutarch, *Placita* (numbering according to Dox).
S = Aëtius in Stobaeus, *Anthologium* (numbering according to Wachsmuth and Hense).

Untersteiner, M., *Senofane: Testimonianze e frammenti*. Milano: Bompiani 2008 (repr. of Florence: La Nuova Italia 1956).
Tannery, P., Histoire du concept de l'infini au VIe siècle avant J.-C. *Revue philosophique de la France et de l'étranger* 15 (1882), 618–636.
West, M.L., *Early Greek Philosophy and the Orient*. Oxford: Clarendon Press 1971.

Chapter 9
Anaxagoras on the Milky Way and Lunar Eclipses

Contents

Introduction

In this chapter, I will explore the interrelations between three astronomical theories that are attributed to Anaxagoras. The first theory is the explanation of the Milky Way as effectuated by the shadow of the earth. The second is the explanation of eclipses of the moon as caused by the earth's shadow. The third is the explanation of eclipses of the moon as due to invisible heavenly bodies below the moon. I will examine how well these theories are attested, to what extent they are mutually compatible, and whether or not they harmonize with Anaxagoras' other astronomical conceptions, particularly that of a flat earth. In Chap. 10, some consequences will be drawn regarding the light and phases of the moon. Chap. 11 will address the question of how Anaxagoras could have measured the distance of the sun.

D. L. Couprie, *When the Earth Was Flat*, Historical & Cultural Astronomy,
https://doi.org/10.1007/978-3-319-97052-3_9

The Milky Way

The Milky Way is visible as a band of varying angular width (roughly 30°) in the night sky.[1] An ancient legend has it that the Milky Way derives its name from the flow of milk that poured from the breast of Juno, heaven's queen. Another story tells us that it is the path through which the souls of heroes pass to heaven. Some people assumed that the Milky Way was the seam where the two hemispheres of heaven were sewn together. Others feared that the firmament was about to split in two.[2] The Milky Way also worried several Presocratics. Metrodorus is said to have identified the Milky Way as the sun's path among the stars. A similar theory, which says it is the former path of the sun, is ascribed to the Pythagoreans and Oenopides. The Pythagoreans seem to have linked the Milky Way with the fall of Phaëton, while Oenopides adds that the direction of the sun's course reversed on that occasion. Others are said to hold that the Milky Way is a reflection of our vision to the sun. Parmenides maintains that a mixture of dense and thin produces the milky color. Anaxagoras explains the Milky Way as a band of stars that light up in the earth's shadow.[3] These ideas unmistakably illustrate how little was understood of the heavenly phenomena at that time. This should be a warning sign for those scholars who are inclined to attribute to the ancient Greek thinkers—in this case, to Anaxagoras—all kinds of astronomical knowledge they did not possess.

Anaxagoras on the Milky Way

As we shall see, Anaxagoras' idea, strange and wrong as it was, can be regarded as one of the first attempts at a rational explanation of the Milky Way in natural terms, supported by an optical theory. Moreover, I will argue that it implied a new answer to the question of why it is dark at night. Anaxagoras' explanation of the Milky Way is well documented: Gershenson and Greenberg, who classify it into their first category "Reliable Traditions," count seven testimonies,[4] the first of which is by

---

[1]For a general review, see Aristotle, *Meteorologica* 345a11–345b32; see also Jaki (1973), 1–32.

[2]Cf. Manilius, *Astronomica* 1.718–761.

[3]For Metrodorus, see P 3.1.3 = S 1.27.1 = DK 70A13; not in LM, Gr, and KRS. For the Pythagoreans, see Aristotle, *Meteorologica* 345a17 = DK 41A10 = LM PYTHS.ANON. D44, not in Gr and KRS; see also P 3.1.2 = S 1.27.1 = DK 58B37c = LM PYTHS.ANON. D45, not in Gr and KRS; see also Manilius, *Astronomica* 1.735–744. For Oenopides, see Achilles Tatius, *Introductio in Aratum* 1.24 = DK 41A10, not in LM, Gr and KRS. For "some others" (perhaps Hippocrates and Aeschylus), see Aristotle, *Meteorologica* 345b9 = DK 42A6; not in LM, Gr, and KRS. For Parmenides, see P 3.1.4 = S 1.27.1 = DK 28A43a = LM PARM. D24 = Gr Prm38; not in KRS. For Anaxagoras, see texts below.

[4]GG, p. 333. Modern handbooks, textbooks, and monographs, however, are rather reticent in providing information about this topic.

Aristotle, who attributes this explanation of the Milky Way not only to Anaxagoras but also to Democritus:

9.1. (1) The schools of Anaxagoras and Democritus posit that the Milky Way is the light of certain stars, (2) for the sun, in its course beneath the earth, does not see (οὐχ ὁρᾶν) [i.e. does not shine upon] some (ἔνια) of the stars. (3) The light of the (stars) (ὅσα μὲν) upon which the sun does shine in the round (περιορᾶται) is of course (μὲν οὖν) not visible (οὐ φαίνεσθαι), for it is prevented (κωλύεσθαι) by the rays of the sun. (4) But those (ὅσοις δ') which are screened (ἀντιφράττει) from the sun by the interposed earth so that it does not shine (μὴ ὁρᾶσθαι) upon them, the light proper to these (οἰκεῖον φῶς), they say, is the Milky Way.[5]

The optical theory behind this explanation of the Milky Way is that lights are more visible in the dark. This is why the stars that lie in the band of the earth's shadow—the Milky Way—are seen to glow more brightly (see also the last lines of text 9.2). Aëtius (in text 9.5) only mentions Anaxagoras and ascribes to Democritus the theory that the Milky Way is the combined light (συναυγασμός) of many stars that are close to one another (διὰ τὴν πύκνωσιν).[6] So it seems that Aristotle's attribution of this theory not only to Anaxagoras but also to Democritus was less accurate, as Diels already remarked.[7]

I have divided Aristotle's text into four clauses, in order to make it easier for the reader to follow the complicated discussion. The usual reading of Aristotle's text is that it describes the situation at night and that the theory of the Milky Way is expressed in clauses (1), (2), and (4). The problem is, then, the third clause: "The light of those (stars) upon which the sun does shine in the round is of course not visible, for it is prevented by the rays of the sun," which is usually understood as having a bearing on the stars at night on either side of the Milky Way. This leads to the strange consequence that most stars would not be visible at night. Lee, for instance, comments on this interpretation: "what is not easy to understand is why, on Anaxagoras' theory, we see any stars outside the Milky Way."[8] The originator of this weird interpretation seems to have been Alexander of Aphrodisias:

9.2. Anaxagoras and Democritus say that the Milky Way is the light of certain stars. They say that at night, when the sun goes under the earth, its rays shine upon some of the stars above the earth (ὅσα περιλάμπει τῶν ὑπὲρ γῆς ὄντων ἄστρων), mask their light, and prevent them from being seen. The stars shielded (ἐμποδιζόμενον) by the earth's shadow (ἡ σκιὰ τῆς γῆς) are hidden from the

[5]Aristotle, *Meteorologica* 345a25–31 = DK 59A80 = LM ANAXAG. D49 = GG 37 = Gr Dmc69; not in KRS. Graham translates οἰκεῖον φῶς as "natural light." Gemelli Marciano, Anaxagoras 55, reads μὲν νῦν instead of μὲν οὖν.

[6]For Aëtius on Democritus and the Milky Way, see P 3.1.6 = S 1.27.1 = DK 68A91 = LM ATOM. D98a; not in Gr and KRS.

[7]Cf. Dox 230.

[8]Lee (1962), 59, note d.

light of the sun and are not illuminated by it. These stars are visible, and their light is the Milky Way.[9]

Unfortunately, in DK, Alexander's text (9.2) is placed between those on Democritus, just behind a reference to Aristotle's text (9.1.). Most authors on Anaxagoras do not even mention it, and if so, they could easily get the impression that the two texts say the same thing. However, while Aristotle (in 9.1, second clause) says that the sun at night, when it is under the earth, does *not* shine upon some stars, Alexander says that the sun, when it is under the earth, *does shine* upon some stars, and then he construes a nonsensical theory that at night the light of these stars is outshone by the sun. According to Alexander, "the sun's rays mask their light, and prevent them from being seen." As already remarked by Tannery, Gomperz, and Heath, this idea could easily be disproved by simple observation.[10] Actually, Alexander in a very confusing way combines Aristotle's second and third clauses.

Cleve has argued that Aristotle meant to say that the sun, by shining on the stars outside the Milky Way, causes their own innate light not to be seen, but instead the reflection of the sun's light from the stars, while the stars in the Milky Way shine with their own light.[11] However, this is not what Aristotle's (9.1) or Alexander's text (9.2) says, but it is taken from Olympiodorus' attempt to provide clarification, which is not very helpful either:

9.3. A third view is that of Anaxagoras and Democritus. They say the Milky Way is the proper light of stars not illuminated (μὴ φωτιζομένων) by the sun. For they say that the stars have their own light on the one hand and the light obtained (ἐπίκτητον) from the Sun on the other. And the Moon proves this. For its own light is of one sort, the light [that it receives] from the Sun is of another; for its own light is coal-like, as it is evident from its eclipse (ἔλλειψις). But, they say not all stars receive light [from the Sun]. The [stars] which do not receive [light from the Sun] produce the circle of the Milky Way.[12]

Olympiodorus (a late source, sixth century A.D.) introduced the confusing idea, which is not in Aristotle's text, that the stars, in addition to their own light, also get light from the sun. This is the opposite of what Aristotle was saying when he spoke of the sun's light preventing us from seeing stars. Olympiodorus' explanation has the strange consequence that the stars outside the Milky Way, having both their own light and additional light from the sun, would be brighter than those of the Milky Way, which only have their own light. Moreover, Olympiodorus uses the example of

[9]Alexander Aphrodisiensis, *In Aristotelis Meteorologica*, 37.28 (on 345a11) = DK 68A91 = GG 287; not in LM, Gr, and KRS. Also not in Gemelli Marciano (2007–2013); Mansfeld (1986); and Curd (2010).

[10]Cf. Tannery (1887), 279; Gomperz (1896), 179; Heath (1913), 84.

[11]Cf. Cleve (1949), 70.

[12]Olympiodorus, *In Aristotelis Meteora*, 67.32 = Gr Axg46 = GG 607; not in LM, Gr, and KRS. GG translate the word ἔλλειψις as "that part of it [sc. the moon] that does not shine," whereas here obviously the equivalent of ἔκλειψις, "eclipse" is meant; see LSJ, s.v. ἔλλειψις.

an eclipse of the moon caused by the shadow of the earth, without noticing that this is at odds with Anaxagoras' explanation of the Milky Way, as we will see. Graham, after having quoted both Aristotle's text (9.1) and that of Olympiodorus (9.3), follows Olympiodorus as if this were the right and only interpretation, and then he comments: "Aristotle distinguishes between the natural light of certain stars and reflected light."[13] In Aristotle's text, however, there is not a single word on reflected light and this distinction.

These attempts to make sense of Aristotle's rendition of Anaxagoras' theory of the Milky Way are not very successful, to say the least. If we try to read Aristotle's text with an eye, unbiased by these confusing suggestions, I think it makes sense to assume that the second clause of text 9.1, "The sun, in its course beneath the earth, does not shine upon (literally: "does not see"—οὐχ ὁρᾶν) some of the stars," explains in a general way why the stars shine at night. During the night, the sun is under the earth, which prevents it from "seeing" the stars. More precisely, Aristotle says that at night the sun "does not shine upon *some* (ἔνια) of the stars", because it shines upon ("sees") the other half of the stars that are under the earth. My suggestion is to read the words "the sun does not see some of the stars" as meaning that at night the sun is so far away from the stars that its light becomes too weak to reach them, and thus the sun does not "see" them. In other words, in the second clause, Aristotle refers to Anaxagoras' theory why it is dark at night. When the shadow of the earth does not cover the whole sky at night, another explication must be found for the darkness of the night sky. We might add that, because the earth is flat, it takes some time for the sun to come under the earth: this is the time of day that we call twilight, when increasingly stars become visible because they are no longer "seen" by the sun (and the opposite phenomenon in the morning is called dawn).

Further, I think it makes sense to assume that Aristotle's third clause, "The light of the stars upon which the sun does shine all around (literally "sees all around," περιορᾶται) is of course not visible," has nothing to do with the stars at night, but should be read as an explanation of why we do not see the stars by day, namely because their light is overpowered by that of the sun above the earth. During the day, the light of the sun is everywhere in the sky above us; this is what the metaphor of "seeing all around" says. The interjection "of course" underlines that the third clause formulates something obvious and not some strange theory.

In this interpretation, the second and third clauses of text 9.1 do not deal specifically with the Milky Way, but outline the general background as to why the stars shine at night and not by day, against which the theory of the Milky Way must be understood. The first and fourth clauses of text 9.1 are about the stars visible in the shadow of the earth (the Milky Way). The stars at night shine in the dark because they are not shone upon by the sun, but the stars of the Milky Way shine in the even deeper dark of the earth's shadow. To sum up: according to Aristotle, the theory of Anaxagoras is that the sun during the day, when it is above the earth, outshines the stars (clause 3); when the sun is under the earth, the stars are visible because the sun

[13] Graham (2013a), 131.

does not "see" them (clause 2). A special category of stars that are not "seen" (μὴ ὁρᾶσθαι) by the sun are those that are in the shadow of the earth and that appear as what we call the Milky Way (clauses 1 and 4). Of course, for us it remains difficult to understand that, according to Anaxagoras, distinction must be made between the general idea of stars that shine because they are not seen by the sun under the earth and the more specific conception of stars that shine even more brightly because they are in the shadow of the earth.

Puzzling as Aristotle's text (9.1) is, and even more enigmatic as it has become through the intervention of its commentators, it does not interfere with the main argument of this chapter. Whatever the interpretation of the second and third clause of text 9.1, its core remains that the Milky Way results from the shadow of the earth. Given that The Milky Way is visible as a band of roughly 30° in the night sky, the shadow of the earth should, in Anaxagoras' conception, cover about 30° of the sphere of the stars. This implies that the sun must be smaller than the earth (and relatively nearby). This observation is confirmed by Aristotle' argument that Anaxagoras' theory of the Milky Way cannot be right because, in fact, the opposite is the case:

9.4. Astronomical researches have now shown that the size of the sun is greater than that of the earth (. . .) therefore the vertex of the cone formed by the rays of the sun will not fall very far from the earth, nor will the earth's shadow (. . .) reach the stars.[14]

When Aristotle claims that the sun is larger than the earth (and thus relatively far away), casting a conical shadow beyond the earth, he implies that in Anaxagoras' theory the earth's shadow must be widening to cover the width of the Milky Way, and that the sun must therefore be relatively close and smaller than the earth.[15] Figure 9.1 gives an impression of how, in Anaxagoras' conception, the Milky Way is dependent on the shadow of the earth. I also tried to imitate the diminution of the sun's rays.

Aristotle's testimony (in text 9.1) is repeated by several authors. Aëtius seems to confirm the interpretation given above:

9.5. Anaxagoras (holds) that the shadow of the earth rests upon this section of the heaven [viz. where the Milky Way is visible] when the sun, having arrived under the earth, no longer illuminates everything (μὴ πάντα περιφωτίζῃ).[16]

I read the phrase "when the sun, having arrived under the earth, no longer illuminates everything" as simply referring to the sky at night (unlike during the day, when the sun "illuminates everything"), while the shadow of the earth is supposed to rest on a special section of the night sky, namely the Milky Way.

[14] Aristotle, *Meteorologica* 345b1–8.

[15] Cf. Guthrie (1965), 309: "He (sc. Aristotle) attacks it from the standpoint of greater astronomical knowledge, for it demands that the sun be smaller than the earth, whereas he knew it to be greater."

[16] P 3.1.5 (= S 1.27.1) = DK 59A80 = GG 173; not in LM, Gr and KRS.

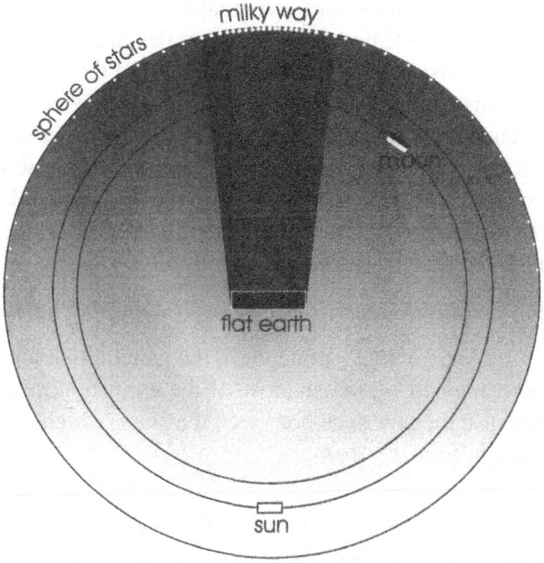

Fig. 9.1 The Milky Way caused by the shadow of the earth, according to Anaxagoras (approximately to scale) (I have drawn not only the earth but also the sun and the moon as flat disks, as was probably Anaxagoras' understanding)

Hippolytus also mentions Anaxagoras' explanation of the Milky Way, with almost the same words as Diogenes Laërtius. Apparently, they drew on the same source.

9.6. The Milky Way is the reflection (ἀνάκλασις) of the light of stars that are not illuminated by the sun.[17]

9.7. The Milky Way is the reflection of the light of stars that are not illuminated by the sun.[18]

The word "reflection" is inaptly chosen for stars that are not illuminated by the sun. Mansfeld rightly remarks that Aëtius' text (9.5) does *not* speak of reflection,[19] and we might add: neither does Aristotle's (9.1). In all these texts (9.2, 9.3, 9.5–9.7) we find the same kernel as in Aristotle (text 9.1): the Milky Way is the result of stars shining more brightly in the shadow of the earth. Anaxagoras' theory was already criticized by Aristotle, who argued that the position of the Milky Way among the stars is always the same but that, if it were the result of the earth's shadow, it would change with changes in the sun's position.[20] Moreover, the shadow of Anaxagoras' flat earth would not be a band across the sky but would show the shape of a circular disk high in the sky at midnight, moving during the night and changing its shape into

[17]Hippolytus, *Refutatio Omnium Haeresium* 1.8.10 = DK 59A42(10) = LM ANAXAG. D4 (10) = Gr Axg38(10) = GG 271; not in KRS.

[18]Diogenes Laërtius 2.9 = DK 59A1(9) = Gr Axg37(9) = GG 340; not in LM and KRS.

[19]See Mansfeld (2010), 489 n. 40. Cf. Ferguson (1968), 100: "This cannot mean 'reflection' unless the doxographers have wholly misunderstood Anaxagoras."

[20]Cf. Aristotle, *Meteorologica* 345a33–38.

an elliptical disk and eventually into a straight stripe at dawn.[21] Anaxagoras' explanation of the Milky Way implies that he has no idea of where the sun actually stands during the night. In the context of this chapter, however, it is not our concern that his explanation of the Milky Way is strange or wrong, but that it is well documented and attributed to Anaxagoras by a witness as early as Aristotle. Anaxagoras' conception of the Milky Way as caused by the earth's shadow can be conceived of as a bold break with the traditional idea that darkness at night is caused by the earth blocking the light of the sun. This idea was worded, perhaps meant as polemical against Anaxagoras, by Empedocles:

9.8. Earth produces night by obstructing the light [of the sun].[22]

In Anaxagoras' conception, the general phenomenon of darkness at night cannot be explained by the shadow of the earth but must be explained otherwise, namely because the sun does not "see" the stars, which means that its rays are not powerful enough to reach them.

Introductory Remarks on Eclipses

Heavenly bodies sometimes disappear from sight. These disappearances can be subdivided in regular and irregular disappearances as well as in partly and totally disappearances. The sun, the moon, the planets, and the stars regularly set and then disappear completely out of sight. There is some regularity in eclipses, because solar eclipses always occur when the moon is new and lunar eclipses when the moon is full. For the Presocratic Greeks, however, the exact date and the magnitude of an eclipse remained unpredictable. The moon can also occult stars and planets, but since we do not possess reports of such occultations from the ancient Greeks, we can leave them out of account.[23] During the month, the moon shows phases, in which it gradually disappears, is out of sight for a few days, and appears increasingly again until it is completely visible.

The first attempts to understand these phenomena of disappearing heavenly bodies had the intention to give one uniform explanation of as many kinds of disappearances as possible. Anaximander regarded the eclipses of the sun and moon and the moon's phases as the complete or partial closure of the aperture in their celestial wheels. As he imagined the celestial wheels of sun, moon, and stars to

[21]Another theoretical possibility would be to imagine Anaxagoras' earth not as a disk but as oblong, which would fit better the shape of the Milky Way (cf. Heath 1913, 84). This would however not affect the argument of this chapter.

[22]Plutarch, *Platonic Questions* 1006e = DK 31B48 = LM EMP. D131 = Gr Emp76.

[23]Perhaps star occultations should be added to the list, but, as far as I know there are no reports of star occultations in Greece from this early period. According to Stephenson (1997), 47, "tens of observations of this kind are described in Babylonian history, but East Asian history is replete with such reports."

turn around the earth, we can assume that he considered their settings to be their becoming invisible under the earth. In this regard, Xenophanes, who explained settings, eclipses, and phases all alike as quenchings, was the most consequent thinker. Another tendency was to make eclipses more like occultations. Some unnamed thinkers explained solar eclipses by invisible condensations of clouds passing in front of the sun (text 9.13). A similar explanation of lunar eclipses is also attributed to Anaxagoras (texts 9.9 and 9.11). The Pythagoreans seem to have been the first to state that an eclipse of the sun occurs when the moon is between the earth and the sun. In these explanations of eclipses, some celestial body (an invisible body, or the moon) comes between the observer and the eclipsed body. The Pythagoreans considered the earth as another celestial body (and not as the central body *sui generis*) and thus could argue that in the case of settings the earth is the celestial body between the observer and the setting sun, moon, planet, or star.

Before we start the investigation of Anaxagoras' theory of lunar eclipses, we must pay attention to yet another phenomenological distinction between two kinds of disappearances of heavenly bodies. The first kind comprises eclipses of the sun, but also occultations of stars or planets, and the settings of the sun, moon, stars or planets. In solar eclipses, occultations, and settings, there is a heavenly body, usually the moon but in settings the earth, between the observer and the eclipsed, occulted or setting body, blocking the sight of the observer.[24] In solar eclipses, occultations, and settings, the order is always: observer—blocking body—eclipsed or occulted body, all three of which must be aligned. Shadow does not play an explanatory role in these phenomena. The second kind of disappearances consists of only one species, namely that of lunar eclipses. During lunar eclipses, it is not a heavenly body between the observer and the eclipsed body that blocks his sight of the eclipsed moon, but the shadow of the earth on the moon, when the earth blocks the light of the sun. The order is also different and requires four instead of three items: light source (the sun)—shadow-throwing body (the earth)—observer—eclipsed body (the moon). Moreover, in this case only the three heavenly bodies must be aligned, but there is no direct need for alignment of the observer. This can be easily demonstrated by ordinary shadows that fall on objects. When I observe the shadow of a tree, I do not have to be in line with the sun, the tree, and the object on which the shadow falls, and usually I am not. In the same way, a lunar eclipse can be observed from outside the alignment of sun, shadow-throwing body, and moon. Pythagoreans used this argument when they argued that lunar eclipses could also be caused by the counter-earth.[25] From this analysis we learn that in ancient times the understanding of the true cause of lunar eclipses must have been much more complicated than that of solar

[24]Similarly, the sight of a heavenly object and, for that matter, any other object, can be blocked by another object, for instance a bird, a tower, our own hand, or whatever. We do not usually call these events "eclipses" or "occultations," although we may say, for example, that the sun is obscured by a cloud or by volcanic dust.

[25]See Aristotle, *De caelo* 293b25–29.

eclipses.[26] It helps us also understand why, as we will see, Anaxagoras tried to explain lunar eclipses in the same way as solar eclipses and occultations, by imagining invisible heavenly bodies between us and the moon. However, let us not anticipate the conclusions of this chapter.

Anaxagoras' Alleged Explanation of Lunar Eclipses

The most often quoted report about Anaxagoras and lunar eclipses is in a text of Hippolytus, just before he mentions Anaxagoras' explanation of the Milky Way:

9.9. (Anaxagoras says) there are below the stars certain bodies invisible to us which are carried around with the sun and moon. (…) *The moon is eclipsed when the earth blocks it*, or sometimes (ἐνίοτε) one of the bodies below the moon. (…) He was the first to determine (ἀφώρισε πρῶτος) what is involved in eclipses and illuminations.[27]

Graham's translation, "He first *correctly* explained eclipses and illuminations" (my italics) says more than the text expresses because, according to Hippolytus, Anaxagoras had two explanations of lunar eclipses. In this section, I will discuss what Hippolytus, in the italicized line above, presents as Anaxagoras' main theory about eclipses of the moon. This resembles the well-known explanation that we still adhere to: the moon is eclipsed when it enters wholly or partly into the shadow of the earth, because at that time the earth is between the sun and the moon, as shown in Fig. 9.2.

We must, however, make seven provisos from which it can be concluded that the drawing in Fig. 9.2 do not reflect the explanation attributed to Anaxagoras:

1. In ancient Greek writings there are no reports of the earth's penumbra or of penumbral eclipses.
2. Anaxagoras believed that the earth is flat. The shadow of a spherical earth on the eclipsed moon will always show the curve of a part of a circle, while the shadow of a flat earth would show a variety of shapes, depending on the positions of the sun and the moon relative to the earth's surface: a part of a circle high in the sky, a part of an ellipse halfway the horizon, and a straight line at the horizon.

---

[26]Bakker (2013), 686, points to the fact that Aristotle and Aëtius used the Greek term ἔκλειψις for eclipses and not for the waning of the moon. The difficulty is, however, in how far their knowledge of the difference between the two (and other phenomena of occultation) prevented them to understand and render truthfully the opinions of the Presocratics who were not yet able to make these differences and even tried to explain as many as possible of these events by the same theory. The confusion between eclipses and moon phases in Aëtius 2.29, which Bakker analyses in his paper, are also due to this misunderstanding.

[27]Hippolytus, *Refutatio Omnium Haeresium* 1.8.6, 1.8.9 and 1.8.10 = DK 59A42(6, 9, and 10) = LM ANAXAG. D4(6, 9, and 10) = Gr Axg38(6, 9, and 10) = GG 270 = KRS 502 (except the here relevant part of 1.8.10).

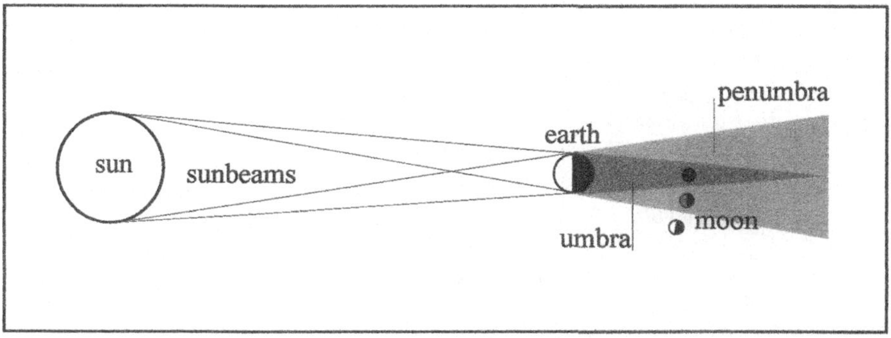

Fig. 9.2 The standard explanation of a lunar eclipse (not to scale)

3. When the earth is considered to be flat, one implication is that the heavenly bodies are nearby and thus relatively small. The belief that the sun is near is implicit in the report, attributed to Anaxagoras, that the stone that fell from heaven in Aegospotami had broken off from the sun.[28] The reports of Anaxagoras claiming that the sun is greater than the Peloponnesus suggest that the sun is smaller than the earth.[29] With a flat earth, the fact that the sun is close by (and thus smaller than the earth) can be easily demonstrated by extending Thales' measurement of the height of a pyramid to the measurement of the sun's distance: In Athens, at noon on the summer solstice, the length of a gnomon is roughly four times its shadow. Accordingly, on a flat earth, the distance from the sun to the sub-solar point (on the Tropic of Cancer) is calculated as roughly four times the distance from Athens to the Tropic of Cancer. This method has already been explained in Fig. 3.9, but see especially Fig. 11.13.
4. When the sun is smaller than the earth, the flat earth's shadow will be widening and not conical. As shown in the previous section, Anaxagoras' explanation of the Milky Way presupposes that the earth's shadow is widening and thus that the sun is smaller than the earth.
5. A widening shadow would produce other shadow lines (sections of a larger circle) on the partly eclipsed moon than a conical shadow.
6. A widening shadow would imply that the moon is eclipsed more often and over a longer period than in the case of a conical shadow.
7. Anaxagoras believed, probably, that the sun and the moon were flat disks.

---

[28]See Diogenes Laërtius, *Vitae Philosophorum* 2.10 = DK 59A1(10) = Gr Axg1(10) = GG 340 = KRS 503; not in LM.

[29]See P 2.21.3 (not in S) = DK59A72 = GG 169 = MR 534, not in LM, Gr, and KRS; Hippolytus, *Refutatio Omnium Haeresium*, 1.8.8 = DK 59A42(8) = LM ANAXAG. D4(8) = Gr Axg38 (8) = GG 270 = KRS 502(8); Diogenes Laërtius, *Vitae Philosophorum* 2.8 = DK 59A1(8) = Gr Axg37(8) = GG 340, not in LM and KRS. Cf. Dreyer (1953), 31: "the sun (...) greater than the Peloponnesus, and therefore not at a very great distance from the earth."

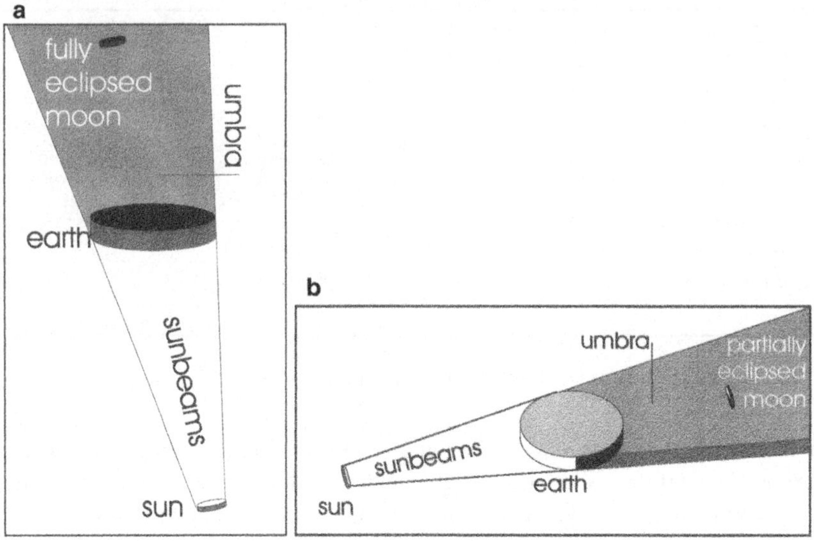

Fig. 9.3 (**a**) Flat earth causing a (total) lunar eclipse at night (approximately to scale). (**b**) Flat earth causing a (partial) lunar eclipse at dawn (approximately to scale)

I have tried to draw what eclipses of the moon would look like in this interpretation of Anaxagoras' explanation of them.[30] Figure 9.3a shows the situation of a totally eclipsed moon in the widening shadow of the earth. Figure 9.3b shows that the shadow line on the partly eclipsed moon on the horizon should be a straight line.

In pseudo-Plutarch's version of Aëtius we read the same explanation of lunar eclipses as in Hippolytus. It is noteworthy that Anaxagoras is not mentioned in this report:

9.10. Plato, Aristotle, and the Stoics agree with the astronomers (. . .) that eclipses of the moon occur when it enters the earth's shadow, when the earth comes between the two heavenly bodies.[31]

In the version of Stobaeus, however, the name of Aristotle has disappeared and been replaced by those of Thales and Anaxagoras. Moreover, Stobaeus attributes a second explanation of lunar eclipses to Anaxagoras, which we will discuss later:

[30]This picture is inspired by Graham 2013a, 130, Figure 4.2. Graham draws parallel instead of widening shadow lines, even though he draws the sun close by and smaller than the earth. Elsewhere, however, when he discusses a solar eclipse, he argues (wrongly) that "Anaxagoras must presume that (. . .) the sun (is) relatively far away" (Graham 2013a 148 and 151). See also Graham and Hintz 2007, 321: "Assuming that the sun was far distant from the earth." But when the sun is far away it must be much larger than the earth and the shadow of the earth must be conical.

[31]P 2.29.6 = MR 614; not in DK, LM, Gr, and KRS, but see Dox 360. Bakker (2013), 685, n.5 mentions a minor difference between the versions of pseudo-Plutarch and Stobaeus, but overlooks the major difference in the names mentioned in the two versions.

9.11. Thales, Anaxagoras, Plato and the Stoics agree with the astronomers (...) that eclipses of the moon occur when it enters the earth's shadow, when the earth comes between the two heavenly bodies. Theophrastus says that Anaxagoras held that eclipses also occur when bodies below the moon happen to obstruct it.[32]

Given these two versions, it is debatable whether Anaxagoras was mentioned at all in Aëtius' original text in relation to the theory that the shadow of the earth produces eclipses of the moon. When we look at the matter in the context of Anaxagoras' other astronomical ideas, pseudo-Plutarch's version (text 9.10) makes more sense.[33] He mentions three schools that were convinced that the earth is spherical and in which Anaxagoras is understandably not included. The concept of a spherical earth fits very well with the standard explanation of eclipses of the moon. The curved shape of the earth's shadow can thus be easily explained, which is not the case with Anaxagoras' assumption of a flat earth. Aristotle builds the question of the shadow lines into one of his proofs that the earth is a sphere.[34] Moreover, if the earth is spherical, the sun must be much larger than the earth and at a great distance, which results in the earth's shadow being conical, as Aristotle already concluded, and not widening (see text 9.4 and Fig. 9.2). The shadow of the earth on the moon shows a width of about 1.5°. This is at odds with the widening shadow of a flat earth, which would cover roughly 30° of the night sky and cause the Milky Way, as was the view of Anaxagoras.

Theon of Smyrna says that it was Anaximenes who discovered how the moon is eclipsed:

9.12. [Eudemus reports that] Anaximenes [was the first] to discover that the moon has its light from the sun and how it eclipses.[35]

Several scholars, and more recently Panchenko, have argued that we should read "Anaxagoras" instead of "Anaximenes."[36] I prefer to follow O'Brien, who suggests that "Eudemus said simply that Anaximenes gave *an* interpretation of the moon's

---

[32]S 1.26.3 = DK 59A77 = LM ANAXAG. D45a and b = GG 486 = MR 614; not in Gr and KRS.

[33]My conclusion here is different from that of MR 617, where it is stated that "Not only (...) does P delete the names of Thales and Anaxagoras (perhaps to avoid the doublet), but he also adds that of Aristotle" and where, finally, a text is offered with all the names mentioned together by pseudo-Plutarch and Stobaeus (MR 621–622). The reasons they adduce have to do with text-critical considerations about the usual methods of the doxographers. My attempt tries to see which of the two versions makes more sense in the context of what we know about Anaxagoras' other astronomical opinions and intends to show that it is not "somewhat unexpectedly" that "the first two names Thales and Anaxagoras are dropped" (MR 615). Moreover, the "standard explanation" of a lunar eclipse as caused by "the moon sink into the *conical* shadow of the earth (MR 616, my italics) cannot be said of Anaxagoras, nor from any other Presocratic flat earth cosmologist.

[34]Aristotle, *De Caelo* 297b23–31. See Chap. 12 below.

[35]Theon of Smyrna, *Expositio* 198.14-199.3 = DK 13A16 = LM ANAXIMEN. R9; not in Gr and KRS, but see p. 156, n.1.

[36]See Panchenko (2002a), 324–326. He mentions others scholars in n. 6.

eclipse," perhaps caused by invisible bodies, which would be compatible with his idea of a fiery moon, and that Theon inaccurately converted this into the suggestion that Anaximenes gave the correct explanation of a lunar eclipse.[37]

The Incompatibility of Anaxagoras' Theory of the Milky Way with His Alleged Explanation of Lunar Eclipses

The Milky Way is a permanent phenomenon, visible every night. Lunar eclipses, on the contrary, are rare phenomena. During Anaxagoras' lifetime, 31 of them were visible in Athens. The inevitable conclusion of the combination of the theories that the moon is eclipsed by the shadow of the earth on the one hand and that the Milky Way is caused by the earth's shadow on the other hand is that the eclipsed moon would always be seen against the background of the Milky Way. In reality, this is not the case, as evidenced by simply observing lunar eclipses. Of the 31 eclipses of the moon that took place during Anaxagoras' lifetime, only eight took place when the full moon was in conjunction with the Milky Way.[38] In other words, the theory of the Milky Way as caused by the shadow of the earth is incompatible with the theory that eclipses of the moon are caused by the shadow of the earth.

I am not the first to note that these two theories about the earth's shadow are incompatible.[39] More than a century ago, several scholars noted that the two theories about the earth's shadow cannot coexist. Tannery remarks: "la lune aurait dû s'éclipser toutes les fois qu'elle traverse la voie lactée, conséquence dont il était également facile de verifier la fausseté."[40] Gomperz writes: "und warum tritt nicht eine Verfinsterung des Mondes ein so oft dieser über die Milchstraße hingeht?"[41] And Heath comments: "if the theory were true, an eclipse of the moon would have been bound to occur whenever the moon passed over the Milky Way and it would have been easy to verify that this is not so."[42] In more recent times, Fehling also concludes, that "seine (...) Erklärung der Milchstraße (...) mit der richtigen Erklärung der Mondfinsternisse (...) unvereinbar ist."[43] Panchenko remarks about the attribution of this theory of the Milky Way to Anaxagoras: "But this is incompatible with other evidence on Anaxagoras' views."[44]

[37]O'Brien (1968), 117.

[38]I used the computer program RedShift 8 Premium.

[39]Nevertheless, there are still scholars who do not recognize the discrepancy. Rechenauer (2013), 777, for instance, mentions the two explanations without noticing any problem.

[40]Tannery (1887), 279.

[41]Gomperz (1896), 179.

[42]Heath (1913), 84.

[43]Fehling (1985), 211. I thank Dmitri Panchenko for drawing my attention to this text.

[44]Panchenko (2013).

As far as I know, no one has yet drawn the conclusion that we must try to determine which of the two theories of the earth's shadow has actually been proposed by Anaxagoras. It is hard to imagine that he would have defended the two conflicting theories at the same time, unless we wish to portray him as a confused fool. If we refuse to accept that Anaxagoras was completely confused, only two options remain: either Anaxagoras was not the author of the idea that the phenomenon of the Milky Way is caused by the shadow of the earth, or he was not the author of the explanation of eclipses of the moon caused by the shadow of the earth.[45]

The results of the textual arguments indicate that Anaxagoras' explanation of the Milky Way is well documented, but that the attribution to him of the right explanation of lunar eclipses depends mainly on the report of Hippolytus. The result of the contextual and observational arguments is that Anaxagoras' theory of the Milky Way harmonizes with his astronomy, but that his alleged theory that lunar eclipses are caused by the earth's shadow is difficult reconcile with the rest of his astronomical ideas, and especially with the concept of a flat earth. If these considerations are correct, it seems plausible that pseudo-Plutarch's version (text 9.10), in which Anaxagoras is not named, represents Aëtius' original text. In that case, Hippolytus (text 9.9) remains the only authority to rely on for the attribution of the generally accepted theory of lunar eclipses to Anaxagoras.[46] We may wonder how reliable Hippolytus' report is, since he mentions it in the same breath together with Anaxagoras' explanation of the Milky Way, without noticing that the two are mutually exclusive. My conclusion is that the correct explanation of lunar eclipses must have been erroneously attributed to Anaxagoras.

Two questions remain, the first of which is whether we can trace the origin of this misallocation. The other question is, how to understand the totally different explanation of lunar eclipses that is also attributed to Anaxagoras. These two questions will appear to be intertwined. We will start our discussion with the second question.

Invisible Heavenly Bodies Below the Moon

In text 9.11, Stobaeus introduces bodies below the moon that can bring about eclipses when they move in front of the moon. Hippolytus (text 9.9) also refers to a theory of lunar eclipses caused by bodies below the moon, in which the shadow of

[45]Perhaps someone would argue that another possible solution for this dilemma could be that, in his youth, Anaxagoras defended the idea of the Milky Way as the shadow of the earth and that, at a later stage, he discovered the true cause of lunar eclipses and abandoned his former idea of the earth's shadow. However, the sources do not give any indication of such a scenario. Even so, the right explanation of lunar eclipses would conflict with his conception of a flat earth.

[46]Cf. Guthrie (1965), 308, n. 1: "For Anaxagoras on the cause of eclipses the authority is Hippolytus."

the earth does not play a role. The word "invisible" obviously means that such an object is invisible until it betrays itself when it covers all or part of the moon.

The idea of invisible heavenly bodies was not new in Presocratic cosmology. Anaximander conceived of the celestial bodies as huge wheels of condensed air filled with fire that we see through an opening. The wheels themselves we do not see because they are made of air, just like the medium in which they orbit around the earth.[47] Another kind of invisible heavenly body is mentioned in the doxography on Anaximenes. He is said to believe that the heavenly bodies are of a fiery nature but that some of them are earthy (γεώδη) and invisible (ἀόρατα).[48] Since this is all that is said about them, it is difficult to understand how earthy bodies could be invisible, and impossible to decide whether they were thought to play a role in lunar eclipses.

According to Anaxagoras, the heavenly bodies are fiery stones.[49] This makes it difficult to imagine how the invisible bodies below the moon could remain invisible during the time in which they were not causing an eclipse. Moreover, the invisible bodies that were able to eclipse the moon must have been much bigger than the stone of Aegospotami, and probably larger than the moon itself, to produce the size of the eclipses we observe on the moon, which makes it even more difficult to understand how they could remain unnoticed. Furthermore, the moon is sometimes faintly visible during an eclipse, which would be impossible if a huge stone were to block its light.

A hypothetical explanation, which could cope with these difficulties and which I consider plausible, is that Anaxagoras' invisible bodies are made of an airy substance and thus form an exception to his theory that the celestial bodies are fiery stones. After all, the very idea of fiery stones being invisible seems to be a *contradictio in terminis*. An indication might be that in text 9.9, Hippolytus distinguishes between invisible bodies below the stars and invisible bodies below the moon. We can imagine that invisible heavenly bodies *above* the moon are fiery stones, which become temporarily visible when they are kindled (such as comets and shooting stars), or when they are driven off course and fall on earth (such as meteorites, for example the stone of Aegospotami), but that the invisible heavenly bodies *below* the moon were conceived of as a kind of meteorological objects that consisted of condensed air, which temporarily became visible during lunar eclipses. However, both when they are eclipsing the moon and when they are not eclipsing the moon, they cannot be fiery, for otherwise we would see them as fiery objects. Their airy or condensed cloudy nature explains why they are only visible when they

---

[47]Hippolytus, *Refutatio Omnium Haeresium* 1.6.4 and 5 = DK 12A11(4 and 5) = LM ANAXIMAND. D7(4 and 5) = Gr Axr20(4 and 5) = KRS 125(4 and 5); S 1.24.1 (not in P) = DK 12A18 = LM ANAXIMAND. D20 = MR 455, not in Gr and KRS.

[48]S 1.24.1 (not in P) = DK 13A14 = LM ANAXIMEN. D13 = Gr Axs16 = MR 455 = KRS 152.

[49]P 2.13.3 = S 1.24.1 = DK 59A71 = LM ANAXAG. D36 = GG 166 and 482 = MR 453, not in Gr and KRS; Hippolytus, *Refutatio Omnium Haeresium* 1.8.6 = DK 59A42(6) = LM ANAXAG, D4(6) = Gr Axg38(6) = GG 271= KRS 502(6). This item will be discussed thoroughly in the second part of this chapter.

eclipse the moon. Aëtius mentions a similar explanation in an anonymous account of eclipses of the *sun*:

9.13. Some (thinkers declare that it is) a condensation of clouds invisibly passing in front of the (sun's) disk (τῶν ἀοράτως ἐπερχομένων τῷ δίσκῳ νεφῶν).[50]

The expression "invisibly passing" (ἀοράτως ἐπερχομένων) is somewhat unfortunate, because these invisible cloudy objects make themselves visible when passing the solar disk. Although Aëtius' item falls under the heading "eclipses of the sun," Bicknell rightly states that "the cloud theory of eclipses is as applicable to lunar as it is to solar eclipses.[51] As regards Anaxagoras, there is no cogent reason to doubt the reports that say the sun is eclipsed when the moon blocks it.[52] But in the case of lunar eclipses, the hypothesis of invisible bodies of an airy nature, which become visible in a cloud-like way during an eclipse, would fit his ideas very well. These airy bodies must become sufficiently condensed to cause an eclipse of the moon, which also would explain why the eclipsed part of the moon has the same color as the surrounding sky (see Fig. 5.6a and b). Sometimes, however, when a blood moon occurs, they are so thin that they allow the moon's own light to shine through. This would also explain why, when moving in front of the moon, do not produce a sharp dividing line, as a stone body would.

Attempts to Understand the Invisible Bodies as an Additional Cause of Lunar Eclipses

The actual difficulty with the theory of invisible objects is that both Hippolytus and Stobaeus (texts 9.9 and 9.11) tell us that it was *in addition to* the right explanation of lunar eclipses. Some authors have tried to argue that the invisible bodies as additional causes of lunar eclipses were introduced to explain specific events. More than a century ago, Schaefer, Boll and Heath suggested that Anaxagoras' invisible bodies would explain the phenomenon of both the sun and moon being visible, opposite one another, during a (total) solar eclipse, a so-called "selenelion."[53] Graham states that the invisible bodies were introduced to explain not only some but all lunar eclipses at the horizon, or as he calls them, "crepuscular eclipses."[54]

---

[50]P 2.24.5 (not in S) = MR 563, DK, LM, Gr, and KRS, but see Dox 354.

[51]Bicknell (1969), 65.

[52]Hippolytus, *Refutatio Omnium Haeresium* 1.8.9 = DK59A42(9) = LM ANAXAG. D4(9) = Gr Axg38(9) = GG 271 = KRS 502(9). Cf. Valerius Maximus, *Facta et Dicta Memorabilia* 8.11, text 1 (not in DK), where it is told how Pericles, quoting what he had learned from Anaxagoras about the orbits of the sun and moon, tried to appease the citizens of Athens who panicked because of an eclipse of the sun.

[53]See Schaefer (1873), 19 n. 1; Boll (1909), 2351; Heath (1913), 80.

[54]Graham (2013a), 128–130.

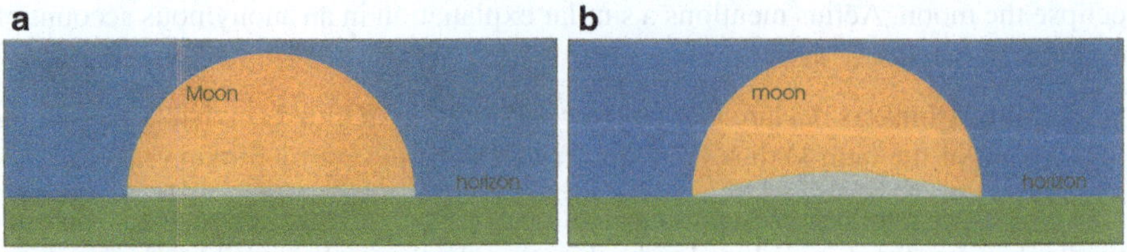

Fig. 9.4 (**a**) The straight shadow line a flat earth should cast during a "crepuscular" eclipse. (**b**) The actual shadow line during a "crepuscular" eclipse

These attempts are suffering from two fundamental mistakes. In the first place, they turn eclipses at the horizon into a special type of lunar eclipses that can be distinguished from other eclipses and thus considered to originate from another cause (invisible heavenly bodies instead of the earth's shadow). So-called "selenelions" and "crepuscular eclipses" are just normal eclipses that have started higher up in the sky, to reach the horizon at a later stage of their existence.[55] For example, the eclipse of March 25, 542 BC started at 5.30 am at an altitude of about 23°. At about 7.06 am it was almost full (altitude about 5°), and when it set at 7.36 am, the moon was still partially eclipsed, with most of the eclipsed part already below the horizon. It would have been very strange indeed if we had to assume that Anaxagoras believed that when the eclipsed moon had reached the horizon, all of a sudden an invisible heavenly body would have taken over the role of the earth's shadow.

In the second place, none of these authors seems to be aware of the discrepancy between the idea that the moon is lit by the sun and the idea of invisible bodies as an additional cause of lunar eclipses. If the moon is lit by the sun, it is hard to understand why the bodies that cover all or part of the full moon, especially when they are supposed to be stony as these authors do, should be invisible or dark and not lit by the sun, just like the moon before which they move.[56]

Graham's idea that Anaxagoras could have proposed invisible bodies as an additional cause of lunar eclipses because of the existence of what he could have observed at the horizon during "crepuscular" eclipses, deserves some further investigation. I quote: "Crepuscular eclipses cannot be explained by the interposition of the earth. For the earth would offer only a thin straight line in profile of the sun. In fact the shadow that falls on the moon in a lunar eclipse is always circular in shape."[57] Graham's problem is illustrated in Fig. 9.4a and b, which show that

[55] Although lunar eclipses that are visible at the horizon can occur either around sunrise or around sunset, all those visible during Anaxagoras' lifetime were at dawn and none at dusk.

[56] This problem does not rise with the counter-earth of the Pythagoreans, which is also said to be an additional cause of lunar eclipses. Of course, there are other problems with the counter-earth as a cause of lunar eclipses. However, the counter-earth does not dwell between the earth and the moon, but is invisible because it orbits between the earth and the central fire, while the part of the earth on which we live is supposed to be always turned away from it.

[57] Graham (2013a), 130.

Furley's remark, repeated by Bakker, that the shape of the earth's shadow on the moon "could be accounted for by a disk-shaped earth as well as by a sphere," is theoretically incorrect.[58] On the other hand, eclipses at the horizon are difficult to observe, because they require a horizon completely without clouds or fog, or interfering buildings, hills and the like. In addition, the light of the sun at the other side of the horizon will often prevent a good view of the eclipsed moon, making it difficult to discern whether the shadow line at the horizon is straight or curved.

Solar eclipses are ephemeral phenomena. The eclipse of August 23, 2017 that was visible in the U.S. lasted about 40 s. The longest calculated solar eclipse (AD 2186) will last 7 min. Lunar eclipses, on the contrary, are usually long-lasting events (depending on several factors, of which the magnitude of the eclipse is the most important) and can last for several hours. During these hours, the moon travels a considerable track along its path in the sky. Sometimes the moon reaches the horizon while it is still eclipsed. It has already been said above that "crepuscular" lunar eclipses are not a special kind of eclipses, but normal lunar eclipses that happen to be visible at the horizon during some time of their existence. Therefore, I write the word "crepuscular" permanently between quotation marks. Such "crepuscular" eclipses are by no means exceptional: of the 31 lunar eclipses during Anaxagoras' lifetime, 13 were visible at the horizon for some time of their existence and thus "crepuscular," which is more than 40% of the total number of 31 eclipses. To me it seems improbable that someone like Anaxagoras would have held that lunar eclipses were caused by the earth's shadow and at the same time would have made an exception for almost half of its occurrences.

In an Addendum, I have generated pictures of all 13 lunar eclipses that were visible at the horizon during Anaxagoras' lifetime. These pictures illustrate that the shadow of the earth showed a variety of shadow lines, at the upper or lower side of the moon, horizontal, vertical, or slanted (or none, in the case of a total eclipse), and that in only one eclipse did the shadow line run parallel to and just above the horizon, as Graham seems to assume[59] (cf. Fig. 9.4b). It seems likely to me that if a defender of the flat earth like Anaxagoras had seen the phenomena of the shapes of the eclipsed moon during so-called crepuscular eclipses, he would have argued: "the earth is flat, and therefore the shapes of the eclipses that can be seen at the horizon cannot be caused by the shadow of the earth; and since these eclipses were just normal lunar eclipses a few hours ago, this indicates that, generally speaking, eclipses of the moon are not caused by the shadow of the earth." And if Anaxagoras had seen the phenomenon of a selenelion in which both the sun and the eclipsed moon were to be seen, he would have been convinced once more that the earth's shadow could not be the cause of a lunar eclipse.

[58] Furley (1996), 121. Cf. Bakker (2016), 171.

[59] See his remark regarding "crepuscular" eclipses: "and these do occur" (Graham 2013a, 129–130).

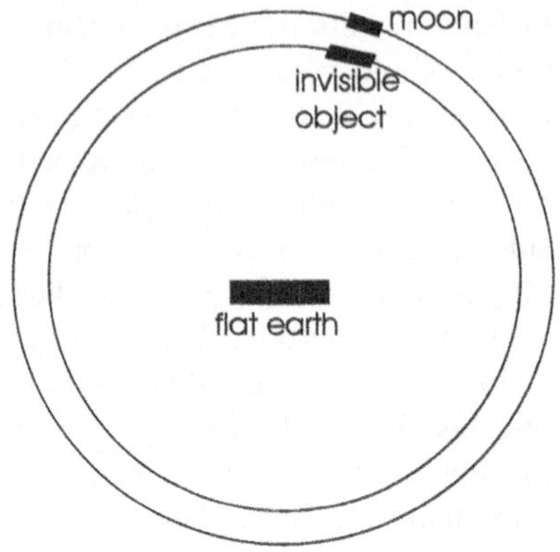

Fig. 9.5 Lunar eclipse caused by an invisible object (approximately to scale) (In this picture, I did not draw the sun because, as said earlier, the shadow of the earth does not play a role in this explanation of lunar eclipses, and Anaxagoras' explanation of the Milky Way implies that he had no idea of the actual position of the sun during the night)

Invisible Bodies as Anaxagoras' Only Theory of Lunar Eclipses

Earlier, I concluded that Anaxagoras could not have been the discoverer of the right explanation of lunar eclipses or could have adhered to this explanation, because this is incompatible with his well-documented theory of the Milky Way. From the previous section, we may conclude that there does not seem to be a reasonable explanation of how the invisible bodies could function as a *supplementary* cause of lunar eclipses. If this analysis is correct and the right explanation of lunar eclipses was wrongly ascribed to Anaxagoras, there is no reason to call his explanation by means of invisible objects "additional." My proposal is therefore that the invisible bodies must be considered as Anaxagoras' one and only way to explain eclipses of the moon. We might assume that this explication was part of a universal theory that also applied to solar eclipses, star occultations, and risings or settings, where in all cases a body (the moon, or the earth, or an invisible body), lying between an observer and the celestial object, blocks the sight of that object and in which no shadow was involved. This explanation of lunar eclipses is compatible with Anaxagoras' conception of a flat earth and would solve the problems that arise, as we have seen, with the interpretation of lunar eclipses as caused by the earth's shadow.[60] This explanation of lunar eclipses is visualized in Fig. 9.5.

[60]Interestingly, Neugebauer (1975), 550. wrote: "One could invent the existence of a special object, a dark 'disk' that obscures the moon, moving always at 180° elongation from the Sun. The mathematics of the ephemerides would allow for this interpretation." Neugebauer is commenting here on the Pythagorean counter-earth, but his remark makes more sense when applied to Anaxagoras.

The Possible Origin of a Misunderstanding

I think the origin of the misunderstanding of calling the explanation of lunar eclipses by means of invisible bodies "additional" is a cryptic text of Aristotle on lunar eclipses in the Pythagorean cosmological system. The Pythagorean system counts one invisible heavenly body, called the counter-earth, which is considered to be another earth, which opposite to the earth orbits around the central fire. But next to this, Aristotle states that some people think that there are invisible *bodies* (in the plural), that cause eclipses of the moon:

9.14. Some even think it possible that there are a number of such bodies [like the counter-earth] carried round the center, invisible to us owing to the interposition of the earth. This serves them too as a reason why eclipses of the moon are more frequent than eclipses of the sun, namely that it [sc. the light of the sun] is blocked by each of these moving bodies, not only by the earth.[61]

Aristotle speaks of heavenly bodies that are "invisible to us owing to the interposition of the earth." This excludes Anaxagoras' invisible bodies, which are said to be below the moon, meaning between the earth and the moon, as Burkert rightly remarks.[62] Yet, some scholars could not resist the temptation to think that Aristotle was not hinting at some unknown Pythagoreans but at Anaxagoras. In modern times, this suggestion has been made several times, and recently by Graham.[63] It is my guess that already in ancient times, Theophrastus, and in his footsteps Stobaeus and Hippolytus, misunderstood Aristotle's words as referring to Anaxagoras' invisible heavenly bodies.

Let us look once more at the relevant texts. Pseudo-Plutarch (text 9.10) has nothing at all to say about Anaxagoras concerning lunar eclipses. Stobaeus (text 9.11) invokes the authority of Theophrastus to attribute the explanation of lunar eclipses by means of invisible bodies to Anaxagoras. Stobaeus may have found the theory of invisible bodies in Aëtius, but I think it is more plausible that he found it in another source that referred to Theophrastus. Hippolytus probably consulted the same source, for both he and Stobaeus use the same words when they mention that, according to Anaxagoras, the moon is eclipsed "by invisible bodies below the moon" (τῶν ὑποκάτω τῆς σελήνης σωμάτων). And since Theophrastus, in misunderstanding of Aristotle's words in text 9.14, spoke of *Anaxagoras'* additional explanation, both Stobaeus and Hippolytus also presented it as his explanation, in addition to the right explanation of lunar eclipses. Consequently, Stobaeus inserted Anaxagoras into the list of names of supporters of the right explanation. Finally, Aristotle's suggestion that invisible bodies were introduced to explain why lunar eclipses occur more

[61]Aristotle, *De Caelo* 293b21–25, not in DK.

[62]Cf. Burkert (1972), 344 n. 34.

[63]See DK, Zweiter Band, 16, note at line 18; Guthrie (1962), 286; Dicks (1970), 66; Graham (2015), 226; Bakker (2013), 693, all of whom overlook the crucial point.

frequently than solar eclipses is probably his own interpretation. Neither Aëtius nor Hippolytus mention it in their reports on the Pythagoreans and Anaxagoras.

Concluding Remarks

If my analysis in this chapter is right, Anaxagoras was not the revolutionary astronomer as which he is presented by modern scholars, but, in several respects, a defender of ancient views. Of course, the results of my investigation are less spectacular than those of scholars who think they can attribute to Anaxagoras the discovery of the real cause of eclipses of the moon, and who think they can explain why he needed an additional theory for some special eclipses. However, they should at least justify why they neglect the conflicting theory of the Milky Way, or they should show that it is compatible with the right theory of lunar eclipses. It has been suggested that this kind of discrepancy is due to the state of astronomical theorizing that was still in its infancy. I would rather say that some ideas and theories of Presocratic astronomy seem strange or even weird to us, but that they often, when we look more carefully, make sense within their contemporaneous context. What is at stake here is not that the ideas in question are strange, but that they are overtly conflicting. I am convinced that some Presocratic thinker who discovered the right cause of lunar eclipses must have thoroughly studied the shadow of the earth on the moon. Therefore, he cannot have defended at the same time the completely different and conflicting theory of the shadow of the earth as the cause of the phenomenon of the Milky Way. To the best of my knowledge, these two theories are not compatible.

In my opinion, the textual, conceptual, and observational evidence does not support the conclusion that Anaxagoras discovered or adhered to the correct explanation of lunar eclipses. Anaxagoras was a great cosmologist, who ingenuously defended conceptions that have since become obsolete, such as the earth being flat and the Milky Way resulting from the earth's shadow, conceptions which did not allow him to discover or accept the true theory of lunar eclipses. His also erroneous solution was to explain eclipses of the moon as analogous to eclipses of the sun and occultations of a star or planet, by assuming that invisible heavenly bodies come between us and the moon. I started my investigation by stipulating that we must be cautious in attributing too much astronomical knowledge to the ancient Greek thinkers. This applies in particular to those Presocratics who, like Anaxagoras, adhered to the concept of a flat earth. Flat earth cosmology often leads to consequences that look surprising and even strange to us, who believe that the earth is a sphere.

A serious problem remains that has to do with the question of what happens to the moon when it is in conjunction with the Milky Way. The belt of the Milky Way is inclined by about 60° in relation to the ecliptic. This means that the moon sometimes passes the Milky Way and thus, according to Anaxagoras' theory that the Milky Way is caused by the earth's shadow, the moon there cannot receive its light from the sun. Nevertheless, the moon is still visible and shows its phases when it passes in front of the Milky Way. We already met this problem in the quotations from Tannery and others, who wondered why the moon was not eclipsed whenever the moon passed over the Milky Way. This leads to the question of what could be meant with the moon receiving its light from the sun, or in other words, what could have been, according to Anaxagoras, the origin of the moon's light. In that context, the question of the invisible bodies must be addressed once again. I will discuss the problem of the origin of the light and the phases of the moon according to Anaxagoras in Chap. 10.

Addendum: "Crepuscular" Lunar Eclipses During Anaxagoras' Lifetime

In order to reconstruct what Anaxagoras could have seen or heard from reliable informants, and to evaluate what he could have concluded from these observations, I have collected all (partial and total) lunar eclipses that were visible at the horizon during Anaxagoras' lifetime. I took Graham's dates for Anaxagoras, ca. 500–428 BC[64] and started in 485 BC, when Anaxagoras was about 15 years old. Taking other dates for the lifetime of Anaxagoras would lead to similar results. I took Athens as observation site (we may imagine Anaxagoras climbing the Acropolis to get a better view). Other places in Greece would have made little difference, but could have provided a better view of the horizon. The observations from Clazomenae, his birthplace, would have been almost identical. In one case (September 7, 460 BC) the eclipse was not visible in Clazomenae because there the moon had already set. During this period, 31 partial or total lunar eclipses were visible in Athens, in 13 of which the wholly or partially eclipsed moon was visible at the horizon (a "crepuscular eclipse" in Graham's terminology). The pictures were generated by the computer program RedShift. For these pictures, I used an older version (1.0), because the newest version does not generate pictures like these. Although lunar eclipses that are visible at the horizon can occur both at sunrise and at sunset, all crepuscular eclipses during Anaxagoras' lifetime occurred at sunrise, and none at sunset.

Three eclipses at the horizon were total (Figs. 9.6, 9.7, and 9.8).

In three eclipses, the line of the shadow on the moon showed the underside of the earth (Figs. 9.9, 9.10, and 9.11).

Three eclipses were not visible when the moon was halfway the horizon, so that there was no shadow line at all to be studied. Therefore, I have rendered the situation when the setting moon was not yet halfway the horizon, but when the moon "sat" upon the horizon (Figs. 9.12, 9.13, and 9.14).

Of three eclipses, the shadow line at the horizon was (almost) perpendicular to the horizon (Figs. 9.15, 9.16, and 9.17).

[64]See Graham (2013a), 85.

Fig. 9.6 Lunar eclipse of November 19, 483 BC

Fig. 9.7 Lunar eclipse of June 26, 475 BC

Fig. 9.8 Lunar eclipse of December 21, 429 BC

Fig. 9.9 Lunar eclipse of September 7, 460 BC

Fig. 9.10 Lunar eclipse of June 26, 456 BC

Fig. 9.11 Lunar eclipse of June 17, 447 BC

Fig. 9.12 Lunar eclipse of November 30, 484 BC

Fig. 9.13 Lunar eclipse of January 2, 429 BC

Fig. 9.14 Lunar eclipse of June 17, 428 BC

Fig. 9.15 Lunar eclipse of May 16, 482 BC

Fig. 9.16 Lunar eclipse of November 30, 446 BC

Fig. 9.17 Lunar eclipse of July 28, 440 BC

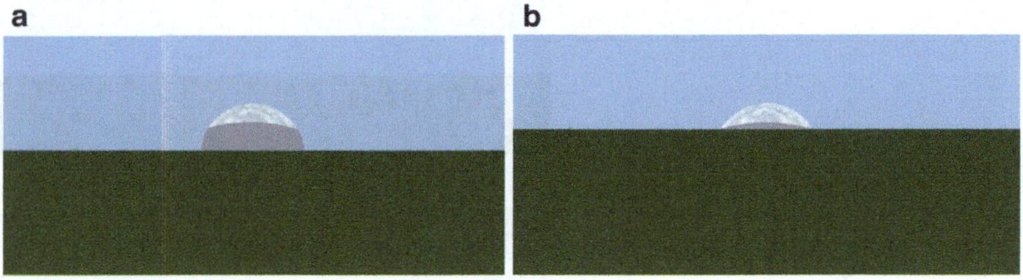

Fig. 9.18 (**a**) Lunar eclipse of March 25, 442 BC. (**b**) The same eclipse, 45 seconds later

In only one eclipse does the shadow line of the upper side of the earth run parallel to the horizon (cf. Fig. 7.14b). It is displayed here in two phases, 45 seconds apart (Fig. 9.18).

References[65]

Bakker, F.A., Aëtius, Achilles, Epicurus and Lucretius on the Phases and Eclipses of the Moon. *Mnemosyne* 66 (2013), 682–707.

—, *Epicurean Meteorology*. Leiden: Brill 2016.

[65] P = Aëtius in pseudo-Plutarch, *Placita* (numbering according to Dox).
S = Aëtius in Stobaeus, *Anthologium* (numbering according to Wachsmuth and Hense).

Bicknell, P. J., Anaximenes' Astronomy. *Acta Classica* 12 (1969) 53–85.

Boll, F., Finsternisse, in A.F. Pauly and G. Wissowa, *Real-Encyclopädie der classischen Altertumswissenschaften* VI.2, Stuttgart 1909, 2330–2363.

Burkert, W., *Lore and Science in Ancient Pythagoreanism*. Cambridge, MA: Harvard University Press. 1972.

Cleve, F.M., *The Philosophy of Anaxagoras*. New York 1949.

Curd, P., *Anaxagoras of Clazomenae*. Toronto: University of Toronto Press 2007 (repr. 2010).

Dicks, D.R., *Early Greek Astronomy to Aristotle*. Ithaca N.Y.: Thames and Hudson 1970.

Diels, H., *Doxographi Graeci*. Berlin: Walter de Gruyter 1879 (= Dox).

Diels, H. and W. Kranz, *Die Fragmente der Vorsokratiker*. Zürich/Hildesheim: Weidmann 1951[6] (= DK).

Dreyer, J.L.E., *A History of Astronomy from Thales to Kepler*. Mineola, NY: Dover Publications 1953.

Fehling, D., Das Problem der Geschichte des griechischen Weltmodells vor Aristoteles. *Rheinisches Museum für Philologie* 128 (1985), 195–231.

Ferguson, J., Two Notes on the Preplatonics. *Phronesis* 9 (1968), 98–106.

Furley, D., The Earth in Epicurean and Contemporary Astronomy. In: G. Giannantoni and M. Gigante (eds.), *Epicurismo Greco e romano: Atti del congresso internazionale, Napoli, 19–26 maggio 1993*, Naples: Bibliopolis 1996, vol 1, 119–125.

Gemelli Marciano, M.L., *Die Vorsokratiker* (3 Bände). Düsseldorf: Artemis & Winkler. 2007–2013.

Gershenson, D.E. and D.A. Greenberg, *Anaxagoras and the Birth of Physics*. New York: Blaisdell Publishing Company 1964 (= GG).

Gomperz, T., *Griechische Denker I*, Leipzig 1896.

Graham, D.W., *The Texts of Early Greek Philosophy*. Cambridge: Cambridge University Press 2010 (= Gr).

—, *Science Before Socrates*. Oxford: University Press 2013a.

Graham, D.W. and E. Hintz, Anaxagoras and the Solar Eclipse of 478 BC *Apeiron* 40.2007, 319–344.

—, Philolaus' Astronomy. *Archive for History of Exact Sciences* 69 (2015), 217–230.

Guthrie, W.K.C., *A History of Greek Philosophy. Volume I. The Earlier Presocratics and Pythagoras*. Cambridge: Cambridge University Press 1962.

—, *A History of Greek Philosophy. Volume II. The Presocratic Tradition from Parmenides to Democritus.* Cambridge: Cambridge University Press 1965.

Heath, T., *Aristarchus of Samos. The Ancient Copernicus*. Oxford: Clarendon Press 1913.

Jaki, S.L., *The Milky Way: An Elusive Road for Science*. Newton Abbot 1973.

Kirk, G.S., J.E. Raven and M. Schofield, *The Presocratic Philosophers*. Cambridge: Cambridge University Press 2009 (= KRS).

Laks, A. and G.W. Most *Early Greek Philosophy VI*, Cambridge, MA. 2016.

Lee, H.D.P. (ed. and transl.), *Aristotle. Meteorologica*, Cambridge, MA 1962.

Liddell, H.G., R. Scott H.S. Jones and R. McKenzie, *Greek-English Dictionary*. Oxford: Clarendon Press 1996. Also available on the internet, e.g.: http://stephanus.tlg.uci.edu/lsj/#eid=1714&context=search (= LSJ).

Mansfeld, J., *Die Vorsokratiker II*. Stuttgart: Reclam 1986.

—, From Milky Way to Halo, in: J. Mansfeld and D.T. Runia, *Aëtiana. Volume 3*. Leiden 2010, 477–514.

Mansfeld, J. and D.T. Runia, Aëtiana. *The Method and Intellectual Context of a Doxographer, Volume Two, Part Two*. Leiden: Brill 2009 (= MR).

Neugebauer, O.E., *A History of Ancient Mathematical Astronomy*. Berlin 1975.

O'Brien, D., Derived Light and Eclipses in the Fifth Century, *JHS* 88 (1968), 114–127.

Panchenko, D., Eudemus Fr. 145 Wehrli and the Ancient Theories of Lunar Light." In: I. Bodnár and W.W. Fortenbaugh (eds.), *Eudemus of Rhodes*, New Brunswick 2002, 323–336.

—, Anaximandros von Milet. In H.J. Gehrke (ed.), *Die Fragmente der griechischen Historiker Part V*, Die Geographen. Published by Brill online 2013b, http://referenceworks.brillonline.com/browse/fragmente-der-griechischen-historiker-v.

Rechenauer, G., Anaxagoras. In the completely revised edition of F. Überweg's *Die Philosophie der Antike* in *Band I. Zweiter Halbband: Frühgriechische Philosophie*, Basel 2013.

Schaefer, H.W., *Die astronomische Geographie der Griechen bis auf Eratosthenes*. Flensburg 1873.

Stephenson, F.R. 1997. *Historical Eclipses and Earth's Rotation*. Cambridge: Cambridge University Press.

Tannery, P., *Pour l'histoire de science Hellène: de Thalès à Empédocle*. Paris: Félix Alcan 1887.

Chapter 10
Anaxagoras on the Light and Phases of the Moon

Contents

Introduction

In Chap. 9, I stated that two different theories about the shadow of the earth have been attributed to Anaxagoras. According to the first theory, the shadow of the earth was responsible for the phenomenon of the Milky Way, while according to the second, the shadow of the earth caused eclipses of the moon. I argued that these two theories are incompatible. I also argued that Anaxagoras' explanation of the Milky Way, which was underpinned by the notion that lights shine brighter in the dark, is better attested than his alleged adoption of the correct explanation of lunar eclipses. Finally, I argued that Anaxagoras' alleged adoption of the correct explanation of lunar eclipses does not harmonize with the rest of his astronomical ideas, especially that of a flat earth. My first conclusion was that Anaxagoras could not have discovered or embraced the theory that lunar eclipses were caused by the shadow of the earth. My second conclusion was that the idea of one or more invisible bodies between the moon and the earth, which according to the doxography was merely additional to the true explanation, in fact constituted Anaxagoras' one and only explanation of lunar eclipses. I suggested that the source of the misunderstanding

D. L. Couprie, *When the Earth Was Flat*, Historical & Cultural Astronomy,
https://doi.org/10.1007/978-3-319-97052-3_10

was a text in Aristotle that mentions some Pythagoreans and the notion of invisible bodies causing lunar eclipses. My interpretation did not, however, address one serious remaining problem, which did not concern eclipses but the light and phases of the moon. During the month, the moon exhibits phases, from new moon to waxing crescent, first quarter, waxing gibbous, full moon, and then back to waning gibbous, last quarter, waning crescent, and new moon. In Chap. 10, I will investigate how Anaxagoras could have explained these phenomena.

My method of investigation will again be to start with the most reliably documented aspects of Anaxagoras' astronomy and to see whether it is possible, from that basis, to interpret the rest of the relevant doxography and to achieve a coherent overall understanding of his astronomical thoughts. The most important certainty we have on Anaxagoras' astronomical thinking is that he believed the earth to be flat. Another of his best documented astronomical ideas is that the Milky Way was the belt of stars not illuminated by the sun. Finally, it is well documented that he thought that the heavens were inclined in relation to the flat earth's surface, that the heavenly bodies were relatively close and smaller than the earth, and that the sun and the stars were of a fiery, stony nature.[1] As already said in Chap. 2, a main presupposition of my method is the conviction that the ideas of Presocratic thinkers like Anaxagoras are not a mere collection of notions that might be overtly contradictory, but that they form a consistent whole. Some ancient ideas that may look strange to our eyes may nonetheless have made sense within the contemporary context. We must always be aware of the anachronistic trap to read into the ancient records notions to which we are accustomed. In this chapter, we will encounter a typical example in expressions such as "the moon receives its light from the sun." A special kind of this mistake, which the Greek doxographers were fond of, is to accredit the ancient Greek philosophers with being the first to offer a given theory. This attitude is still not absent in the interpretive work of some modern scholars. Take, for example, the recent claims that Parmenides and Anaxagoras were the first advocates of "heliophotism"—the idea that the moon is illuminated by the sun—and that Anaxagoras was the discoverer of the true cause of lunar eclipses. The danger of such interpretations is that they easily tend to disregard data that do not concur with them. I must confess that I made this kind of mistake in what I wrote some years ago about Anaxagoras, eclipses and the moon's light. This means that I must withdraw most of what I wrote on page 177 of my *Heaven and Earth in Ancient Greek Cosmology*, New York 2011. The Chaps. 9 and 10 contain my current ideas on these subjects. The studies that most provoked my thinking about Anaxagoras' astronomy were Dennis O'Brien's 50-year-old paper "Derived Light and Eclipses in the Fifth Century"[2] and Daniel Graham's recent and innovative book *Science*

[1]The moon is also stony, but whether or not (and to what degree) it has a fiery nature is one of the topics investigated in this chapter. As stated in Chap. 9, I think an exception must be made for the invisible bodies below the moon; they are obviously not fiery, and it can be argued that they are not stony either.

[2]O'Brien (1968).

Before Socrates,[3] even and especially when I disagree (from time to time fundamentally) with them.

Could Anaxagoras Could Have Given the Correct Explanation of the Moon's Phases?

The standard interpretation of Anaxagoras' explanation of the phases of the moon is that they display the shapes of the portion of the moon that is illuminated by the sun, as seen by an observer on earth. The moon's phases are usually illustrated with the help of a diagram such as Fig. 10.1.

There are at least two reasons for doubting whether Anaxagoras could have understood the phases of the moon as we do. The first is that our understanding of the shapes of the moon's phases requires that the moon is spherical, while Anaxagoras, in all probability, thought of the heavenly bodies as flat disks like the earth.[4] Several texts that refer to his ideas state that he thought the moon had hills, and ravines, just like the earth, which he considered to be flat.[5] Plato says that, according to Anaxagoras, the moon is earth.[6] Another report clearly states:

10.1. This same Anaxagoras says that the moon is a flat place (χώρα πλατεῖα) (. . .).[7]

Graham calls this text a "testimony of uncertain pedigree and value."[8] It is, though, the only explicit text we have on Anaxagoras and the shape of the moon.

If the phases were caused by the light of the sun, the moon as a flat disk would always be full, except at new moon, as Cleomedes already remarked: "So if the moon's shape were flat, it would be full as soon as it passed by the sun after conjunction, and would remain full until [the next] conjunction."[9] This can be demonstrated by means of a picture (Fig. 10.2).

In much more recent times, Heath wrote, "Whether Anaxagoras reached the true explanation of the phases of the moon is doubtful. (. . .) it required that the moon should be spherical in shape; Anaxagoras, however, held that the earth, *and doubtless the other heavenly bodies also*, were flat. And accordingly, his explanation of

[3]Graham (2013a).

[4]An indication could be that Empedocles still believed that the moon does not have the form of a sphere, but that of a disk, as is reported by Plutarch, *Roman Questions* 288b = DK 31A60 = LM EMP. D135a = Gr Emp85, not in KRS; Diogenes Laërtius, *Vitae Philosophorum* 8.77 = DK 31A1 (77) = LM EMP. D135b, not in Gr and KRS.

[5]Cf. text 10.2 below and Diogenes Laërtius, *Vitae Philosophorum* 2.8 = DK 59A1(8) = Gr Axg37 = GG 340; not in LM and KRS.

[6]Plato, *Apologia* 26D1 = DK 59A35 = LM ANAXAG. R4 = Gr Axg9 = GG 1 = KRS 465.

[7]*Scholium on Apollonius of Rhodes* 1.498 = DK59A77 = LM ANAXAG. D43; not in Gr, GG, and KRS.

[8]Graham (2013a), 251 n. 21.

[9]Cleomedes in Bowen and Todd (2004), II.5.37–40.

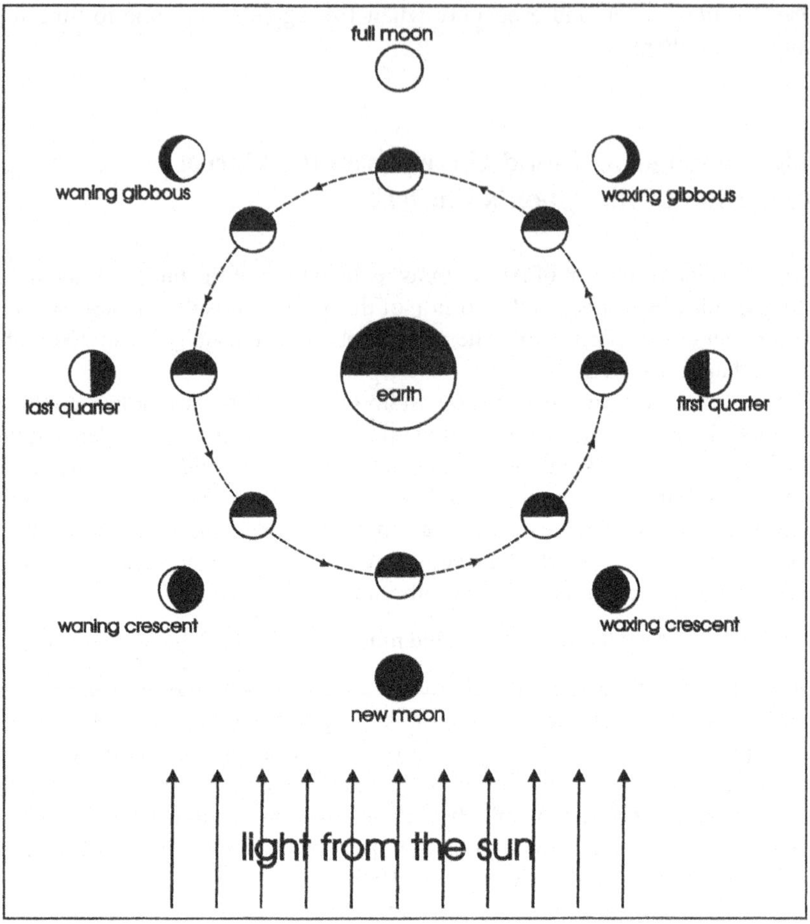

Fig. 10.1 The standard explanation of the phases of the moon (A similar diagram in Graham 2013a, 98, figure 3.1)

the phases could hardly have been correct."[10] In other words, considering the moon as flat, Anaxagoras could not have explained the phases of the moon as caused by the light of the sun.

Graham, being convinced that Anaxagoras believed that the moon was illuminated by the sun, argues the other way around and claims that Anaxagoras must have held that the moon was spherical because, otherwise, his understanding of the phases of the moon would have been impossible.[11] Yet there exists no report that confirms

[10]Heath (1913), 80–81, my italics. See also Tannery (1887), 278.

[11]See Graham (2013a), 99: "the moon's shape is a *function* of its angular distance to the sun. This is what heliophotism, taken as a hypothesis, predicts." Graham's argument is not always consistent. He states that "if Parmenides fully understood heliophotism, he would see that the moon provides a

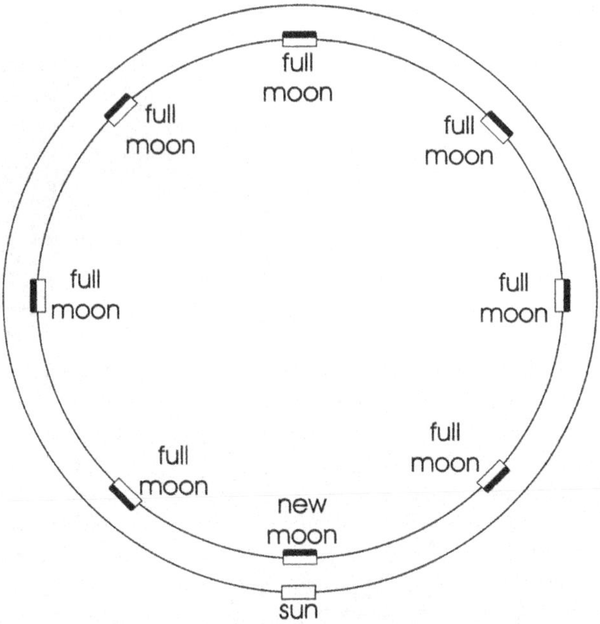

Fig. 10.2 The moon as a flat disk does not show phases (approximately to scale)

that Anaxagoras considered the moon to be spherical. As far as I know, Aristotle was the first to claim that the moon's spherical shape could be deduced from its phases.[12] Although I think Anaxagoras believed that the moon was a flat disk like the earth, and although I think he could not have explained the phases of the moon the way we do, this does not mean that he was not able to offer an explanation that would fit into the context of his flat earth astronomy. However, the two possible explanations given at the end of this chapter for the moon's phases in Anaxagoras' astronomy are independent of the moon's shape.

The second reason why Anaxagoras could not have explained the phases of the moon as we do is because of his explanation of the Milky Way. Aristotle and several other sources assert that, according to Anaxagoras, the phenomenon of the Milky Way results from the shadow of the earth, cast upon the stars by the sun. The optical theory behind this is that lights shine brighter in the dark. This explanation of the Milky Way is strange and definitely wrong, but it is one of the best attested of Anaxagoras' astronomical theories and do not I know of any author who questions its authenticity or has tried to argue it away. The belt of the Milky Way is inclined by about 60° in relation to the ecliptic. The moon's monthly path among the stars, in its

model for all the heavenly bodies. (...) Heavenly bodies, including *the earth*, *must*, by parity of reasoning, *be spherical*" (Graham (2013a), 114, my italics). Elsewhere, he declares that "it is important to notice that Anaxagoras seems to grasp *all the implications* of heliophotism" (id. 124, my italics). However, Anaxagoras does not seem to have grasped *all* the implications of heliophotism, because he believed that the earth is *flat*.

[12]Aristotle, *De Caelo* 291b18–23 and *Analitica Posteriora* 78b4–12.

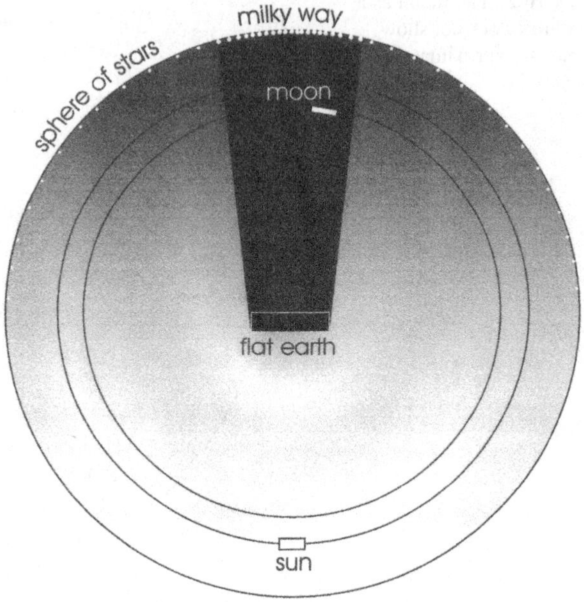

Fig. 10.3 The full moon in the shadow of the earth (approximately to scale)

turn, is inclined by about five degrees relative to the ecliptic. This means that the moon regularly passes through the Milky Way, where it is visible and shows phases. If Anaxagoras really believed that the moon's light is reflected light from the sun, it is difficult to see how he could have explained the visibility of the moon and its phases when the moon is in the Milky Way, where it does not receive light from the sun (see Fig. 10.3).

O'Brien underestimates the problem when he writes that "the shadow of the earth must therefore be *a fairly narrow band*, which would *occasionally* obscure the light of the moon," but on the other hand he overestimates the problem when he writes that "the moon would be eclipsed *night after night*."[13] In the case of Anaxagoras it is not the night sky as a whole that is dark because the earth blocks the light of the sun, but only that part of the night sky that is called the Milky Way (see Fig. 10.3). The rest of the darkness at night is due to the fact that the sun's rays do not reach the stars, or, as text 9.1 says it more poetically, do not "see" the stars. On the other hand, the Milky Way is not a fairly narrow band in the night sky. The width of the Milky Way is roughly 30°, through which the moon passes twice a month for several nights. The suggestion that this problem may have escaped Anaxagoras' attention is barely convincing, because it concerns a frequently recurring and easily observable phenomenon. Moreover, since the shadow of the earth, according to Anaxagoras' theory of the Milky Way, cannot be conical but must be widening (see again Fig. 10.3), the

[13]See O'Brien (1968), 125 and 124; my italics.

Moon should remain in the shadow of the earth much longer than it actually does during eclipses.

Anaxagoras' views on the phases of the moon must, of course, have been closely linked to his ideas about the nature of the moon's light, of which we have several reports. Aëtius' statements on the subject of the moon's light are scattered over four chapters. We will discuss them, together with analogous texts, in the next sections and return to the moon's phases at the end of this chapter.

Anaxagoras on the Light of the Moon in Aëtius 2.25 and Analogous Texts

Aëtius' first relevant chapter is the particularly well-attested[14] chapter 2.25, called "On the substance (περί οὐσίας) of the moon." The item about Anaxagoras says:

10.2. Anaxagoras and Democritus [declare that it is] an inflamed solid mass (στερέωμα διάπυρον), which has in it plains and mountains and ravines.[15]

Anaxagoras' conception of the moon's substance was not exceptional. Almost all philosophers mentioned in Aëtius 2.25 were of the opinion that the moon was, in one way or another, fiery.[16] Anaximander believed that it was "a wheel with a hollow rim and full of fire (πυρὸς πλήρη);" Anaximenes, Parmenides, and Heraclitus that it was "fiery (πυρίνη);" Xenophanes, "an inflamed condensed cloud (νέφος πεπυρωμένον);" Posidonius and most of the Stoics, "combined out of fire and air (μικτὴ ἐκ πυρὸς καὶ ἀέρος);" Cleanthes, "fire-like (πυροειδῆ);" Empedocles, "compacted air, fixed by fire (πεπηγότα ὑπὸ πυρός);" Plato, "formed for the most part from fiery material (τοῦ πυρώδους);" Diogenes, "a sponge-like ignited mass (ἄναμμα);" and Berosus, "half-inflamed (ἡμιπύρωτος)." The only exceptions are Thales ("earthy"), Aristotle ("formed from the fifth body"), Ion ("partly glass-like and transparent, partly opaque"), and Pythagoras ("mirror-like").[17] It should be noted that, unlike Posidonius, Cleanthes, Empedocles, Plato, and Berosus, the entry on Anaxagoras does not contain any restrictions or further qualifications. That the moon, according to Anaxagoras, consisted of inflamed material is confirmed by Origen:

10.3. (...) nor will we call the sun, moon, and stars inflamed clumps (μύδρον διάπυρον) as Anaxagoras did.[18]

---

[14]For this qualification, see MR 572.

[15]P 2.25.9 = S 1.26.1 = DK 59A77 = LM ANAXAG. D42 = MR 573 = GG 171; not in Gr and KRS.

[16]See Dox 355–357 and MR 572–587.

[17]Assuming that pseudo-Plutarch's κατὰ τὸ πυροειδὲς σῶμα must be replaced by Stobaeus' κατοπτροειδὲς σῶμα. See Dox 357 n.1 and MR 381(c).

[18]Origenes, *Contra Celsum* 5.11 = GG 268; not in DK, LM, Gr, and KRS.

Achilles Tatius' chapter "About the Moon" does not mention specific names, but one statement is equivalent to that of pseudo-Plutarch about Anaxagoras and Democritus:

10.4. Some (say the moon is) a solid ignited earth containing fire (ἕτεροι δὲ γῆν πεπυρομένην στερέμνιον ἔχουσαν πῦρ).[19]

In the same sense, Hippolytus relates Anaxagoras' beliefs as follows:

10.5. The sun and moon and all the heavenly bodies are fiery stones (λίθους ἐμπύρους) carried around by the revolution of the aether.[20]

It is striking that in Aëtius' chapter 2.20 "On the substance of the sun" the same or similar words are used with respect to the sun. In the case of Anaxagoras, almost the same characterizations are used in relation to the moon ("an inflamed solid mass," στερέωμα διάπυρον) and to the sun ("an inflamed clump or rock," μύδρος ἢ πέτρος διάπυρος).[21] Hippolytus calls both the sun and the moon "inflamed stones" (λίθοι ἐμπύροι) (text 10.5). These texts leave no doubt that, according to Anaxagoras, the moon was an inflamed solid body like the sun and the stars. The most obvious interpretation is that these qualifications also describe the light of the moon: the moon is fiery and shines with its own light. This seems to rule out the option that Anaxagoras considered the moon's light to be the reflection of the light of the sun. If we take seriously the proposition that, for Anaxagoras, the moon was a fiery, inflamed body—and I see no reason we should not—this is another argument why Anaxagoras could not have understood the phases of the moon as we do. If these were the only texts about Anaxagoras and the light of the moon, I think nobody would ever have thought of attributing to him "heliophotism" in the sense of light reflected from the sun. But let us see what the other texts have to say.

Anaxagoras on the Light of the Moon in Aëtius 2.28 and Analogous Texts

Aëtius' second relevant chapter is 2.28, "On the lights (φωτισμῶν) of the moon."[22] In Stobaeus' version, Anaxagoras is mentioned as one of the successors of Thales:

10.6. Thales was the first to say that it is illuminated by the sun (ὑπὸ τοῦ ἡλίου φωτίζεσθαι).

---

[19]Achilles Tatius, *Introductio in Aratum* 21 = DK 59A77 = MR 578; not in LM, Gr, GG and KRS.

[20]Hippolytus, *Refutatio Omnium Haeresium* 1.8.6 = DK 59A42(6) = LM ANAXAG. D4(6) = Gr Axg38(6) = GG 270; not in KRS, but see p. 344, n.1.

[21]P 2.20.6 = S 1.25.1 = DK 59A11 = MR 515 = GG 168; not in LM, Gr and KRS.

[22]See Dox 358–359 and MR 601–612, who translate: "On the illuminations of the moon."

Pythagoras, Parmenides, Empedocles, Anaxagoras and Metrodorus (declare) likewise.[23]

Instead of these lines pseudo-Plutarch writes this:

10.7. Thales and his successors (οἱ ἀπ' αὐτοῦ) (declare that) it is illuminated by the sun.[24]

Mansfeld and Runia suppose that pseudo-Plutarch shortened the original series of names that Stobaeus has preserved.[25] Assuming that they are right, the phrase "the moon is illuminated by the sun" seems to contradict what we found in Aëtius' chapter 2.25: the moon is of a fiery substance. Another possibility is that Stobaeus felt obliged to offer his own instances of "Thales' followers." Be that as it may, Hippolytus also reports on Anaxagoras, a few lines after his remark that the sun and moon are fiery bodies:

10.8. The moon does not have its own (μὴ ἴδιον ἔχειν) light, but [gets it] from the sun.[26]

And Plutarch writes:

10.9. A favorable reception was given to our friend's exposition, which presented the Anaxagorean theory that the sun imparts (ἐντίθησι) brightness (τὸ λαμπρόν) to the moon.[27]

The oldest and at the same time most enigmatic account of Anaxagoras' ideas about the moon's light is in Plato's dialogue *Cratylus*, when he discusses a curious etymology of the word σελήνη:

10.10. Socr.: It seems to show that the view he has recently advocated—that the moon gets (ἔχει) its light from the sun—is quite ancient (παλαιότερον).
(. . .)
Socr.: This light (φῶς) around (περί) the moon is always (ἀεί) new (νέον) and old (ἔνον), if the followers of Anaxagoras are right. For as the sun is always traveling around the moon in a circle, presumably (που) it always sheds (ἐπιβάλλει) new light (νέον) on it, while the old (ἔνον) of the previous month persists (ὑπάρχει).[28]

---

[23]S 1.26.2 = DK 59A77 = MR 602 = GG 485; not in LM, Gr, and KRS.

[24]P 2.28.5 = MR 602; not in DK, but see Dox 358; not in Gr and KRS.

[25]Cf. MR 603.

[26]Hippolytus, *Refutatio Omnium Haeresium* 1.8.8 = DK 59A42(8) = LM ANAXAG. D4(8) = Gr Axs38(8) = GG 270 = KRS 502(8). I put the words "gets it" between brackets, because there is no verb in this clause.

[27]Plutarch, *Face on the moon* 929b = DK 59B18 = LM ANAXAG. D40 = Gr Axg44 = GG 188 = KRS 500. Curd (2010), 27, translates "the sun places the light in the moon."

[28]Plato, *Cratylus* 409A7–B10 = DK 59A76 = LM ANAXAG. D41 (both leaving out that Anaxagoras' view is called "ancient") Gr Axg45 = GG 5; not in KRS.

The words "the sun is always traveling around the moon in a circle" are a somewhat strange way of saying that the sun and moon are in opposition once a month and are in conjunction half a month later. The words "the old light of the previous month persists" seem to have to do with the moon's phases. But why is "the moon *always* new and old"? Even more interesting is the question of the precise meaning of "the moon gets its light from the sun." This is usually assumed to mean that the moon reflects the light of the sun, which seems to contradict the content of texts 10.2–10.5. One also wonders, why the view Anaxagoras recently advocated is called "ancient." These problems will be discussed in later sections of this chapter. Plutarch refers to Plato's text:

10.11. (...) he said that Anaxagoras was embarrassed by the name of the moon, since he tried to claim as his own some very ancient opinion in regard to its illumination (περὶ τῶν φοτισμῶν). Has not Plato said this in the *Cratylus*?[29]

At first sight, these texts (10.6–10.11) seem to contradict what was said in the previous section (texts 10.2–10.5). It is especially difficult to understand how Hippolytus can state both that the moon is a fiery stone (text 10.5) and that the moon has no light of its own (text 10.8).

Anaxagoras on the Light of the Moon in Aëtius 2.29 and Analogous Texts

The third relevant chapter of Aëtius is 2.29, "On the eclipse (περί ἐκλείψεως) of the moon."[30] Surprisingly, four items in this chapter also contain opinions (of Anaximander, some unnamed youngers, Xenophanes, and Anaxagoras) on the phases of the moon. Anaxagoras is mentioned in Stobaeus' version of an item, part of which I have already discussed in Chap. 9. The relevant lines read as follows:

10.12. Thales, Anaxagoras, Plato, and the Stoics agree with the astronomers that it (the moon) produces the monthly concealments (τὰς μηνιαίους ἀποκρύψεις) by following the sun's path and being illuminated (περιλαμπομένην) by it (...).[31]

[29]Plutarch, *De E apud Delphos* 15 = GG 153; not in DK, LM, Gr, and KRS.

[30]See Dox359–360; MR 613–623.

[31]S 1.26.3 = DK 59A77 = LM ANAXAG. D45a = GG 486 = MR 614; not in Gr and KRS.

In pseudo-Plutarch's version, however, Anaxagoras is not mentioned:

10.13. Plato, Aristotle, and the Stoics agree with the astronomers that it produces the monthly concealments by following the sun's path and being illuminated by it (. . .).[32]

In their reconstructed text, Mansfeld and Runia insert Aristotle, who only appears in pseudo-Plutarch's version of this passage.[33] In Chap. 9, I argued that, from the viewpoint of astronomical conceptions, pseudo-Plutarch's enumeration, "Plato, Aristotle, the Stoics and the astronomers," all of whom were advocates of a spherical earth, makes more sense than Stobaeus' version. Strictly speaking, the words "monthly concealments" in this text only allude to the new moon, but implicitly the moon's phases are also meant.

Similarly, with the term "illuminations," Hippolytus means the moon's phases,[34] when he states:

10.14. He was the first to determine (ἀφώρισε πρῶτος) what is involved in (τὰ περὶ) eclipses and illuminations (φωτισμούς).[35]

As already remarked at text 9.9, Graham's translation, "He first *correctly* explained eclipses and illuminations" says more than Hippolytus' words express. The verb ἀφορίζω means "to determine" and this determination is not necessarily a correct explanation After all, Hippolytus indicates that Anaxagoras offered two explanations for the moon's eclipses and this can hardly be called "correct." Hippolytus' text says more: the words τὰ περὶ, which Graham leaves untranslated, are an interesting restriction, meaning something like "the circumstances under which" eclipses and phases of the moon may occur. As we have seen (text 10.5), Hippolytus said that, according to Anaxagoras, the moon was a fiery stone, and (in text 10.8) that the moon did not have its own light but got it from the sun. Gershenson and Greenberg rightly comment, "He nowhere explains how (. . .) these statements [in texts 10.5, 10.8, and 10.14] are to be reconciled."[36] This comment can be generalized as the question of how to reconcile what is said in Aëtius' chapters 2.28 and 2.29 with what is said in his chapter 2.25.

Two other items in Aëtius' chapter 2.29 deserve our attention. One of them is interesting in the context of our enquiry, although Anaxagoras is n
ot mentioned. In pseudo-Plutarch's version, it reads as follows:

---

[32] P 2.29.6 = MR 614 (P2.29.5, in their numbering); not in DK, but see Dox 360; not in LM, Gr, and KRS.

[33] MR 622.

[34] Curd (2010), 95 translates "the phases of the moon"; LM ANAXAG. D4(10) add between square brackets, "[i.e. the lunar phases]."

[35] Hippolytus, *Refutatio Omnium Haeresium* 1.8.10 = DK 59A42(10) = LM ANAXAG. D4 (10) = Gr Axg38(10) = GG 270; this line not in KRS 502(10).

[36] GG, p. 339.

10.15. The youngers (οἱ δὲ νεώτεροι) [say that the phases of the moon appear] in accordance with the spreading of a flame (κατ' ἐπινέμησιν φλογὸς) that is kindled little by little in an orderly manner (κατὰ μικρὸν ἐξαπτομένης τεταγμένως),[37] until it produces the complete full moon, and analogously diminishes (μειουμένης) again until the conjunction [of the sun and the moon], when it is completely quenched (σβέννυται).[38]

Where in the version of pseudo-Plutarch is written "the youngers," Stobaeus' version has "there are some of the youngers in whose opinion..." (τῶν δὲ νεωτέρων εἰσι τινες οἷς ἔδοξε). After the words "the youngers," Mansfeld and Runia, who follow Stobaeus' version, put "members of the school" between brackets, and Huffman adds "Pythagoreans," but Dumont notes, "il n'est pas sûre que ses modernes soient eux aussi des pythagoriens."[39] Mansfeld and Runia read, "in whose opinion (an eclipse takes place)," but remark a few pages earlier, "note again the confusion between eclipses and phases."[40] Huffman reads, "who thought that [the phases of the moon?]" and Dumont adds, "La seconde explication (i.e. that in text 10.15) rend compte des phases de la lune." In my opinion, this text is clearly not about eclipses but about the phases of the moon, as indicated by the sequence "full moon—until the conjunction." I added, between square brackets, "of the sun and the moon." According to Graham, "the most important feature of this account is that it seems confused: what the sentence describes is not a lunar eclipse—which happens in hours, not in the course of a month—but rather the phases of the moon."[41] In my view, the sentence is not confused but placed under the wrong heading. At the end of this chapter, I will come back to its interpretation. "The conjunction" means the conjunction of the new moon with the sun.

Anaxagoras on the Light of the Moon in Aëtius 2.30 and Analogous Texts

The fourth relevant chapter is 2.30, "On its [sc. the moon's] appearance (περὶ ἐμφάσεως) and why it appears to be earthy."[42] In Stobaeus' version, the item on Anaxagoras reads as follows:

[37]MR 2009, 622 translates "that slowly catches alight", which says pretty much the same.

[38]P 2.29.4 ≈ S 1.26.3 = DK 58B36 (cf. Dox 360) = MR 613 and 614 = LM PYTHAGOREAN DOCTRINES D42, not in Gr.

[39]MR 622; Huffman (1993), 237; Dumont (1988), 581 and 1405, n. 5 at 581.

[40]MR 618.

[41]Graham (2013a), 196–197.

[42]See Dox 361–362; MR 624–634. GG 172 translates: "Concerning the reflection of light from the moon (...)," which is certainly not right.

10.16. Anaxagoras (declares the appearance of the moon is caused by) the unevenness of its composition on account of cold being mixed together with the earthy, the moon having some parts that are high, others that are low, and others that are hollow. Moreover, (he declares that) the dark (τὸ ζοφῶδες) has been mixed in with the fire-like (παραμεμῖχθαι τῷ πυροειδεῖ), the effect of which causes the shadowy (τὸ σκιερόν) to appear; for this reason, the heavenly body is called "falsely appearing" (ψευδοφανῆ).[43]

Pseudo-Plutarch's version is much shorter:

10.17. Anaxagoras (declares the appearance of the moon is caused by) the unevenness of its composition on account of cold being mixed together with the earthy, because (γάρ) the dark has been mixed in with the fire-like. For this reason, the heavenly body is called "falsely appearing" (ψευδοφανῆ λέγεσθαι).[44]

Mansfeld and Runia say that in pseudo-Plutarch's version, "the information about the unevenness of its surface is deleted."[45] I think it is also possible that Stobaeus inserted some clarifying text, freely borrowed from Aëtius' chapter 2.25 (cf. text 10.2). Pseudo-Plutarch's text makes it clear, by means of the word γάρ, that the words "the cold is mixed with the earthy" are intended to mean the same as "the dark is mixed with the fire-like." Apparently, the dark spots on the moon should be regarded as places that are less hot; this is a kind of mitigation of the fiery moon in Aëtius' chapter 2.25.9 (text 10.2). As far as I can see, the issue of texts 10.16 and 10.17 is the light and dark spots on the moon, or "the face on the moon." The same applies to the other texts in Aëtius' chapter 2.30, as its title, "On its appearance and why it appears to be earthy," indicates.

The manuscripts of pseudo-Plutarch have the variants ψευδοφαῆ and ψευδοφανῆ. I followed Mansfeld and Runia's reading ψευδοφανῆ and their translation "falsely appearing."[46] The dictionary has for both terms "shining with false, i.e. borrowed, light,"[47] but in texts 10.16 and 10.17, the issue is not whether the moon borrows its light from the sun, but what the surface of the moon looks like.[48] Whatever this word may indicate, it does not have to do with the phases of the moon, but with "the face on the moon." The last lines are a duplicate with the text on Parmenides, two items further on.[49] Although ψευδοφανής fits nicely in a

[43]S 1.26.4 = DK 59A77 = LM ANAXAG. D44 (only the first line) = GG 487 = MR 624; not in Gr and KRS

[44]P 2.30.2 = GG 172 = MR 625; not in DK, but see Dox 361; not in LM, Gr, and KRS.

[45]MR 626.

[46]See MR 628, n. 514; LSJ, s.v. ψευδοφαής.

[47]LSJ, s.v. ψευδοφαής.

[48]The term ψευδοφαῆ is used by Diogenes Laërtius 2.1 in his account on Anaximander (DK12A1 (1), but DK (81 note at lines 11 and 12) comments: "das Theophrastexcerpt wohl von Anaxagoras fälschlich übertragen.".

[49]Cf. MR 627–628.

hexameter,[50] from Parmenides' poem we only know the word νυκτιφαές (shining by night).[51] While Diels has argued that the word ψευδοφανῆ was wrongly attributed to Parmenides, Mansfeld and Runia argue that it makes sense to reserve the last line of text 10.16 for Parmenides. Nevertheless, they include it in their reconstructed text of Anaxagoras.[52]

Finally, a passage in Plutarch's biography of *Nicias* deserves our attention:

10.18. Anaxagoras first put in writing in the clearest and boldest terms of all a theory (λόγος) concerning the radiant and shadowy (places) of the moon (περὶ σελήνης καταυγασμῶν καὶ σκιᾶς). This theory, which was not ancient (παλαιός) or generally accepted, at this time still went about whispered in secret with caution rather than confidence among a few men.[53]

The interpretation of this cryptic text encounters several problems. In the first place, Plutarch speaks rather vaguely about "a theory" (λόγος), and when he describes it, he uses the word καταυγασμός that is not attested elsewhere, but is a verbal noun, derived from καταυγάζω and translated in LSJ as "shining brightly". He then emphasizes that this theory is new and not generally accepted, using the words οὔτε παλαιός, which seems to be intended as a polemic against Plato (text 10.10), who calls "quite ancient" (παλαιότερον) the view that the moon gets its light from the sun. Sometimes, the second sentence of text 10.18 is taken as a reference to Anaxagoras instead of to a theory: "Anaxagoras himself was not venerated (παλαιός), nor was his doctrine the best known".[54] And finally, Plutarch calls this theory, whatever it was, both "written in the clearest and boldest terms" and "whispered in secret", which looks contradictory.

Plutarch's text can be interpreted in at least three different ways, two of which can be found in the translations and commentaries. Gershenson and Greenberg translate περὶ σελήνης καταυγασμῶν καὶ σκιᾶς as "of the phases of the moon" or "about the waxing and the waning of the moon."[55] Similarly, Gilardoni and Giugnoli translate: "una teoria sui periodi di illuminazione e di oscuramento della luna" and comment that the text is about "fasi lunari".[56] Curd translates it as "about the changing phases of the moon", but elsewhere, she explains that the text is about eclipses.[57] Graham

[50] Cf. MR 628.

[51] DK 28B14.

[52] Cf. Diels (1897), 110–112; MR 628 and 632.

[53] Plutarch, *Nicias* 23.2 = DK 59A18 = LM ANAXAG. D38 and P25b = Gr Axg6 = GG 197; not in KRS.

[54] Curd (2010), 85.

[55] GG 197 and Panchenko (2002), 326, who adds that this is also Perrin's translation in the Loeb edition.

[56] Gilardoni and Giugnoli (2002), 61 and 254.

[57] Curd (2010), 85 and 211.

writes that Plutarch's text is "concerning the illumination and shadow of the moon"[58] and adds: "Hippolytus agrees: He [Anaxagoras] first correctly explained eclipses and illuminations".[59] Laks and Most write, "concerning the illuminations and darkenings of the moon", and summarize elsewhere that this text is about the light of the moon.[60] According to Guthrie, the text is about lunar eclipses.[61]

We can conclude that these recent commentators hesitate whether Plutarch is talking about Anaxagoras' explanation of the phases of the moon or about his (alleged) theory of eclipses. In favor of the former interpretation, one can argue that the most natural translation of περὶ σελήνης καταυγασμῶν καὶ σκιᾶς seems to be that the theory was about the changing phases of the moon. In favor of the latter interpretation, one can point to the context, in which Plutarch is speaking about eclipses. On the other hand, it sounds somewhat strange to introduce a theory of eclipses with the word καταυγασμός ("shining brightly"). Moreover, the text does not seem to speak about the shadow of the earth, as would be the case in an explanation of lunar eclipses, but about shadows (on the surface) of the moon. I would like to add a third possible interpretation, namely that what is at issue is the light and dark spots on the moon or "the face on the moon" (compare the word σκιᾶς in text 10.18 and τὸ σκιερόν in text 10.16, which is clearly about the appearance of the moon). This interpretation would explain why the theory had to be "whispered in secret with caution": it had to do with Anaxagoras' blasphemous conception of the heavenly bodies as (fiery) stones, for which he was condemned.[62] It is not clear to me, whether Laks and Most's interpretation that the text is about the light of the moon fits into one of these three interpretations, or whether it is meant as a separate one. In the end, I think we must conclude that Plutarch's text does not help us very much, because whatever interpretation we prefer, it remains unclear what precisely the content of the "theory" in question should have been supposed to have been.

Problems and Past Suggestions to Solve Them

The texts collected in the previous sections show that the question of Anaxagoras' conception of the moon's light and phases is quite complicated. Sometimes, information can be found in a chapter of Aëtius in which we would not expect it. It is not

[58]Graham (2013a), 138. Graham quotes this text first in a discussion about the relative ages of Empedocles and Anaxagoras and a second time when he summarizes the thesis of his book—that Parmenides and Anaxagoras were the heroes of early Greek astronomy (Graham 2013a, 138 and 247)—but not when he discusses Anaxagoras' alleged heliophotism and states that he "seems to grasp all the implications of heliophotism" (id. 124).

[59]Graham (2013a), 138.

[60]LM, ANAXAG. D38 and P25b.

[61]Guthrie (1965), 306.

[62]Cf. Diogenes Laërtius, *Vitae Philosophorum* 2.12 = DK 59A1(12) = LM ANAXAG. P23 = Gr Axs 1(12) = GG 340, not in KRS.

always immediately clear whether a text is about eclipses, about the waning and waxing of the moon, or about the light and dark spots on the moon.[63] The Presocratics did not always distinguish clearly between phenomena like the waning and waxing of the moon, eclipses, and the risings and settings of the heavenly bodies, in all of which a celestial body disappears wholly or partially for some time, to appear again at a later time.[64] In the case of Anaxagoras, we may especially wonder how far he had advanced on the path of distinguishing between lunar eclipses and the waning and waxing of the moon.

With respect to the question of whether the moon has its own light or receives its light from the sun (in whatever sense), there seems to be a crucial divergence between the accounts in Aëtius' chapter on the substance of the moon (*Placita* 2.25) and those in his chapter on the illuminations of the moon (*Placita* 2.28). In *Placita* 2.25, most Presocratics are said to hold that the moon is fiery in one way or another. Apart from the dubious testimonies on Thales (the moon is earthy) and Ion (the moon is partly glass-like and transparent, partly opaque) the only exception in this chapter is Pythagoras, who is said to have declared that the moon is a mirror-like body (κατοπτροειδὲς σῶμα).[65] From this, we would expect that *Placita* 2.28 would tell us that almost all Presocratics believed that the moon has its own light and that only Pythagoras thought that the moon is illuminated by the sun, but this is not the case. Not only Pythagoras, but also Thales, Parmenides, Empedocles, Anaxagoras, and Metrodorus are mentioned as thinkers who said that the moon is illuminated by the sun (ὑπὸ τοῦ ἡλίου φωτίζεσθαι) (text 10.6), while only Anaximander, Xenophanes, and the sophist Antiphon are said to have held that the moon has its own light (ἴδιον φῶς, ἰδιοφεγγής). Apparently, there is no consistent correlation between the notions of the moon "being fiery" and "having its own light." And in Stobaeus' version of chapter 2.29,[66] not Pythagoras but Thales and Anaxagoras are mentioned as saying that the monthly concealments of the moon result from its being illuminated (περιλαμπομένην) by the sun (text 10.12). With regard to Anaxagoras, this means that we must investigate whether the apparent contradiction between texts 10.2–10.5 (the moon is an inflamed solid mass) and texts 10.6 and 10.8–10.14 (the moon is illuminated by the sun) can be resolved within the context of his astronomy.

The simplest solution, which is widely adopted, seems to be that the moon not only has its own light, which is sometimes visible as earthshine or as a blood moon, but that the moon is also, except during new moon, illuminated by the sun, whose light normally overpowers the much fainter light of the moon. This interpretation has been defended, with some slight variations, of O'Brien, Wöhrle, Panchenko, and

[63]Cf. MR 661.

[64]See Chap. 8, Sect. *The Nature and Movements of the Celestial Bodies*, and Chap. 9, Sect. *Introductory Remarks on Eclipses*.

[65]Cf. P 2.25.14 ≈ S 1.26.1 = MR 573 and 574; not in DK, but cf. Dox 357; not in Gr and KRS. For the reading κατοπτροειδὲς σῶμα also in pseudo-Plutarch's corrupted text, see MR 581.

[66]See Dox 359–360; MR 613–623.

Graham, and also by myself some years ago.[67] The text that is usually referred to as evidence is that of Olympiodorus, of which I showed in Chap. 9 how confused it is:

10.19. A third view is that of Anaxagoras and Democritus. They say the Milky Way is the proper light of stars not illuminated (μὴ φωτιζομένων) by the sun. For the stars (τὰ ἄστρα), he [sc. Aristotle] says, have their own light as well additional (ἐπίκτητον) light acquired from the sun. And the case of the moon makes this clear. For this has one kind of light of its own and another from the sun. *Its own light is coal-like, which the moon's eclipse shows us.* However, they say, not all the stars receive additional light from the sun and those which do not, compose the band of the Milky Way.[68]

O'Brien rightly comments that "the parallel with the moon seems to be Olympiodorus' own illustration (...). It would be wrong therefore to take Olympiodorus' words as positive evidence for Anaxagoras." Yet he suggests that "in this instance, Olympiodorus' idea seems to have a good chance of representing Anaxagoras' view."[69] Panchenko sees in this text "direct evidence that Anaxagoras assigned a double nature to lunar light."[70] He translates τὰ ἄστρα as "the luminaries,"[71] which is not right here because the reference is to the explanation of the behavior of the stars within and outside of the Milky Way. Graham also reads this text as a confirmation that Anaxagoras believed in the double nature of the moon's light. He comments: "Anaxagoras (...) wanted to account for the light that is emanating from the moon even during its complete eclipse. The moon must have a natural source of light that is normally overpowered by its reflection of the sun's light."[72]

Not only is it dangerous to take Olympiodorus' own illustration in the last sentence of text 10.19 as an exemplification of Anaxagoras' intentions, but what these authors (and Olympiodorus in the first place) also overlook is that, if the moon has its own source of light, this must also be visible when the moon is in conjunction with the Milky Way. When this happens, the rays of the sun cannot shine upon the moon, because then it is, according to Anaxagoras, in the earth's shadow, which implies that the moon's own light would shine brightly in the dark, just like the stars of the Milky Way. Similarly, when the moon is in conjunction with the Milky Way, its phases could not be caused by the sun's light. As mentioned earlier, is hardly credible that this problem had escaped Anaxagoras' attention. The assumption that Anaxagoras' moon had a mixed light, one reflected from the sun and another of its

[67]Cf. Dreyer (1953), 32, n.1. O'Brien (1968), 126–127. Wöhrle (1995), 245. Panchenko (2002), 329–331. Graham (2013a), 131. Couprie (2011), 177.

[68]Olympiodorus, *In Aristotelis Meteora* 67.32 = Gr Axg46 (except the last sentence) = GG 607, not in DK, LM, and KRS; my italics.

[69]O'Brien (1968), 126.

[70]Panchenko (2002), 329.

[71]Ibid.

[72]Graham (2013a), 131.

own, does not, therefore, solve the problem of the explanation of the moon's light and phases in Anaxagoras' astronomy.

Most authors also bring up Plato's words in the *Cratylus* (text 10.10) as evidence for this interpretation of Anaxagoras' ideas about the light and phases of the moon. In Panchenko's words: "If we take the Platonic words seriously, it follows that the moon not only shines by reflection, but also in some way absorbs and stores the light received from the sun".[73] Again, this does not solve the problem of the moon's phases twice a month for several nights when it is in conjunction with the Milky Way. Moreover, Plato's text does not speak of "reflection" but successively says, that the moon gets (ἔχει) its light from the sun, that the light is around (περί) the moon, and that the sun always sheds (ἐπιβάλλει) new light on the moon. Ferguson explicitly states, "This is a theory of borrowed light, but it is not a theory of reflection".[74] This brings us to the fundamental ambiguity to be discussed in the next section.

The Ambiguity of "Received Light"

The question is, then, whether there might not be another explanation for the light and phases of the moon that would be compatible with Anaxagoras' other astronomical ideas and that would reconcile the texts attributing to him the view that the moon is an inflamed solid body with the texts that mention him as saying that the moon gets its light from the sun.

In a commentary on Empedocles, Ferguson wrote, "'the moon has its light from the sun'. This apparently simple statement bristles with difficulties. (...) The actual words do not necessarily mean that the moon shines with *reflected* light; they are compatible with the idea that the moon is *kindled* by the sun."[75] O'Brien adopted this idea more specifically with regard to Anaxagoras: "The proper solution, I suggest, lies in breaking the (...) assumption: that derived light means reflected light. This is in fact a modern assumption, which was not shared in later antiquity."[76] The problem is, that we are easily tempted to interpret the words "the moon receives its light from the sun" in conformity with our modern conception of the moon reflecting the light of the sun, but we may question whether this was as evident to the ancient Greeks as it is to us. In other words, the very use of the word "reflection" has strong anachronistic overtones and should be better avoided in understanding ancient Greek explanations of where the moon gets its light from. One possible exception

[73]Panchenko (2002), 329. See O'Brien (1968), 127; Wöhrle (1995), 246; Couprie (2011), 177; Graham (2013a), 132.

[74]Ferguson (1968), 100.

[75]Ferguson (1968), 99. Cf. DK 31A30 (Plutarch, *Stromata* 10) = LM EMP. D134c = Gr Emp64 = KRS 300.

[76]O'Brien (1968), 122.

could have been Pythagoras (or the Pythagoreans?), according to whom the moon is "a mirror-like body" (κατοπτροειδὲς σῶμα).[77]

One might even wonder whether such bias already affected the accounts of Presocratic conceptions in the doxography. In other words, in conformity with their acquaintance with the right explanation of the moon's phases, the authors of these texts could have understood expressions such as "the moon receives its light from the sun" to mean "the moon reflects the light of the sun." In addition, it is important to note that the expression "the moon has its own light" is also ambiguous. It could imply that the light of the moon does not reflect the light of the sun, but it is not at odds with theories according to which the moon is ignited by the sun. Once the moon has received its light by being kindled by the sun, this light could be said to be the moon's own light. In the same sense, we say that a candle it is lit by a match but, once lit, has its own light.

In the context of Anaxagoras' astronomical ideas, it is highly plausible that expressions such as "the moon receives its light from the sun" must be read as meaning that the moon is in one way or another ignited or kindled by the sun. To quote O'Brien again, "It is not explicitly stated that Anaxagoras' moon shines by reflection. Plutarch's (...) sentence shows that the moon's light is derived light, but not whether it is derived by kindling or by reflection."[78] Elsewhere, O'Brien writes, "A fiery moon, even a partially fiery one, would seem to be inconsistent with the moon's deriving her light from the sun, if derived light means reflected light."[79] To quote O'Brien once more, "the simple theory of a moon whose light is kindled from the sun will at once resolve the difficulties in the evidence for the fifth century. For derivation by kindling, as distinct from reflection, is not inconsistent with, in fact it demands, a fiery moon."[80] Unfortunately, as we have seen, O'Brien, did not grasp the full impact of his own words because he did not consider the implications of Anaxagoras' explanation of the Milky Way. Graham ignores the ambiguity of the expression "The moon receives its light from the sun."[81] In his book, "illumination," "derived light" and "reflected light" are synonymous, as his definition of heliophotism shows: "Heliophotism makes a causal connection between the phases of the moon and the sun: the sun's light is *reflected* from the surface of the moon."[82] Significantly, Graham, who advocates that Anaxagoras defended heliophotism,[83]

---

[77] S 1.26.1 = MR 381(c), not in DK, but see Dox 357, not in LM, Gr, and KRS. See also note 17.

[78] O'Brien (1968), 125, referring to Plutarch, *Face on the Moon* 929b = DK 59B18 (see text 10.9).

[79] O'Brien (1968), 121.

[80] See O'Brien (1968), 123.

[81] In an earlier paper, he discusses this ambiguity. See Graham (2002), 364, where he concludes: "L'ensemble de l'explication n'est pas nécessaire. Car, quoi que puisse être la physique de la lumière de la lune, il s'avère que l'éclairage de la surface de la lune par le soleil est toujours une condition nécessaire pour que la lune émette de la lumière." In this and the next section of this chapter it will be argued that in Anaxagoras' explication of the light and phases of the moon, "éclairage" does not mean that the moon reflects the light of the sun.

[82] Graham (2013a), 109–110 (my italics).

[83] See Graham (2013a), 87–88.

almost completely ignores the texts that say the moon is fiery, just as he almost completely ignores the texts that say the Milky Way is caused by the shadow of the earth.[84]

The Moon's Light and Phases According to Anaxagoras; Suggestions for a New Interpretation

Parmenides said that the lit side of the moon is always turned towards the sun.[85] It is hard to believe that he was the first to discover this. We can read it as a statement of a well-known fact, because it is a primary observational datum. Thales had already studied and tried to predict eclipses of the sun. He could not have done this without being aware of the observational fact that a solar eclipse occurs during new moon and a lunar eclipse during full moon and that, between these two events, the phases of the moon appear, in which the bright side of the moon is always turned towards the sun. As the cases of Anaximander and Xenophanes show, this knowledge did not automatically lead to a correct explanation of the light and the phases of the moon. There is no reason to doubt that Anaxagoras was also familiar with this observational fact. However, as we have seen, its correct explanation would have been incompatible with the rest of his astronomical ideas. As defended above and in Chap. 9, pseudo-Plutarch's version (texts 9.10 and 10.13) of Aëtius' text on the right explanation of the moon, which does not mention Anaxagoras, is preferable to the version of Stobaeus (texts 9.11 and 10.12).[86] This means that we do not possess information on Anaxagoras' explanation of the moon's phases, except for Hippolytus dubious words that he "was the first to determine what is involved in phases of the moon.[87] Nevertheless, given our knowledge of his other astronomical ideas and taking into account the ambiguity of expressions such as "the moon receives its light from the sun" and its equivalents (in texts 10.6–10.13), we can make a reasonable guess. As far as I can see, two options deserve serious consideration, the first of which I personally prefer.

---

[84]Graham mentions text 10.2 once, in a footnote, but only in relation to the claim that the moon has plains, mountains, and ravines. And his only and disputable comment on text 10.5 is this: "the sun, moon, and stars are fiery stones, hence solid, massive bodies *of presumably spherical shape*" (my italics). See Graham (2013a), 123, n. 14, and 124. He does not mention texts 10.3 and 10.4.

[85]See Plutarch, *Face on the Moon* 929b = DK 28B15 = LM PARM. D28 = Gr Prm33. A lot has been written about Parmenides' alleged discovery of heliophotism. Even after the recent thorough studies on this subject (e.g., Mourelatos 2013), I remain skeptical as to whether someone who called the moon νυκτιφαές (or νυκτὶ φάος) and who reportedly called it fiery (πυρίνη), could have developed the theory that the moon reflects the light of the sun. But a discussion of this issue would be far beyond the scope of this book.

[86]Even Graham (2013a) does not use Stobaeus' version as an argument for his interpretation of Anaxagoras.

[87]Cf. text 10.14.

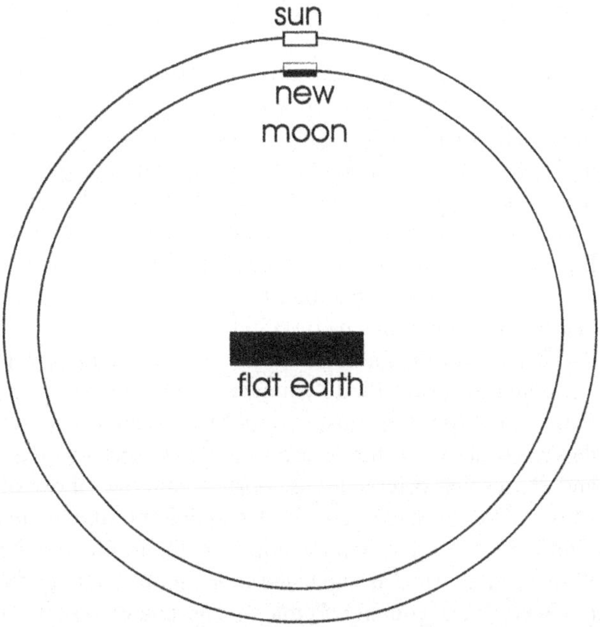

Fig. 10.4 During new moon, the sun is very close to the moon (approximately to scale)

O'Brien and Panchenko questioned whether a pure theory of derived light, kindled by the sun, ever existed.[88] They overlooked text 10.15, according to which unnamed "youngers" defended a full-fledged theory of a fiery moon and its phases. If my analysis is correct, Anaxagoras may have been one of the advocates this theory. His conception of the earth as flat and his explanation of the Milky Way implied that the heavenly bodies must be relatively near and smaller than the earth. This means that, when the moon and the sun during new moon are in conjunction, the two luminaries must be very close to each other, as shown in Fig. 10.4. At this point, the heat of the sun on the back of the moon—the side that is turned away from the earth—would necessarily be very intense, enabling the sun to ignite the moon.[89] However, during new moon, we do not see this light of the heated moon because the side that is kindled is the one that is turned away from us.

Subsequently, this light, which is actually the glowing stony surface of the moon, expands. We see the first glimpse of fire creeping over the rim of the moon when we observe the small crescent a few days after new moon. As the moon goes through the phases of waxing crescent, first quarter, waxing gibbous, and finally full moon, the glow spreads gradually, covering an ever-growing part of the moon and finally its entire surface. We can compare this process with a fireplace that is lit on one side with a small fire that grows bigger and bigger until the whole fireplace is burning.

[88]Cf. O'Brien (1968), 123; Panchenko (2002), 328.

[89]Cf. Panchenko (2002), 333: "At the time of conjunction (...), the side of the moon turned to the sun is turned *from* us, while the side which is not affected by heating is turned towards us."

However, because the moon is stony, it is not ignited with a raging fire but with the quiet glow we perceive, like glowing coals. After full moon, when the sun is farthest away from the moon, the glow shrinks again, gradually decreasing as the moon passes through the phases of waning gibbous, last quarter, and waning crescent, until it is finally extinguished at new moon and then kindled anew. With this explanation of the phases of the moon there is no question of reflected light. The light we see on the moon is not the reflection of the light of the sun, but the glow of the heated surface of the moon. In this explanation, expressions such as "the moon receives its light from the sun" are understood literally: the moon is kindled by the sun. Although it must be reignited every month, it can be said that, once kindled, it has its own light, just like a lit lamp has its own light.

This is the explanation of the moon's light and phases that in text 10.15 is attributed to unspecified "youngers". Although the text does not mention how the flame is kindled, the most natural reading is that the moon is kindled by the sun as described above. It might even be argued that this explanation of the moon's light and phases was offered as an improvement over those of Anaximander and Xenophanes, who did not explain why the opening of the vents in the celestial wheels or the kindling started during new moon and then followed the rhythm of the lunar month. Usually, text 10.15 is supposed to be about younger Pythagoreans, but it is hard to tell who these younger Pythagoreans could have been,[90] who allegedly rebelled against the Pythagorean theory that the moon, functioning like a mirror (κατοπτροειδής), has its light by reflection (ἀνταυγείᾳ).[91] Moreover, text 10.15 is about the phases of the moon, while the immediately preceding text is about the Pythagorean (Philolaic) theory of lunar eclipses. If we assume that, in text 10.15, not Pythagoreans but others are meant, the most likely candidate would be Anaxagoras (and his followers), in whose system this explanation of the phases of the moon would fit very well.

This explanation also makes sense with respect to Plato's text in the *Cratylus* (text 10.10). Socrates can call this explanation "ancient" because it presupposes a fiery moon as did other Presocratic thinkers (cf. the remarks after text 10.2). The light of the moon can be called "always new" because it is kindled anew every month. We can easily imagine that what we see during the month as the dark part of the moon has a faint afterglow, comparable with a peat-moor fire that spreads underground as the remnant of an earlier ignition. Usually, we do not see this faint afterglow because it is outshone by the light part (in this theory: the burning part) of the moon. Only when the light of the crescent moon is very small can we perceive it as what we now call earthshine. Because it is the faint afterglow of the extinguished fire, this light can also be called "old." Socrates uses the words "the followers of Anaxagoras" (οἱ Ἀναξαγόρειοι), which can be compared with "the youngers" in text 10.15.

---

[90] Cf. Dumont (1988), 1405, n.5 at p. 581: "Il n'est pas sûre que ces modernes soient eux aussi des pythagoriens."

[91] Cf. = S 1.26.1 ≈ P 2.25.14 = MR 574 and 573 not in DK, but see Dox 357, not in LM and Gr.

An explanation like the one suggested above has been proposed by Sider in his interpretation of Anaxagoras' fragment B18 (text 10.9). I quote: "The sun actually gives up some of its λαμπρόν (in the form of bright aither), which becomes part of the moon during and, to a lesser extent, after the time of direct illumination." And somewhat further: "Only if some light was physically absorbed could the moon glow from the light of the sun when the sun no longer shines directly on it." And again: "(...) the sun had physical substance which would penetrate into the moon's surface."[92] In Sider's interpretation, too, the moon's light is not reflected light from the sun, but in a way kindled by the sun, although according to him in the form of bright aither, while in the interpretation suggested above it is the sun's fire that starts the moon's glow.

The other possibility that deserves to be mentioned is an extrapolation of the conception of invisible heavenly bodies, which must have been, as argued in Chap. 9, Anaxagoras' one and only explanation for lunar eclipses. As argued above, these invisible bodies must be conceived of as consisting of a kind of condensed air. An argument in favor of this theory could have been that the not illuminated part of the moon has the same color as the surrounding sky, black at night and blue by day, as shown in Figs. 5.1a and 5.2. The phenomenon of "earthshine" during the crescent waxing or waning moon could, by analogy with the explanation of the "blood moon" during lunar eclipses, be explained, by the temporary transparency of the air-like invisible heavenly body. Previous thinkers, such as Anaximander and Xenophanes, also made no distinction between the explanations of eclipses and phases of the moon. Anaximander said they were both due to the closing of the apertures of the moon wheel. Xenophanes considered them as quenchings. Anaxagoras may have found it satisfying to propose a uniform explanation for eclipses, occultations, settings, and phases, explaining them all by means of a body that obstructs our vision of another celestial body: the moon (in solar eclipses and star occultations), the earth (in the settings of sun, moon, and stars), or an invisible body (in the case of lunar eclipses and phases). In this scenario, too, the moon must be a fiery stone ignited by the sun's heat, but its phases are not explained by the gradually spreading and extinguishing fire but by the intervening bodies. This second explanation of the moon's phases would not explain why the cycle starts during new moon and follows the rhythm of the lunar month.

Conclusion

According to Graham, "Anaxagoras profoundly changed the understanding of the heavens irreversibly and forever."[93] In my opinion, on the contrary, Anaxagoras inventively defended ideas that were already outdated when he wrote them down—

[92] Sider (2005), 158–159 (= Sider 1981, 122–123).

[93] Graham (2013a), 242.

about the shapes of the earth and of the other heavenly bodies, the Milky Way, lunar eclipses, and the light of the moon—contrary to what we would now consider to be more progressive ideas. Taken together, however, his ideas formed a coherent whole. Anaxagoras' main achievement in astronomy was his acknowledgement that the heavenly bodies are fiery stones, and for this idea he had to go into exile. But as far as his general understanding of the heavenly phenomena is concerned, perhaps in the end he is best described as a tragic figure.

References[94]

Bowen, A.C. and R.B. Todd: *Cleomedes' Lectures on Astronomy*. Berkeley etc.: University of California Press 2004.

Couprie, D.L., *Heaven and Earth in Ancient Greek Cosmology*. New York: Springer 2011.

Curd, P., *Anaxagoras of Clazomenae*. Toronto: University of Toronto Press 2007 (repr. 2010).

Diels, H., *Parmenides, Lehrgedicht*. Berlin: Georg Reimer 1897.

Diels, H. and W. Kranz, *Die Fragmente der Vorsokratiker*. Zürich/Hildesheim: Weidmann 1951[6] (= DK).

Dreyer, J.L.E., *A History of Astronomy from Thales to Kepler*. Mineola, NY: Dover Publications 1953.

Dumont, J.-P., *Les Présocratiques*. Paris: Gallimard 1988.

Ferguson, J., Two Notes on the Preplatonics. *Phronesis* 9 (1968), 98–106.

Gershenson, D.E. and D.A. Greenberg, *Anaxagoras and the Birth of Physics*. New York: Blaisdell Publishing Company 1964 (= GG).

Gilardoni, G. and Giugnoli, G., *Anassagora: Frammenti e testimonianze*. Milano: Bompiani 2002.

Graham, D.W., La lumière de la lune dans la pensée grecque archaïque. In: A. Laks and C. Louguet (eds.), *Qu'est-ce que la Philosophie Présocratique?* Villeneuve d'Ascq 2002, 351–380.

—, *The Texts of Early Greek Philosophy*. Cambridge: Cambridge University Press 2010 (= Gr).

—, *Science Before Socrates*. Oxford: University Press 2013.

Guthrie, W.K.C., *A History of Greek Philosophy. Volume II. The Presocratic Tradition from Parmenides to Democritus*. Cambridge: Cambridge University Press 1965.

Heath, T., *Aristarchus of Samos. The Ancient Copernicus*. Oxford: Clarendon Press 1913.

Huffman, C.A., *Philolaus of Croton*. Cambridge: Cambridge University Press 1993.

Kirk, G.S., J.E. Raven and M. Schofield, *The Presocratic Philosophers*. Cambridge: Cambridge University Press 2009 (= KRS).

Laks, A. and G.W. Most *Early Greek Philosophy VI*, Cambridge, MA. 2016.

Liddell, H.G., R. Scott H.S. Jones and R. McKenzie, *Greek-English Dictionary*. Oxford: Clarendon Press 1996. Also available on the internet, e.g.: http://stephanus.tlg.uci.edu/lsj/#eid=1714&context=search (= LSJ).

Mansfeld, J. and D.T. Runia, Aëtiana. *The Method and Intellectual Context of a Doxographer, Volume Two, Part Two*. Leiden: Brill 2009 (= MR).

O'Brien, D., Derived Light and Eclipses in the Fifth Century, *The Journal of Hellenic Studies* 88 (1968), 114–127.

Mourelatos, A.P.D., Parmenides, Early Greek Astronomy, and Modern Scientific Realism. In: J. McCoy, ed., *Early Greek Philosophy*, Washington, D.C.: The Catholic University of America Press 2013, 91–112.

[94]P = Aëtius in pseudo-Plutarch, *Placita* (numbering according to Dox).
S = Aëtius in Stobaeus, *Anthologium* (numbering according to Wachsmuth and Hense).

Panchenko, D., "Eudemus Fr. 145 Wehrli and the Ancient Theories of Lunar Light." In: I. Bodnár and W.W. Fortenbaugh (eds.), *Eudemus of Rhodes*, New Brunswick 2002, 323–336.
Sider, D., *The Fragments of Anaxagoras*. Sankt Augustin: Academia Verlag 2005 (first edition Meisenheim am Glan: Verlag Anton Heim 1981).
Tannery, P., *Pour l'histoire de science Hellène : de Thalès à Empédocle*. Paris : Félix Alcan 1887.
Wöhrle, G., "Wer entdeckte die Quelle des Mondlichts?," *Hermes* 123 (1995), 244–247.

Chapter 11
Anaxagoras and the Measurement of the Sun and Moon

Contents

The Doxographical Evidence

According to the available sources, Anaxagoras compared the size of the sun and moon with that of the Peloponnesus. The relevant texts are the following:

11.1 The sun is much bigger than (πολλαπλάσιον) the Peloponnesus.[1]
11.2 The sun exceeds (ὑπερέχειν) the Peloponnesus in size.[2]
11.3 The sun is bigger (μείζω) than the Peloponnesus.[3]
11.4 Anaxagoras says that it (sc. the sun) is larger than (μείζονα) the Peloponnesus.[4]
11.5 The moon is as big as (ὅση) the Peloponnesus.[5]
11.6 Anaxagoras says that the sun is 18 times the earth.[6]

---

[1]P 2.21.3 (not in S) = DK 59A72 = GG 169 = MR 535; not in LM, Gr, and KRS.

[2]Hippolytus, *Refutatio Omnium Haeresium* 1.8.8 = DK 59A42(8) = LM ANAXAG. D4 (8) = Gr. Axg38(8) = GG 270 = KRS 502(8).

[3]Diogenes Laërtius, *Vitae Philosophorum* 2.8 = DK 59A1(8) = GG 340, not in Gr, LM, and KRS.

[4]Theodoretus, *Curatio affectionum Graecarum* 1.97 and 4.22 = GG 445 and 453 = MR 536; not in DK, LM, Gr, and KRS.

[5]Plutarch, *The Face on the Moon* 932a = Gr. Axg41 = GG 189; not in DK and KRS.

[6]*Isagoge bis excerpta* 17 = MR 525 (see also 538).

D. L. Couprie, *When the Earth Was Flat*, Historical & Cultural Astronomy,
https://doi.org/10.1007/978-3-319-97052-3_11

Plutarch (text 11.5) is the only one to report on Anaxagoras and the size of the moon; the other testimonies are about the size of the sun. Text 11.6 is a strange anomaly with its specific number for the size of the sun relative to the earth. With the exception of MR, most collections of texts apparently did not consider it worth the mentioning. Perhaps there is a confusion with Anaximander's number 18 for the moon.

Did Anaxagoras Measure the Size of the Sun and Moon with the Help of a Solar Eclipse?

Several scholars have tried to determine how Anaxagoras could have achieved these measurements. Recently, Graham repeated the suggestion he made in 2007 in an article written together with the astronomer Hintz, that Anaxagoras made use of the solar eclipse of February 17, 478 BC (or in astronomical year numbering: −477 February 17)[7] to measure the sizes of the moon and the sun.[8] This idea is not new; Sider and West proposed it earlier, although with the eclipses that occurred on May 19, 557 BC (West) and April 30, 463 BC (Sider).[9] They suggested that "Anaxagoras might have argued: the moon's shadow must be the same size as the moon (...), therefore the moon is as big as the Peloponnese, therefore the sun, which looks the same size but is further away, must be that much bigger than the Peloponnese."[10] Graham convincingly argues that Sider's choice is not very probable because the shadow track of the eclipse of April 30, 463 BC passed over Thessaly, not over the Peloponnesus, and that West's choice is highly improbable because the eclipse of May 19, 557 BC took place 50 years before Anaxagoras was born.[11] In this chapter, I will discuss and evaluate Graham's own attempt. I will also briefly address Fehling's attempt, which was based on quite different data.[12] Finally, I will propose my own suggestion, namely that Anaxagoras has transformed Thales' measurement of the height of a pyramid into a method for making astronomical calculations.

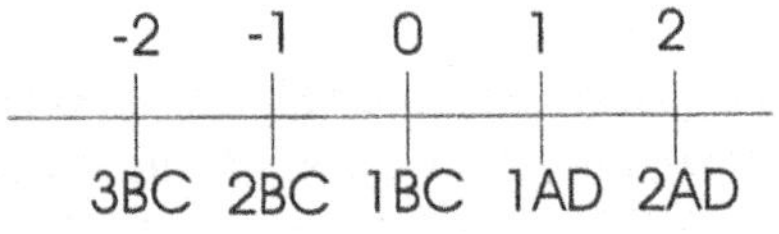

Fig. 11.1 Timeline

[7]The reason for the difference is that the AD–BC year numbering does not have a year 0; this is best illustrated with a small picture of the timeline around the beginning of our era (Fig. 11.1).

[8]See Graham (2013a, 149; 2013b, 4); Graham and Hintz (2007, 324).

[9]Cf. West (1971, 233, n. 1); Sider (1973).

[10]West (1971, 233, n. 1).

[11]Cf. Graham (2013a, 147–148); Graham and Hintz (2007, 322–324).

[12]Fehling (1985, 209–210).

Fig. 11.2 The eclipse of February 17, 478 BC, Athens (37°56′ N, 23°42′ E)

The eclipse of February 17, 478 BC was annular, which means that a ring of sunlight remained visible around the moon. This means that Anaxagoras, who supposedly observed this eclipse in Athens, could easily have concluded that the sun is bigger than the moon.[13] The track of the eclipse of February 17, 478 BC (or more precisely, the track of its annularity, as will be explained below) went across the Peloponnesus, as well as across Clazomenae, Anaxagoras' birth place. It must have looked like Fig. 11.2 (generated with Redshift 9 Premium).

Solar Eclipses; Umbra, Penumbra, and Antumbra

To make it possible to follow the argument, it is necessary to first recapitulate the types and main features of solar eclipses and to illustrate them schematically with images that can be found in any astronomy book. The so-called *umbra* is the shadow cone in which the sun is completely blocked by the moon; an observer in the *umbra* experiences a total eclipse. The so-called *penumbra* is the area in which only a part of the sun is obscured by the moon; an observer in the *penumbra* experiences a partial eclipse (see Fig. 11.3).

An annular eclipse is a special kind of partial solar eclipse, in which the shadow cone of the moon (the *umbra*) does not reach the earth. The so-called *antumbra* is the area from which the moon appears entirely contained within the disc of the sun; an observer in this area experiences an annular eclipse. In Figs. 11.3, 11.4a, b, the sun and moon are depicted as spheres. In all probability, Anaxagoras believed that not only the earth, but also the heavenly bodies were flat. For the argument presented in

[13]Cf. Graham (2013a, 150): "(. . .) he could *see* that the sun was larger, for the moon failed to block the periphery of the sun."

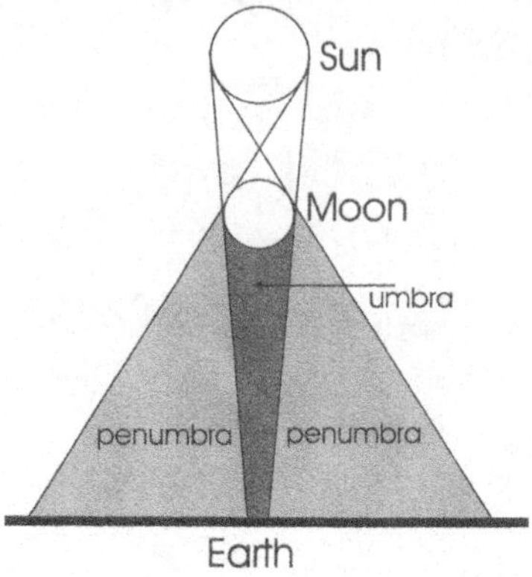

Fig. 11.3 During a total solar eclipse, the conical *umbra* of the moon reaches the earth

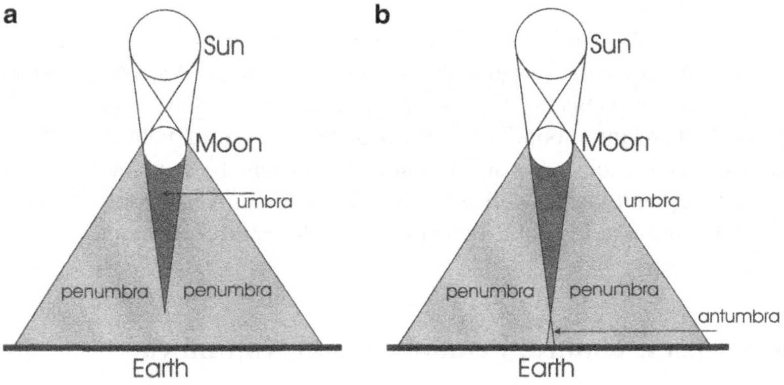

Fig. 11.4 (**a**) During an annular eclipse, the *umbra* of the moon does not reach the earth. (**b**) The area in which an annular eclipse can be observed is called the *antumbra*

this chapter, however, his ideas on the shape of the sun and the moon are of minor importance.

As shown in Fig. 11.4a, b, it is important not confuse the *antumbra* of an annular eclipse with the shadow (*umbra*); the *antumbra* is better understood as part of the *penumbra*. This fact alone was sufficient for Sider to discard the annular eclipse of February 17, 478 by stating that "there would be no *umbra*."[14]

[14]Sider (1973, 129, n. 10, my italics).

Graham and Hintz on the Eclipse of February 17, 478 BC

Graham and Hintz' reference to Sider when they write, "Some scholars seem to dismiss the eclipse of 478 because it was not total and allegedly might have been missed"[15] is incorrect. The only thing Sider says is: "The eclipse of 478 BC was annular (i.e., there would be no umbra),"[16] which does not imply that it could have been missed. Rather, it implies that this eclipse was, according to Sider, irrelevant for Anaxagoras' project. Graham and Hintz argue (or speculate) differently: "Anaxagoras and his generation presumably had no knowledge of this phenomenon and would have treated the *antumbra* as an *umbra*."[17] They state consequently,: "The *shadow* of the eclipse (. . .) covered the whole breadth of the Peloponnesus except for the extreme northwest," and elsewhere: "the *umbra* passed south of Sicily."[18] In his book, Graham writes that "the *umbra* (technically, the '*antumbra*'), obscured almost the whole of the Peloponnesus," and: "To be precise, the shadow of this eclipse was an *antumbra*, formed by an annular eclipse."[19] In the footnote to his text Graham states, "In an annular eclipse the shadow on the earth lies beyond the focal point,"[20] which is completely incomprehensible. During an annular eclipse, the "shadow on the earth" (the *antumbra*) is actually a part of the *penumbra*, while the "focal point" of the *umbra* lies above the earth's surface, as shown in Fig. 11.4a. Moreover, the only diagram of a solar eclipse in Graham's book[21] shows a total eclipse, not an annular eclipse as was the case on February 17, 478 BC. The fundamental issue, however, is not whether Anaxagoras was able to tell the difference between *umbra* and *penumbra*, but that an annular eclipse is visually essentially different from a total eclipse. As we will see, this last difference would have influenced the kind of information Anaxagoras was supposed to gather according to Graham.

In the three sources I consulted (the NASA list of eclipses on the Internet, the computer program RedShift 8 Premium, and Graham and Hintz' article), the renditions of the path of the *antumbra* of the eclipse of February 17, 478 BC are slightly different, but they all agree that it went across the Peloponnesus. Figure 11.5 presents a reproduction of the NASA version. Graham and Hintz draw the path of the *antumbra* somewhat smaller than RedShift and NASA, from the northern border of the Peloponnesus to just below Kythira (the little isle south of the Peloponnesus).

Calculated using the NASA list, the data (which will be only slightly different from the two other calculated paths) of this eclipse were as observed in Athens:

---

[15]Graham and Hintz (2007, 327 and n. 24).

[16]Sider (1973, 129, n. 10).

[17]Graham and Hintz (2007, 327, n. 21, my italics).

[18]Graham and Hintz (2007, 324 and 336, my italics).

[19]Graham (2013a, 149 and 144 n. 17, my italics).

[20]Graham (2013a, 149 n. 25).

[21]Graham (2013a, 152, Fig. 5.2).

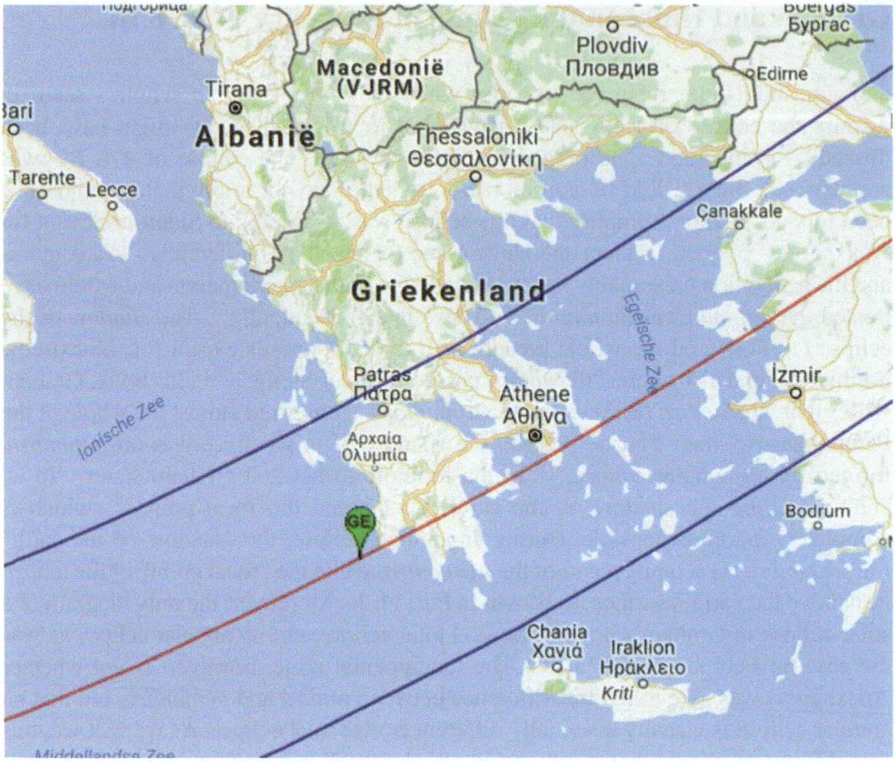

Fig. 11.5 The path of the *antumbra* of the solar eclipse of February 17, 478 BC. Made with the help of https://eclipse.gsfc.nasa.gov/SEsearch/SEsearchmap.php?Ecl=-04770217

Latitude: 37.9554° N
Longitude: 23.7524° E
Magnitude: 0.968
Obscuration: 89.31%

When Graham and Hintz suggest that Anaxagoras was "collecting data on the visibility of the eclipse,"[22] they are again inaccurate. What they should have said is that Anaxagoras tried to collect data of where the eclipse was seen as a special type of a partial solar eclipse known as an *annular* eclipse. Outside the path of annularity, in the wide area of the *penumbra*, the eclipse was a normal partial eclipse. This difference is worded more correctly in Graham's book: "Those who saw (...) the annular eclipse (...) would be duly impressed. (...) Those farther away would see less and would only observe a partial eclipse (...)."[23]

[22]Graham and Hintz (2007, 330).

[23]Graham (2013a, 153).

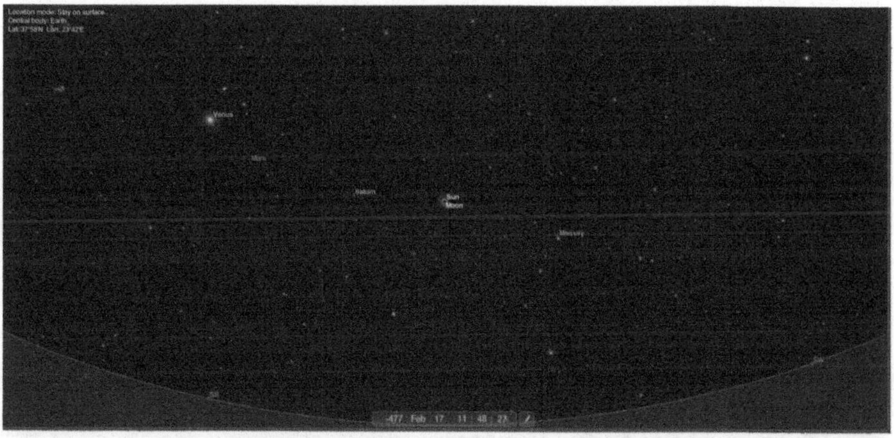

Fig. 11.6 Computer picture of the solar eclipse of February 17, 478 BC after Graham and Hintz. Cf. Graham and Hintz (2007, Fig. 4). Generated with Redshift 9 Premium

These inaccuracies are connected with another one. Graham and Hintz state that "watchers would have seen at least Venus and perhaps all four planets shine forth," and in a footnote, they refer to a picture that looks like Fig. 11.6.[24]

In their picture, Graham and Hintz show the heavens during the eclipse not just with Venus, Mars, Saturn, and Mercury, but also with many stars. This was definitely not what would have been seen in Athens during the eclipse of February 17, 478 BC.[25] My computer program indicates that it is likely that only Venus was visible and no other planets or stars (see Fig. 11.7; Venus faintly visible in the southeast). This picture shows the heavens above Athens at exactly the same moment as in Fig. 11.2

According to Graham, Anaxagoras assumed that "the moon was relatively close to the earth and the sun relatively far away. Thus, the roughly parallel rays of the sun would project onto the earth a shadow roughly the diameter of the moon."[26] This presupposition has two features that deserve special attention: the sun is relatively far away and its rays fall roughly parallel on the earth. The assumption that Anaxagoras thought that the sun was relatively far away cannot be right, because if the earth was flat, as Anaxagoras believed, the sun and the stars would be relatively close and not much farther away than the moon, as extensively argued and shown in Chap. 3, section *Distance of the Heavens*. The presupposition that the sun is far away is typical of what a modern astronomer, who knows that the earth is spherical, would think about what Anaxagoras would have assumed. Characteristically, the same mistake already occurred in the article co-authored with Hintz: "Assuming that the

[24]Graham and Hintz (2007, 327, n. 25), and 328 Fig. 4.

[25]Perhaps they forgot to use the day light vision button of their computer program.

[26]Graham (2013a, 151, 148).

Fig. 11.7 What the annular eclipse of February 17, 478 BC must have looked like in Athens. Generated with Redshift 9 Premium

sun was far distant from the earth (. . .)."[27] What is evident to a modern astronomer, who is familiar with the sphericity of the earth and the actual distance of the sun, was not at all evident to an ancient Greek cosmologist who believed that the earth was flat. This demonstrates how difficult it is for us to imagine what the world looked like for people like Anaxagoras, who thought the earth was flat.[28] For them, the sun and the other celestial bodies could not be far away; they were relatively close.

Let us elaborate on one specific argument for the relative proximity of the sun under the presupposition of a flat earth. Thales is said to have written a *Nautical Star Guide*.[29] Whether or not this report is apocryphal, it shows that there was an ancient tradition of using the stars as a means of orientation at sea. Sailors, who navigated at night by using the stars, and especially those indicating where to find the north, would certainly have noticed that in the northern part of the Mediterranean Sea, e.g., in what is now the Gulf of Venice, the pole was much higher in the sky (at about an altitude of 45°) than in its southern part, e.g., in what is now the Gulf of Sidra (at about an altitude of 31°). Compare Fig. 11.8 with Fig. 11.9.

On a flat earth, the only way to account for these observations is that the heaven of stars cannot be far away, as shown in the Fig. 11.10.

---

[27] Graham and Hintz (2007, 321).

[28] That Anaxagoras believed that the earth is flat is reported by several sources: Diogenes Laërtius, *Vitae Philosophorum* 2.8 = DK 59 A1(8) = Gr Axg32(8) = GG 340, not in LM and KRS; Hippolytus, *Refutatio Omnium Haeresium* 1.8.3 = DK 59A42(3) = LM ANAXAG. D4 (3) = Gr Axg38(3) = GG 271= KRS 502(3); Plato, *Phaedo* 97B = DK 59A47 = LM ANAXAG. R5 = GG3, not in Gr and KRS. He even put forward a proof of its flatness, as is shown by Panchenko (1997) and will be discussed below, in Chap. 12, section *Aristotle on Empirical Arguments for a Flat Earth*.

[29] Diogenes Laërtius, *Vitae Philosophorum* 1.23 = DK 11A1(23) = LM THAL. R6 = Gr Ths12 = TP1 Th237(23).

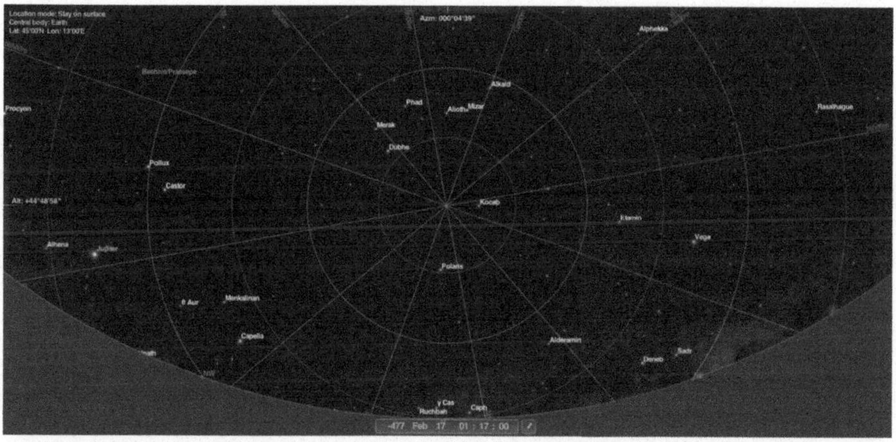

Fig. 11.8 The pole seen from the Gulf of Venice. This and the next picture generated with Redshift 9 Premium. I reproduced some markers for better orientation

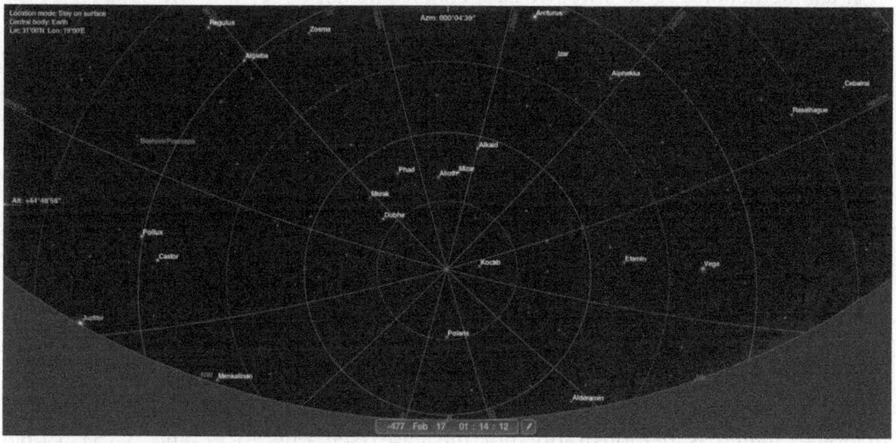

Fig. 11.9 The pole seen from the Gulf of Sidra

Anaxagoras knew that the sun is below the stars, so the sun must be even closer than the stars and thus quite small. This means that a major presupposition for Graham's method, which is that Anaxagoras assumed that the sun was far away, does not hold. However, he also notes: "On all early models of the heavens, the earth is relatively large in relation to the sun and moon,"[30] without explaining how to understand this statement in relation to the other statements quoted above that say that the sun was supposed to be relatively far away.

[30]Graham (2013a, 120).

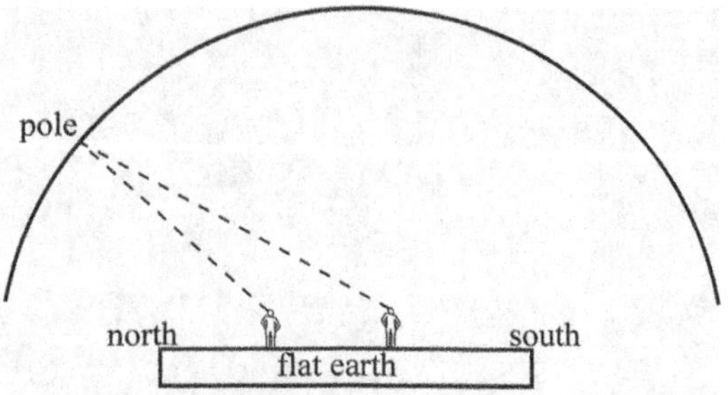

Fig. 11.10 The celestial pole seen from different places on a flat earth

With respect to the alleged parallel rays of the sun, Graham cites Empedocles, who seems to equate the moon's width with that of its shadow during a solar eclipse:

11.7 [the moon] did away with his [the sun's] rays

To the earth from above, and it obscured the earth
As much as was the width of the bright-eyed moon[31]

It can be debated as to whether Empedocles meant to say that the shadow of the moon on the earth is as large as the moon itself, or whether he was only saying that during a total solar eclipse, the shadow of the entire moon falls on the earth. In any case, Empedocles was clearly speaking of a total eclipse and not of an annular eclipse, like the one of February 17, 478 BC, in which the *umbra* does not reach the earth. Even if Anaxagoras mistakenly thought that during an annular eclipse, the shadow (*umbra*) of the moon fell on the earth, it is hard to believe that he thought that the rays of the sun on the earth were approximately parallel and that its shadow was as large as the moon. A ring of the sun was visible around the moon. Therefore, the shadow would have been conical and thus smaller than the moon itself. In passing, Graham and Hintz quote West's critical remark that, if Anaxagoras used the argument that the sun's rays ran parallel, he did it falsely.[32] It has already been shown in Chap. 3 that, if the rays of the sun were to run parallel, all gnomons, wherever they were placed on the surface of a flat earth, would cast identical shadows, which is demonstrably not the case (see Figs. 3.12 and 3.13). Either Graham and Hintz overlooked this fact, or they supposed that Anaxagoras had overlooked it.

Graham's description of what happened during this eclipse is as follows: "The sky in Athens began to grow dark"[33] Compare this with Stephenson's remark, which is worth quoting in full: "The eclipse of 478 BC can be fairly easily eliminated from further consideration. On this occasion, no more than 0.89 of the solar diameter

[31]Empedocles, quoted by Plutarch, *The Face on the Moon* 929c–d = DK 31B42 = LM EMP. D132 = Gr Emp87. The same quotation in West (1971, 233, n. 1).

[32]Graham and Hintz (2007, 322), cf. West (1971, 233 n. 1).

[33]Graham (2013c, 139, my italics).

Fig. 11.11 Total solar eclipse November 13, 2012. Photo Dennis Mammana

would have been obscured (even in the zone of annularity) so that the event could not have caused darkness; it might well have passed unnoticed."[34] Graham and Hintz object that Stephenson "attributes to a non-astronomer, Thucydides, the ability to discern a 'small' eclipse of 0.72 in 424 BC"[35] However, Thucydides does not claim to have observed this eclipse, but probably refers to observations of astronomers. In the same sentence, he reports that an earthquake has occurred, which also does not imply that he himself has felt it.

Elsewhere, Graham distinguishes between the degrees of darkness during an annular and a total eclipse: "(. . .) an annular eclipse is not nearly as obvious as a total eclipse. In a total eclipse direct sunlight is completely blocked and *the sky turns dark as at night*, with stars and planets becoming visible."[36] Contrary to what he claims, even during a total eclipse, the sky does not turn as dark as it does at night. The degree of darkening depends, of course, on the magnitude of the eclipse, but what we really see is more like the splendid photo, taken by Dennis Mammana during a total solar eclipse in Australia on November 13, 2012. Apart from the sun and the moon, only Venus is vaguely visible somewhat to the left of the center at the top of the photo (Fig. 11.11). If this was the case during a total eclipse, it would have been improbable to see more planets (let alone stars) during an annular eclipse like the one of February 17, 478.

[34]Stephenson (1997, 345–346).

[35]Graham and Hintz (2007, 327, n. 24); Thucydides, *Peloponnesian War* 4.52.

[36]Graham (2013a, 149).

Further Critical Remarks on Graham and Hintz' Attempt

In Anaxagoras' time, it was still impossible to predict a solar eclipse. The best thing astronomers of the time could produce was a list of *possible* dates for an eclipse. They certainly knew that solar eclipses occurred during the new moon. Perhaps they were also familiar with the rule that "the interval in months between consecutive visible eclipses is always a multiple of 6 months or a month less than such a multiple."[37] While the eclipse may not have been a complete surprise for Anaxagoras, it certainly was for the majority of his fellow citizens. As the eclipse was annular, and thus partial, it could only be observed with the naked eye under penalty of risking serious eye damage. Graham and Hintz' statement, "This eclipse would have been an impressive sight to all who witnessed it,"[38] is wishful thinking. Nowadays, even during total eclipses (let alone during annular eclipses), spectators wear special eyeglasses or look at the screens of their cameras to prevent eye damage. The usual way for astronomers at the time of Anaxagoras to prevent the danger of eye injury was by indirectly observing an eclipse on the surface of a bowl filled with water or oil. Despite his definite statements about the darkness during total and annular eclipses, Graham acknowledges this, and even adds another gadget: "The eclipse could be seen clearly only in reflection from water or pinhole projections."[39]

On the occasion of the annular eclipse of May 20, 2012, which he observed from Bryce Canyon National Park in Utah, Graham notes: "the sun's light was not noticeably weaker than before the eclipse, [and] *the eclipse could not be observed with the naked eye*, without appropriately darkened glasses."[40] Calculated with the help of the NASA list, the data of this eclipse as observed from Bryce Canyon National Park are:

Latitude: 37.6691° N
Longitude: 112.1539° W
Magnitude: 0.956
Obscuration: 87.22%

When we compare these figures with those of the annular eclipse of February 17, 478 BC as observed in Athens, we note a difference of 2.09% in obscuration. However, in Athens, too, to watch the eclipse with the naked eye.

Graham, having ascertained that the eclipse of February 17, 478 BC could only be observed in a bowl of liquid or by means of a pinhole, concludes nevertheless: "So while an annular eclipse is not an obvious event, its viewing seems to have presented no major difficulties to ancient Greeks."[41] This conclusion formulated in general terms, is confusing. We cannot seriously imagine a lot of inhabitants of Athens and

[37]Britton (1989, 5). See also Steele (2000, 423); Aveni (1993, 47); O'Grady (2002, 139).

[38]Graham and Hintz (2007, 327).

[39]Graham (2013c, 139); cf. Graham (2013a, 149–150).

[40]Graham (2013a, 149 n. 26).

[41]Graham (2013a, 150)

the Peloponnesus looking at the image of the eclipse in bowls of water or by means of pinholes, just as nowadays hosts of people, warned by newspapers, drive to places where they can observe an eclipse using dark glasses. The methods Graham describes where only used by astronomers and learned people who knew the possible dates of coming eclipses. The relevance of this critical remark will become clear when we see how Graham imagines that Anaxagoras was gathering the information he needed.

Graham assumes that Anaxagoras, on the conviction, albeit wrongly, that the shadow of the moon on the earth is about as large as the moon itself, tried to determine how large that shadow was. He presents a vivacious description of Anaxagoras' inquiry: "(...) he went down to the port city of Piraeus day after day for weeks or months and collected information from travelers and correspondents. When he was done, he knew that he had made a first-rate discovery."[42] According to Graham, Anaxagoras had to ask questions such as: "Where do you come from? Did you see the eclipse? Was the sun completely blocked out?"[43] If Anaxagoras had used this approach to gather his information, the last and most essential question would have been wrongly formulated, because during an annular eclipse, the sun is never and nowhere completely blocked out. Instead, Anaxagoras should have asked "Did you see an annular eclipse?" or "Did you see a ring of sunlight all around the eclipsed moon?"

Reliable information, which would have been decisive for Anaxagoras' measurements, could only have been expected from experienced stargazers, who had observed the phenomenon by watching its reflection in a bowl of water or another liquid, or perhaps by means of a pinhole. Modern astronomers can dispose of well-equipped informants all over the world, who know exactly when, where, and what kind of eclipse is to be seen, and who have the right equipment to make their observations. We cannot expect Anaxagoras to have had such a set of reliable informants. What Anaxagoras needed to know was who had seen an annular eclipse and who had been far enough from its central path to see a normal partial eclipse. The eclipse could not have been observed with the naked eye, and the farther an informant would have been from the central line of its path, the greater the risk of eye damage by looking directly at the sun. As for the possibility of the eclipse of February 17, 478 BC having been observed at all, Stephenson holds that most people would have missed it.[44] While this might be too pessimistic an assumption for people that were on or near the central line of the eclipse's path, he certainly has a

[42]Graham (2013c, 139–140). See also Graham and Hintz (2007, 339) and Graham (2013a, 153–154), where he adds: "he could have also sent letters to stargazers in other cities seeking information."

[43]Graham (2013c, 151).

[44]Graham and Hintz (2007, 327 n. 24) refer to Stephenson (1997, 345–346) and Sider (1973, 129, n. 10). They object that Stephenson "attributes to a non-astronomer, Thucydides, the ability to discern a 'small' eclipse of 0.72 in 424 BC" (ibid.). However, in *Peloponnesian War* 4.52, Thucydides does not imply to have observed this eclipse, but probably refers to observations of astronomers. In the same sentence, he reports that there was an earthquake, which also does not imply that he himself has felt it.

point when it concerns Anaxagoras' alleged informants that were outside the central path. In another article, Graham admits: "Unlike a total eclipse, an annular eclipse might not be evident to the casual observer."[45]

Moreover, as the picture of the path of the eclipse shows, much of the crucial information had to come from people at sea, especially those south of the Peloponnesus. Even if they had noticed that something was going on and had nothing better to do, it would have been difficult for them to observe the eclipse in a bowl of water on a rolling ship in order to obtain the information Anaxagoras needed.[46] We may even doubt whether such informants would have been available, because, as Graham himself notes, "Most ships laid up from early November to early March to avoid storms and adverse sailing conditions."[47] Further south, on Crete, for example in Heraklion, it was certainly not possible to observe the eclipse with the naked eye and it will even have passed unnoticed by most people.

Graham and Hintz (2007, 330–331) still note: "To get an accurate report in an age with few written records, and no periodicals or journals, he must have collected information soon after the event. For memories are fleeting (. . .)."[48] In Graham's later publications, this "soon" has become "a few weeks," and even "weeks or months."[49] We may wonder how reliable such information was if it came weeks or even months after the event.

A real possibility that should not be overlooked, but that Graham does not even mention, is that the sky was overcast at the time of the eclipse, which was (and is) not unusual in Greece in February, as the above-mentioned weather conditions for sailors confirm. If this had been the case in Athens on February 17, 478 BC, Anaxagoras would have missed his once-in-a-lifetime opportunity. Perhaps this explains why there is no record of the eclipse, as would be expected anyway, because its path went over Athens, where it "would have been an impressive sight to all who witnessed it,"[50] and where Anaxagoras allegedly made his observations with such spectacular results. This is all the more remarkable since, as Graham points out, three other annular eclipses (although Greece they were only partial eclipses) were recorded, namely October 2, 480 BC, August 3, 431 BC, and March 21, 424 BC[51] Anaxagoras may have been lucky and there were only light clouds in Athens that day, which

[45]Graham (2013b, 5, n. 18).

[46]According to Casson (1995, 48 and 234), sails were generally made of linen patches sewn together. Perhaps they proffered a tool for observing solar eclipses with the naked eye, but one might wonder how accurate those observations could have been.

[47]Graham (2013a, 155, n. 35).

[48]Graham and Hintz (2007, 330–331).

[49]Graham (2013a, 155; 2013c, 139).

[50]Graham and Hintz (2007, 327).

[51]See Herodotus, *Historiae* 7.37.2, Thucydides, *Peloponnesian War* 2.28.1 and 4.52.1, mentioned in Graham (2013a, 150, n. 30). To these, we might add (perhaps) April 16, 1178 BC (Homer's eclipse, *Odyssea* 20.345), May 28, 565 BC (Thales' eclipse, cf. Herodotus, *Historiae* 1.74), April 6, 648 BC (Archilochus' eclipse, cf. S 4.46.10), and May 19, 557 BC (eclipse of the Siege of Larissa, cf. Xenophon, *Hellenica* 2.3.4).

would have made it possible to observe the eclipse even with the naked eye. However, in order to be able to collect relevant information from others, such favorable weather conditions must have been present for a wide range of his informants.

A final problem must also be mentioned. Graham excludes the possibility that, having observed the eclipse in Clazomenae, Anaxagoras started his alleged investigations and calculations there, because, at that time, "Clazomenae was in a war zone. Regular commerce and communication across the Aegean were difficult at best."[52] Graham does not see the fact that Athens was in ruins at that time as a problem for Anaxagoras' alleged investigations.[53] The data of Anaxagoras' life, including the year of his arrival in Athens, are still a matter of discussion among scholars. In a thorough and well-documented article, Mansfeld argued that Anaxagoras must have arrived in Athens in 456/5 BC.[54] If Mansfeld is right, Anaxagoras would not have been able to observe the annular eclipse of February 17, 478 BC in Athens. Graham argues the other way around: Anaxagoras used the eclipse of February 17, 478 BC to do his measurement of the sun and moon, so he must have arrived in Athens before that date. Graham notes: "If the eclipse of 478 provided the data for Anaxagoras' estimate of the size of the sun, we can learn something about both Anaxagoras' life and his contributions to science," and "(...) it is controversial when Anaxagoras spent those 30 years [in Athens]: roughly 480–450, or 460–430? (...) *The fact of the eclipse counts strongly for the former date.*"[55] I have difficulty taking this kind of argument seriously.

The above criticism is not meant as hair-splitting or nitpicking, but rather with the question of how Anaxagoras and others could have observed the eclipse. In my opinion, these objections make the theory that Anaxagoras measured the sizes of the moon and sun with the aid of a solar eclipse highly improbable. It seems to me that Fehling's qualification of Sider's article: "der kaum realistische Versuch von D. Sider,"[56] also applies to Graham's attempt.

Fehling's Attempt

Although Fehling's own suggestion contains some sensible ideas, it is based on a kind of *petitio principii*. He starts by qualifying Anaxagoras' indication of the size of the sun as an estimation based on rational suppositions: "eine Schätzung auf Grund vernünftiger Annahmen."[57] He goes on to say that the largest distance on the

[52]Graham (2013a, 154).

[53]See Graham (2013a, 154–155).

[54]Mansfeld (1979).

[55]Cf. Graham and Hintz (2007, 330, my italics).

[56]Fehling (1985, 209).

[57]Fehling (1985, 209).

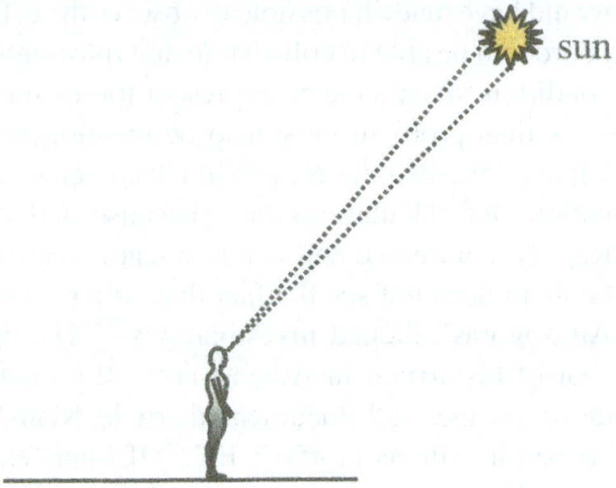

Fig. 11.12 The angular diameter of the sun

Peloponnesus is about 220 km, so Anaxagoras could have estimated the diameter of the sun to be approximately 250 km.[58] Given this diameter, Fehling calculates the distance of the sun, according to Anaxagoras, in the range between 15,000 and 60,000 km, depending on the measurement of the angular diameter of the sun between the values of 0.5° and 2° (the two values known in the tradition of ancient Greece).[59] The angular (or apparent) diameter of the sun indicates how large the sun appears from a given point of view (see Fig. 11.12). If the angular diameter of the sun had been taken to be 0.5° (which is close to its actual value), or 1/720th of a circle, then the ancient Greeks would have expressed this by saying that 720 suns in a row make the full circle of its orbit around the earth. If the angular diameter of the sun had been taken to be 2°, 180 suns in a row make the full circle of its orbit.[60]

However, there is nothing rational ("vernünftig") in this alleged estimate of the sun's size, because Anaxagoras would have had no rational argument beforehand to compare the size of the sun with the Peloponnesus. The real question is: How could Anaxagoras have argued and calculated to find a number for the size of the sun that prompted him to compare it to the size of the Peloponnesus? In other words, Fehling takes as the arbitrarily chosen starting point of his argument what should have been its outcome.

[58] Fehling (1985, 210).

[59] Fehling (1985, 210). The (right) value of 0.5 degrees is attributed to Thales by Diogenes Laërtius, *Vitae Philosophorum* 1.24 = DK 11A1(24) = LM THAL. R14 = Gr Ths1(24) = TP1 Th237(24); not in KRS, but see p. 83. See also Apuleius, *Florida* 18.32 = DK 11A19 = LM THAL. R13 = TP1 Th178; not in Gr and KRS. The (wrong) value of 2 degrees (1/15 of a sign of the zodiac) is, surprisingly, used by Aristarchus, see Heath (1913, 352–355) (Hypothesis 6 for the moon and Proposition 2 for the sun).

[60] This calculation presupposes that it was done from or near the center of the flat earth's surface. Since on a flat earth the sun is not far away, its angular diameter must be greater if an observer was situated at the edge of the earth. The ancient Greeks believed, however, that their land and especially Delphi, was the middle of the earth.

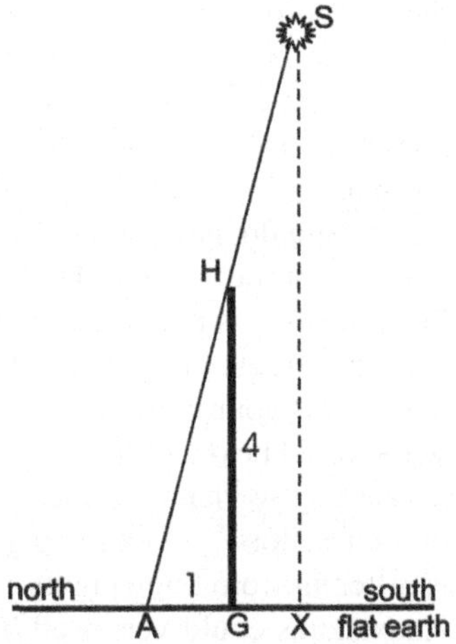

Fig. 11.13 Measuring the height of the sun on a flat earth (not to scale)

An Extrapolation of Thales' Method to Measure the Height of a Pyramid

Unfortunately, we do not have records about how Anaxagoras managed to measure the sizes of the sun and moon. My suggestion is that Anaxagoras must have been familiar with a simple procedure to measure those sizes. The difference between Fehling's approach and mine is that Fehling started with the assumption that Anaxagoras, for whatever reason, thought that the sun must be somewhat larger than the Peloponnesus, whereas in my proposal, Anaxagoras' estimate that the sizes of the sun and moon were in the order of the size of the Peloponnesus was a result of his calculations. This method has already been discussed above in Chap. 3. To enable the reader to follow the argument without looking back at that Chapter, I will show the simple method (see Fig. 11.13), which is based on Thales' measurement of the height of a pyramid.

In Athens, the shadow (AG) at noon at the summer solstice was roughly one-quarter of the length of the gnomon (HG). Accordingly, if X is the place south of Athens where the sun stood in the zenith at noon during the summer solstice, the distance AX must be also one-quarter of the distance SX. A very rough estimate of the distance AX is 2000 km. Then, the height of the sun above the flat earth must be about 8000 km. Even with a mistake of several hundred kilometers in the estimate of the distance (AX) between Athens and the place where the sun stood in the zenith, the result is that the sun must be close by. A similar measurement could have been made each time the moon was at its highest point (altitude about 70°) and due south, by extending the line AH in the direction of the moon instead of the sun.[61] The radius of

[61]This remark is intended to overcome Graham's objection (2013a, 147) that with this method it is the sun, not the moon, that is measured.

the sun's orbit is the line AS in Fig. 11.13, which is about 8250 km, and thus the orbit of the sun $2\pi \times 8250 =$ about 50,000 km (assuming $\pi = 3$, as was usual in ancient times), and the diameter of the sun between $50{,}000 \div 720 =$ about 70 km and $50{,}000 \div 180 =$ about 280 km, depending on the estimated angular diameter of the sun.

We should not forget that Anaxagoras had to estimate not only the angular diameter of the sun and the distance AX, but also the size of the Peloponnesus. The Peloponnesus is not "roughly circular in shape,"[62] but an irregularly shaped piece of land. Its largest distance is approximately 220 km, as Fehling said, but its smallest diameter is approximately 130 km. Moreover, as Graham and Hintz state: "Maps to scale would be out of the question, especially since the Greeks had no reliable way of measuring overland distances."[63] They add that the Peloponnesus was better known to some sailors,[64] but knowing how long it takes to circumnavigate the Peloponnesus is different from knowing its size. Given the uncertainties of the estimates involved, Anaxagoras could very well have calculated a number that matched the size of the Peloponnesus. If Anaxagoras used this method, his estimate of the sun's angular diameter was probably closer to Aristarchus' 2° than to the 0.5° that was ascribed to Thales.

A major advantage of my proposed method is that Anaxagoras was not dependent on one accidental and unpredictable occurrence of a solar eclipse, nor on reports gathered from informants whose reliability he was not able to verify. On the contrary, Anaxagoras could have carried out the calculations of the sun's size all by himself and at his ease any year at (or around) the summer solstice (weather permitting). It is not coincidental that, when measuring the circumference of the earth, Eratosthenes used exactly the same experimental set-up as suggested above for Anaxagoras' measurement of the distance and size of the sun. The only difference was in their main presupposition. While Anaxagoras was convinced that the earth was flat, Eratosthenes knew that the earth is a sphere. In view of these different starting points, both calculations were mathematically correct. Eratosthenes' famous experiment did not arise out of the blue; it was rooted in a tradition that had started with Thales' measurement of the height of a pyramid, which Anaxagoras transformed into a method for making astronomical calculations. It is instructive to place side by side, on the one hand, Eratosthenes' experimental set-up on a spherical earth and, on the other hand, how it would have been interpreted under the assumption of a flat earth (see Fig. 11.14a, b). Actually, Eratosthenes' set-up was the same as used by ancient Chinese flat earth astronomers to calculate the distance to the sun, as will be explained in Chap. 13.

[62]Graham (2013a, 146).

[63]Graham and Hintz (2007, 323).

[64]Graham and Hintz (2007, 324).

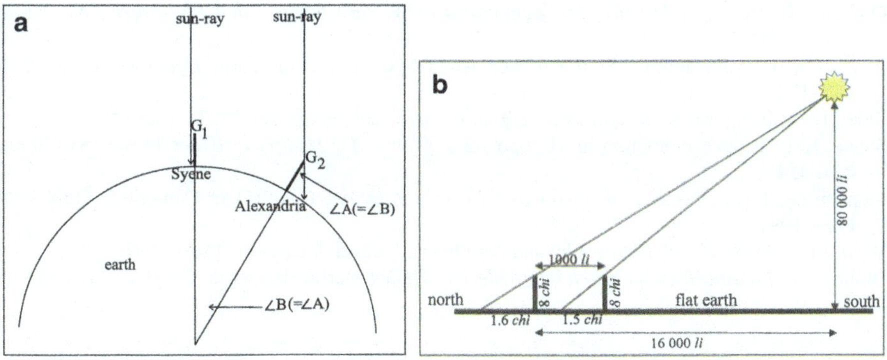

Fig. 11.14 (**a**) Eratosthenes' measurement of the earth's circumference. (**b**) Chinese astronomers used the same experimental set-up to measure the distance to the sun

References[65]

Aveni, A.F., *Ancient Astronomers*, Washington DC: Smithsonian Books 1993.

Britton, J.P., An Early Function for Eclipse Magnitudes in Babylonian Astronomy. *Centaurus* 39 (1989) 1–52.

Casson, L. *Ships and Seamanship in the Ancient World*. Baltimore and London: The John Hopkins University Press 1995 (repr. of 1971).

Diels, H. and W. Kranz, *Die Fragmente der Vorsokratiker*. Zürich/Hildesheim: Weidmann 1951[6] (=DK).

Fehling, D., Das Problem der Geschichte des griechischen Weltmodells vor Aristoteles. *Rheinisches Museum für Philologie* 128 (1985), 195–231.

Gershenson, D.E. and D.A. Greenberg, *Anaxagoras and the Birth of Physics*. New York: Blaisdell Publishing Company 1964 (=GG).

Graham, D.W., *The Texts of Early Greek Philosophy*. Cambridge: Cambridge University Press 2010 (=Gr).

—, *Science Before Socrates*. Oxford: University Press 2013a.

—, Anaxagoras and the Comet. *Ancient Philosophy* 33 (2013b), 1–18.

—, Anaxagoras: Science and Speculation in the Golden Age. In: J. McCoy (ed.), *Early Greek Philosophy. The Presocratics and the Emergence of Reason*. Washington DC: The Catholic University of America Press 2013c, 139–156.

Graham, D.W. and E. Hintz, Anaxagoras and the Solar Eclipse of 478 BC *Apeiron* 40 (2007), 319–344.

Kirk, G.S., J.E. Raven and M. Schofield, *The Presocratic Philosophers*. Cambridge: Cambridge University Press 2009 (=KRS).

Laks, A. and G.W. Most, *Early Greek Philosophy VI*, Cambridge MA. 2016.

Mansfeld, J, The Chronology of Anaxagoras' Athenian Period and the Date of his Trial I. *Mnemosyne*, 4th ser., 32 (1979), 39–69 (reprinted in: J. Mansfeld, *Studies in the Historiography of Greek Philosophy*. Assen: Van Gorcum 1990. 264–294).

Mansfeld, J. and D.T. Runia, Aëtiana. *The Method and Intellectual Context of a Doxographer, Volume Two, Part Two*. Leiden: Brill 2009 (=MR).

[65]P = Aëtius in pseudo-Plutarch, *Placita* (numbering according to Dox).
S = Aëtius in Stobaeus, *Anthologium* (numbering according to Wachsmuth and Hense).

O'Grady, P. *Thales of Miletus: The Beginnings of Western Science and Philosophy*, Aldershot: Ashgate 2002.
Panchenko, D., Anaxagoras' Argument Against the Sphericity of the Earth. *Hyperboreus* 3 (1997), 175–178.
Sider, D., Anaxagoras on the Size of the Sun. *Classical Philology* 68 (1973), 128–129.
Steele, J.M., Eclipse Prediction in Mesopotamia. *Archive for History of Exact Science* 54 (2000), 421–454.
Stephenson, F.R., *Historical Eclipses and Earth's Rotation*. Cambridge: Cambridge University Press 1997.
West, M.L., *Early Greek Philosophy and the Orient*. Oxford: Clarendon Press 1971.
Wöhrle, G., *Traditio Praesocratica. Die Milesier: Thales*. Berlin: Walter de Gruyter 2009 (=TP1).

Chapter 12
Aristotle's Arguments for the Sphericity of the Earth

Contents

Introduction

According to Aëtius, Thales stated that the earth is shaped like a ball.[1] Diogenes Laërtius reports that Hesiod, Anaximander and Pythagoras taught that the earth was spherical.[2] These testimonies are usually considered false.[3] Several scholars have argued that Parmenides was the first to accept the earth's sphericity.[4] In the *Phaedo*, Plato tells us that someone has convinced Socrates that the earth is a sphere, without indicating who this "someone" was or what his arguments were. He then compares the earth with a dodecahedron.[5] There must have been a heated debate among ancient Greek cosmologists about the shape of the earth, some of which can still be followed in the reports of the doxographers. For example, a curious argument,

[1]P 3.9.1 (not in S) = DK 50, 4 = TP1 161; not in LM, Gr, and KRS.

[2]Diogenes Laërtius, *Vitae Philosophorum* 8.48 = DK 28A44 = LM PARM. D33b = Gr Prm40, not in KRS (Pythagoras and Hesiod); Diogenes Laërtius, *Vitae Philosophorum* 2.1.1 = DK 12A1(1) = LM ANAXIMAND. D31 = Gr Axr 1(1) = TP2 Ar92 (Anaximander).

[3]Although O'Grady (2002, 94–100), still broke a lance for Thales. Bakker (2016, 165–166) mentions Pythagoras without a critical note.

[4]Recently, e.g., Panchenko (2008), Graham (2013, 105–106).

[5]Plato, *Phaedo*, 108C–E and 110B.

D. L. Couprie, *When the Earth Was Flat*, Historical & Cultural Astronomy,
https://doi.org/10.1007/978-3-319-97052-3_12

attributed to Anaxagoras, that is intended to prove that the earth is flat.[6] Aristotle, who was familiar with the ins and outs of the debate, sought to terminate it by putting forward arguments for the sphericity of the earth and arguing against alleged evidence that the earth is flat. Since then, hardly any serious philosopher or astronomer has doubted that the earth is a sphere.[7] Aristotle's arguments can be divided into two groups: theoretical and empirical.[8]

Recently, several scholars highly praised Aristotle's empirical arguments. Stephen Hawking admired Aristotle for delivering "two good arguments for believing that the earth was a round sphere rather than a flat plane."[9] Daniel Graham stated, "Aristotle defends and indeed proves the sphericity of the earth in the *De Caelo* with adequate scientific arguments."[10] Carlo Rovelli, in a paper on Aristotle's physics, wrote about Aristotle's most famous argument, which used the shape of the shadow on the moon during a lunar eclipse to prove the earth's sphericity: "This proves empirically, and very solidly indeed, that the earth has a shape that is (approximately) spherical."[11]

In this chapter, I will evaluate Aristotle's empirical arguments and try to find out to what extent they are convincing, especially in the context of contemporaneous thinking. If it can contribute to a better understanding, I will place Aristotle's arguments against a broader historical background, without pretending to present a complete historical survey. Finally, I will argue that Aristotle's empirical arguments for the sphericity of the earth were meant to confirm a conclusion drawn from his physical theories. That is why, throughout this chapter, I shall speak of Aristotle's "empirical arguments" (except where I quote other authors who call them "proofs"), to distinguish them from his theoretical arguments. Of course, Aristotle's theoretical arguments were not shared by his adversaries, for instance the schools of Anaxagoras and Democritus, who claimed that the earth was flat, and not even by the followers of Archelaus, who taught that the earth was spherical but not in the center of the universe. It might be interesting to ask whether Aristotle' empirical arguments could have convinced the adherents of a flat earth. In that context, it is also interesting to see which available arguments Aristotle did not use. One section of Bakker's *Epicurean Meteorology* is also devoted to *Ancient Proofs of the Earth's Sphericity*.[12] I have taken the opportunity to insert discussions, especially on those points where our results diverge.

---

[6]See the section *Aristotle on empirical arguments for a flat earth*.

[7]Bakker's *Epicurean Meteorologica* tries to answer the intriguing question of whether Epicurus (and Lucretius) believed the earth to be flat. A notorious flat-earther was Cosmas Indicopleustes (sixth century A.D.).

[8]Simplicius (*In Aristotelis De Caelo Commentaria* 542.14) counts five arguments, but the last one has to do with measuring the earth's circumference and thus presupposes the sphericity of the earth.

[9]Hawking (1988, 2–3).

[10]Graham (2013, 95).

[11]Rovelli (2015, 35).

[12]See Bakker (2016, 169–175) (Chap. 4.2.2).

The First Empirical Argument

Let us start with what is usually considered to be Aristotle's most convincing empirical argument, which is quite sophisticated:

12.1 If the earth were not spherical, eclipses of the moon would not exhibit segments of the shape which they do. (. . .) in eclipses the boundary [sc. between the bright and dark portions of the moon] is always convex (ἀεὶ κυρτὴν).[13]

We can explain the word "always" as meaning that during eclipses, the boundary line between light and dark on the moon is curved, regardless of whether the eclipse is visible near the horizon or high in the sky, and regardless of where on the moon the shadow line is visible.[14] The word κυρτός, which Aristotle uses here, means "bulging," "curved," "convex," and can be used, for example, to describe a hunchback or a shield.[15] Aristotle uses it elsewhere to indicate the sphericity of the earth.[16] It may be questioned, however, whether astronomers, in the time of Aristotle, were always able to identify this shadow line not only as curved, but also as part of a circle. Therefore, it is a bit tendentious to write, as Rovelli does with reference to this line, "By careful observation, we see that this shadow is circular."[17]

Neugebauer has formulated an objection to Aristotle's argument that sounds both fundamental and relevant: "It is an often-repeated statement—from Aristotle to modern textbooks—that the sphericity of the earth is demonstrated by the fact that the earth's shadow on the moon is always bounded by a convex arc. This, of course, is mathematically inconclusive, quite aside from the fact that nobody ever explains how to establish the accurate nature of the observed curve. But even if we take it for granted that the shadow of one object on another unknown surface appears as a circle one should remember that there exists an unlimited number of shadow casting and shadow receiving bodies which produce identical shadow limits."[18] Elsewhere, he even calls it "Aristotle's pseudo-argument."[19]

Be this as it may, another interesting and typically contemporaneous problem with Aristotle's argument is implied in the sentence that immediately follows the one quoted above:

---

[13]Aristotle, *De Caelo* 297b24ff; the text between square brackets is my addition.

[14]The last is Kepler's interpretation; see note 31.

[15]See LSJ, s.v. κυρτός.

[16]Cf. Aristotle, *Meteorologica* 265a31.

[17]Rovelli (2015, 35).

[18]Neugebauer (1975, 1093–1094). See also, e.g., Dicks (1970, 260, n. 379): "Strictly, this proves only the curvature of the earth's surface," and North (2008, 82): "Not a perfect argument by itself, of course."

[19]Neugebauer (1975, 576).

12.2 Thus, *if the eclipses are due to the interposition of the earth*, the shape must be caused by its circumference, and the earth must be spherical.[20]

In the context of the current discussions, the actual handicap of Aristotle's argument is that it is indirect and depends upon the conviction that the shadow lines during lunar eclipses are caused by the shadow of the earth, and that this belief, in its turn, depends on the idea that the moon's light is reflected light from the sun. Bakker also notes that "the argument depends, of course, on the assumption that the moon receives its light from the sun."[21] In Cleomedes' terminology, the argument is an ἔφοδος ("procedure"). Bowen and Todd remark: "the hallmark of the 'procedures' in the *Caelestia* is the presence, whether explicit or implicit, of an independently identifiable truth or principle that is the foundation, and so, in effect, the axiom, of the argument."[22]

In Aristotle's time, his explanation was not as evident as tend to think today, and his presuppositions were not generally accepted. As we have seen in Chap. 10, section *The Moon's Light and Phases According to Anaxagoras; Suggestions for a New Interpretation* and text 10.15, there were people who seriously defended the theory that the moon had its own light, kindled by the sun at new moon and gradually spreading and extinguishing during the month, as Aëtius reports. We can reasonably assume that these people did not subscribe to the theory that lunar eclipses were caused by the of the shadow earth. There were at least two schools that defended an alternative origin of lunar eclipses. As Aristotle himself and others report, there were people who defended that at least some lunar eclipses were caused by invisible heavenly bodies (see texts 9.9, 9.10, and 9.13). If my analysis in Chap. 9 is right, both Anaxagoras' conception of the Milky Way as caused by the earth's shadow, and his conviction that the earth is flat must have made him reject the possibility that lunar eclipses were caused by the shadow of the earth. I argued that he must have held that lunar eclipses are due to invisible heavenly bodies.

Even though they agreed with Aristotle that the moon receives its light from the sun, the Pythagoreans (and Philolaus in particular), postulated an invisible heavenly body that they called counter-earth. In Pythagorean cosmology, the earth and the other heavenly bodies orbit around a common center, the central fire. The orbit of the counter-earth lies between the earth and the central fire. The counter-earth always remains invisible because we live on the part of the earth that is turned away from it. An eclipse of the moon caused by the counter-earth could be represented as shown in the schematic drawing of Fig. 12.1. Philolaus probably believed that all heavenly bodies were spheres. The shadow on the moon would be the same, then, whether it originated from the shadow of the earth or from the shadow of an alleged counter-earth. Therefore, the theory of the counter-earth could worry Aristotle in that it states that the earth is not the only spherical body that causes lunar eclipses.

---

[20]Aristotle, *De Caelo* 297b29–31, my italics.

[21]Bakker (2016, 171).

[22]Bowen and Todd (2004, 1, 157 n. 8).

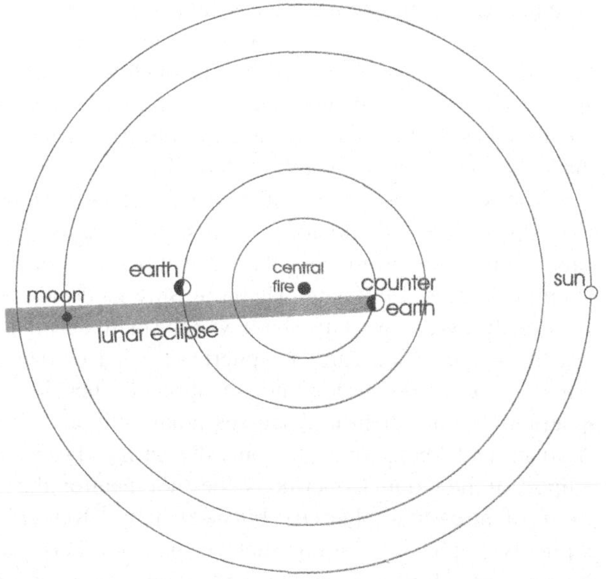

Fig. 12.1 Lunar eclipse caused by the counter-earth. In this representation, the shadow of the earth is not conical. It is not handed down whether the Pythagoreans already realized that when the earth is conceived of as spherical, the sun must be far away and much larger than the earth, and that the earth's shadow must therefore be conical

Obviously, in the representation of Fig. 12.1, an observer on the earth is not aligned with the sun and moon, but the Pythagoreans argued that this is not necessary and is not the case on a central earth either:

12.3 Since the earth's surface is not in any case the center, but distant the whole hemisphere from the center, there is no more difficulty, they (sc. the Pythagoreans) think, in accounting for the observed facts on their view that we do not dwell at the center, than on the common view that the earth is in the middle.[23]

Aristotle, then, had to counter the argument that denied his presupposition that it was the shadow of the earth that produced lunar eclipses. If the shapes of the dividing line between dark and light on the moon during an eclipse were not caused by the shadow of the earth, but by the interposition of an invisible heavenly object, then that line told nothing about the shape of the earth but only about the shape of the intervening object. As long as this possibility was not excluded, the conclusiveness of the argument concerning the shapes of the shadows during lunar eclipses must remain questionable. He should have disproved it as a possible cause of eclipses of the moon, but he kept silent. The point is not that both the Philolaic and the Anaxagorean versions of the theory of invisible bodies causing eclipses are strange to us, but that Aristotle should have countered them in the context of his argument about the role of the earth's shadow during lunar eclipses.

There were also observational facts that could have been used as arguments against the explanation of lunar eclipses as caused by the earth's shadow. For

[23] Aristotle, *De Caelo* 293b25–29.

example, the phenomenon of a selenelion, in which both the sun and the fully eclipsed moon are visible on opposite sides above the horizon, which is difficult to explain if the eclipse is caused by the shadow of the earth.[24] In Chapter 5, section *lunar eclipses*, I also mentioned the strange fact that the eclipsed part of the moon usually takes on the color of the surrounding air, unlike what we usually see with shadows cast on objects (see Fig. 5.6a, b).

Perhaps it is for reasons like these that later authors, who list proofs of the sphericity of the earth, do not give Aristotle's argument of the earth's shadow during lunar eclipses the prominent place that it deserves according to modern scholars.[25] Manilius, Strabo, Pliny, and Theon of Smyrna do not mention it. Ptolemy does not mention the argument of the shadow of the earth during lunar eclipses when he argues that the shape of the earth is spherical,[26] and neither does he when he discusses eclipses.[27] Cleomedes does not mention it when he offers proofs for the earth's sphericity,[28] but when he discusses lunar eclipses.[29] He argues, obviously against Philolaus and Anaxagoras, that only the earth and no other bodies can be the cause of eclipses of the moon. Copernicus does not mention the argument in the chapter with proofs of the earth's sphericity, but elsewhere.[30] Kepler is the first, as far as I know, to explicitly emphasize that on whatever part (northern, southern, eastern, western) of the moon an eclipse is seen, the shadow line is always a part of a perfect circle.[31] His words can be seen as an explication of Aristotle's "always."

The Second Empirical Argument

Aristotle's second empirical argument consists of two parts:

12.4 (...) (1) a small change of position on our part southward or northward visibly alters the circle of the horizon, so that the stars above our heads change their position considerably, (2) and we do not see the same stars as we move to the

[24]This phenomenon has already been mentioned in Chap. 9, section *Attempts to Understand the Invisible Bodies as an Additional Cause of Lunar Eclipses*.

[25]Bakker (2016, 171, 174–175, Table 4.2), also notes this curious fact. His suggestion, however, that this could be due to the idea that the same effect of shadow lines on the moon could be caused by a flat, disk-shaped earth, is not right.

[26]Ptolemy, *Almagest* I.4.

[27]Ptolemy, *Almagest* VI.

[28]Cf. Cleomedes in Bowen and Todd (2004, I.5).

[29]Cleomedes in Bowen and Todd (2004, II.6). This text is not mentioned in Table 4.2 in Bakker (2016, 174–175).

[30]See Copernicus (1543, not in I.2) ("Quod terra quoque sphaerica sit"), but at the end of 1.3 ("Quomodo terra cim aqua unum globum perficiat").

[31]See Kepler (1635, I, 25): "Terminos umbrae terrestris, in corpore lunae deficientis, tam qui sunt ad septentriones, quam qui ad Austrum, tam ad Orientem, quam ad Occidentem [sc. partem Lunae], esse arcus perfecti circuli." The text between square brackets is my addition.

> North or South. Certain stars are seen in Egypt and the neighborhood of Cyprus, which are invisible in more northerly lands, and stars, which are continuously visible in the northern countries are observed to set in the others.[32]

The first part of this argument is not conclusive because, on a flat earth, in which the celestial bodies are not at a great distance, the same stars are also seen at a different angle when you are going north- or southward (see Fig. 3.11). Bakker draws the same conclusion, which he makes dependent on the assumption, suggested by Furley, that on a flat earth the heavens are relatively close.[33] On a flat earth, however, the idea that the heavenly bodies are close by is not an assumption to be made or not to be made, but a necessity, as has been extensively demonstrated in Chaps. 3 and 11, section *An extrapolation of Thales' method to measure the height of a pyramid*. The method to measure the distance of the sun that is suggested in these chapters is essentially the same as Thales' measurement of the height of a pyramid. It is much simpler than Bakker's reconstruction of Furley's calculation,[34] which requires the measurement of angles (up to 1/10 of a degree) and the use of trigonometry, which was not available to the ancient Greeks.

Actually, the ancient Greeks were familiar with only a small part of the earth—roughly the regions around the Mediterranean Sea. Figures 11.8 and 11.9 show how sailors could have observed the differences in height of the stars at the greatest north-south distance in the Mediterranean Sea. Although "a small change of position on our part southward or northward visibly alters the circle of the horizon, so that the stars above our heads change their position considerably," as Aristotle says (text 12.4), we may wonder whether the range of north-south displacement at that time was sufficient to decide, by means of the available instruments, whether the differences in the angles under which the stars were seen were due to a flat or a curved surface of the earth.

An alternative form of Aristotle's argument, mentioned by Pliny, has to do with the observed differences in the length of gnomon shadows at noon on different latitudes, which are directly related to the height of the sun above the observer's horizon.[35] This phenomenon was used by Eratosthenes in his famous calculation of the circumference of the earth, when he saw that at noon during the summer solstice in Alexandria a gnomon cast a small shadow, while in Syene a gnomon cast no shadow at all. Eratosthenes' presupposition was that the earth is a sphere. On the other hand, under the assumption of a flat earth, the observation of these differences in shadow length can be used to demonstrate that the sun must be close by, as explained in Chaps. 3 and 11.

The second part of Aristotle's argument is not conclusive, because when we move southwards, the effect of seeing new stars appear in the south and some circumpolar stars rise and set in the north would also occur if the surface of the earth were not

[32]Aristotle, *De Caelo* 297b31ff, the numbers between brackets are my addition.

[33]Bakker (2016, 172).

[34]Bakker (2016, 238, Fig. 4), cf. Furley (2010, 429, n. 41): "Given the same figures as attributed to Eratosthenes, the sun would have to be only 39,579 stades from the earth (less than 5000 miles)".

[35]Pliny, *Natural History* 2.182–184.

exactly flat but slightly convex and had the shape of a shield.[36] Nobody in Aristotle's time could travel far enough to see the effects all around the earth. According to Panchenko's intriguing hypothesis, by Archelaus, Leucippus and Democritus have advocated the idea of a convex earth.[37] Another objection to the second part of Aristotle's argument could have been that we do not see some stars because of our limited range of vision.[38] When we move around on a flat earth, we see stars appear as they enter our range of vision, and disappear when they are outside our range of vision. According to Dmitri Panchenko,[39] this argument was used by Anaximenes:

12.5 The sun is hidden not by going under the earth, but (...) by being a greater distance away from us.[40]

If Panchenko is right, Aristotle should have countered these objections, and even if he is not right, they show that Aristotle's argument is not conclusive. Aristotle's argument has been repeated by many authors, starting with Manilius, who is the first to mention Canopus which is visible in Egypt, but not in Greece or Italy.[41] Copernicus adds that the argument does not only hold for phenomena on the same meridian, but that the inclination of the pole is the same for observers at the same latitude, which happens in no figure except the sphere.[42]

The Third Empirical Argument

It is not certain whether Aristotle's third empirical argument should be taken seriously. At any rate, he himself is rather cautious and says that it does not seem utterly incredible. The argument, which occurs in the context of his conviction that the periphery of the earth is not large, runs as follows:

12.6 For this reason, those who imagine that the region around the Pillars of Heracles joins on to (συνάπτειν) the regions of India, and that in this way the ocean is one, are not, it would seem, suggesting anything utterly incredible.

---

[36]This is more specific than a "reference to the unevenness of the earth's surface" (Bakker 2016, 172).

[37]Panchenko (1999, 32) dismisses somewhat too easily Cleomedes' remark (in Bowen and Todd 2004, I.5.12) that those who taught that the earth was concave did so because otherwise the water would flow off the earth.

[38]In fact, as will be explained in Part Two of this book, this argument was used by ancient Chinese astronomers of the *gai tian* system to explain why stars appear and disappear when we move around on a flat earth. See also *the section Empirical arguments that Aristotle did not use.*

[39]See Panchenko (2015).

[40]Hippolytus, *Refutatio Omnium Haeresium* 1.7.6 = DK 13A7(6) = LM ANAXIMEN. D3(6) = Gr Axs12 = TP2 Aa 56 = KRS 156.

[41]Manilius, *Astronomica*, 1.215–220.

[42]Copernicus (1543, I.2).

> They produce also in support of their contention the fact that elephants are a species found at the extremities of both lands, arguing that this phenomenon at the extremes is due to communication between the two.[43]

In the time of Aristotle, there were apparently elephants in northern Africa, near the Pillars of Hercules. The suggestion is not only that Africa borders India, but also that there exist, or existed, headlands in the ocean that connected Africa to India, allowing elephants to travel from one country to the other. The argument could have been countered by asking why the elephants could not have migrated from one country to the other, if humans had been able to do the same. Moreover, there is no evidence of the actual or past existence of such a connection between Africa and India. In the *Meteorologica*, Aristotle even asserts that "beyond India and the Pillars of Heracles it is the ocean which severs the habitable land and prevents it forming a continuous belt round the globe."[44] As far as I know, the argument has not been repeated by other authors.

Empirical Arguments that Aristotle Did Not Use

It is interesting to note that Aristotle does not mention several arguments that could have been available to him. The first is that when a ship is at sea, we first see the mast and only when it approaches the coast can we see its hull. Since Strabo,[45] this and similar arguments have been used time and again by later authors to prove the sphericity of the earth. It could have been countered, however, by pointing to the effect of perspective, which makes distant objects appear smaller and even disappear from our sight. In addition, the disappearance of objects also would make sense if the earth were a shield-shaped, convex slice of a sphere. Curiously enough, this can be illustrated by a picture from Kepler's *Epitome of Copernican Astronomy*, which is meant to illustrate the argument, but unintentionally shows its weak spot (a shield-shaped earth) (Fig. 12.2).

The second empirical argument not mentioned by Aristotle is that of the time difference in observed phenomena as seen from places that lie more or less at the same longitude, but far from each other. Cleomedes, for instance, writes: "the Persians who live in the East are said to encounter the onset of the sun four hours earlier than the Iberians who live in the West."[46] A related argument comes from stories of travelers to the north, who reported that the day was shorter there than they were used to. One of the main consequences of the conception of a flat earth is, however, that it is always the same time everywhere. When the sun rises upon a flat

[43]Aristotle, *De Caelo* 298a9–16. Perhaps Aristotle obtained this information from Eudoxus. See Bigwood (1993, 546–547).

[44]Aristotle, *Meteorologica* 362b28–30.

[45]Strabo, *Geographica* 1.1.20.18–27.

[46]Cleomedes in Bowen and Todd (2004, I.5.30).

Fig. 12.2 Kepler's drawing of an empirical proof of the earth's sphericity (Kepler (1635), Chap. 1, figure on p. 1)

earth, it rises for every place on earth, and a day has the same length everywhere. Curiously, this problem seems to have been almost completely neglected in Presocratic cosmology. The only time it is mentioned is in a late source. Hippolytus reports that according to Archelaus, the shape of the earth was concave: "lofty around the edge and hollow in the middle." And then he adds:

12.7 He adduces as a proof of this hollowness the fact that the sun does not rise and set at the same time for all men, as would inevitably happen if the earth were flat.[47]

It has been often remarked that this report is confused, because on a concave earth the sun would be seen earlier by those in the occidental regions than by those in the oriental regions, which is the opposite of the actual state of affairs. Paul Tannery, for example, wondered how Archelaus could have drawn conclusions that were precisely the opposite of those he should have drawn.[48] Perhaps it is because of this almost total neglect in Presocratic cosmology of the problem of time differences, why Aristotle does not mention it among his empirical arguments that the earth is a sphere.

Later on, this subject was often discussed in connection with the observation of eclipses. Ptolemy points out that eclipses were seen in an oriental region at later hours of the day than in an occidental region, which is impossible on a flat earth.[49] Pliny reports more precisely: "The eclipse of the sun which occurred the day before the calends of May, in the consulship of Vipstanus and Fonteius [=April 30, AD 59], not many years ago, was seen in Campania between the seventh and eighth hour of the day; the general Corbulo informs us, that it was seen in Armenia, between the eleventh and twelfth hour."[50] Besides, Pliny tells the nice story of a messenger who ran the same way eastbound and westbound, but although the way was eastbound downhill, this run lasted longer.[51] Pliny is also the first to mention the argument of

---

[47]Hippolytus, *Refutatio Omnium Haeresium* 1.9.4 = DK 60A4(4) = LM ARCH. D2(4), KRS 515 (4); not in Gr.

[48]Cf. Tannery (1887, 279). See for more scholars on this subject: Panchenko (1999, 23).

[49]Ptolemy, *Almagest*, I.4.

[50]Pliny, *Natural History* 2.72.

[51]Pliny, *Natural History*, 2.73.

the shortening of the days in the northern regions.[52] Again, these arguments were not decisive, for the same reason that one could argue that the flat earth was curved like a shield, which no one in Aristotle's time was able to disprove.

It is interesting to note that while the problem of time differences was not solved and even neglected in ancient Greek flat earth cosmology, it had found an ingenious solution elsewhere in the ancient world, in the so-called *gai tian* system of ancient Chinese flat earth cosmology.[53] It can even be said that the solution of this problem was the main *raison d'être* of the *gai tian* model. The main ingredients of this system were: (1) all celestial bodies circle around the celestial pole in a plane, parallel to that of the flat earth; (2) the reach of the sun's light is limited, and the sun in its daily orbit around the pole illuminates successive parts of the earth; (3) the horizon, where the planes of heaven and earth seem to meet, is an illusion; (4) accordingly, the rising and setting of the heavenly bodies are illusions; (5) the range of human vision is also restricted and has the same size as the reach of the sun's light; (6) in winter, the orbit of the sun around the pole is at its smallest, in summer at its widest. Although all the ingredients mentioned were needed for the solution of the problem of time differences on a flat earth, the most important component was the idea sub 2, which says that the sun in its daily orbit illuminates successive parts of the earth.

The third and most important empirical argument not mentioned by Aristotle is that, when sailors crossed the equator in an attempt to circumnavigate Libya, they saw the stars circling around another pole.[54] On a flat earth, on the contrary, and even on a shield-shaped earth, only one pole is visible, wherever one goes. Herodotus tells a story that could have proved the contrary, although he himself is skeptical about its truth:

12.8 One of their claims—which I personally find incredible, although others may not—was that, while sailing round Libya [i.e. Africa] they had the sun on their right hand side [meaning in the northern sky].[55]

Panchenko rightly notes that the Phoenician mariners "would also have been very much impressed to observe the rotation of the stars around another, southern pole."[56] Panchenko hypothesizes that this "made Parmenides formulate his great theory of a spherical earth,"[57] obviously assuming that Parmenides had heard the story and believed it, but this is pure speculation. More likely, Parmenides came to his conception of a spherical earth on metaphysical grounds.

Not in the context of evidence of the earth's sphericity, but in a discussion about the use of sundials, Pliny has presented a similar report from the time Alexander the Great reached India: "In India, in the celebrated seaport Patale, the sun rises to the right hand

52 Pliny, *Natural History*, 2.72.

53 See Part Two of this book.

54 This crucial proof is not mentioned in Bakker's list (2016, 174–175, Table 4.2).

55 Herodotus, *Historiae* 4.42; the remarks between square brackets are my additions.

56 Panchenko (2008, 192).

57 Ibidem.

and the shadows fall towards the south."[58] It is possible, between the Tropic of Cancer and the equator, to see the shadow fall to the south during the summer, but Pliny's story does not imply the observation of the southern celestial pole, because India is in the northern hemisphere. The Phoenician sailors who were said to have circumnavigated Africa, on the other hand, must have crossed the equator and thus seen the stars orbiting around another pole, although Herodotus does not explicitly report this crucial observation. Nevertheless, I think Aristotle missed a manifest opportunity here. He had formulated the reasons for believing the words of the Phoenician sailors when he advocated the existence of an immobile South Pole of the celestial axis that is invisible to us, and the existence of another inhabitable zone beyond the tropics.[59] He also states:

12.9 The lands beyond the tropics are inhabitable, as there the shadow would not fall towards the north, and we know that the earth ceases to be habitable before the shadow disappears or falls towards the south.[60]

For Aristotle, however, this was a theoretical conclusion, drawn from reflection on the sphericity of the earth. He was convinced that nobody could live near the equator and that it was also impossible to cross the equator because of the heat. Therefore, he could not believe the stories of those who saw it with their own eyes, and could not use it as an empirical argument of the earth's sphericity. Curiously enough, not only Aristotle, but also none of the later authors I consulted, mentions the visibility of another celestial pole as an argument for the sphericity of the earth. I call this "the northern hemisphere bias."[61] The final empirical evidence for the sphericity of the earth was presented AD 1522, when the ship Victoria of Magellan's fleet returned from its journey around the earth, mostly in the southern hemisphere, where the sailors saw the stars orbiting around the south celestial pole. Copernicus does not yet mention the circumnavigation of the earth. As far as I know, Kepler was the first astronomer to use it as an argument, but he misses its most important point: the stars circling around another pole.[62]

Aristotle does not use an argument by analogy that he uses for the other celestial bodies. After he has argued that the moon must be spherical, he concludes:

12.10 If, then, one of the heavenly bodies is spherical, the others will clearly be spherical also.[63]

As such, this does not sound like a good argument, but rather like a generalization based one case. But if it could be used for the heavenly bodies, one would expect that it could be applied to the earth as well. But Aristotle avoids this kind of inference,

---

[58] Pliny, *Natural History* 2.75.

[59] See Aristotle, *De Caelo* 284b6–286a2, where Aristotle also argues that the South Pole is the uppermost and the North Pole the lowest. See also *Meteorologica* 362a32–b9.

[60] Aristotle, *Meteorologica* 362b6–9. See also Heidel (1937, 86).

[61] For a survey of the sources until the end of the thirteenth century, see Hamel (1996, 38–109).

[62] Kepler [1635, I.3 (p.19)].

[63] Aristotle, *De Caelo* 291b23.

obviously because he does not conceive of the earth as a heavenly body. The earth is made of the heaviest element, "earth," while the heavenly bodies are made of the fifth element. The earth stands still in the center of the cosmos, while the heavenly bodies circle around the earth, although the movements of the sun, moon and planets are more complicated than that of the fixed stars. According to his physical theory, the heavenly bodies are spherical, because the circular movement and the spherical shape belong to the purest bodies.[64] The arguments for the sphericity of the heavenly bodies do not therefore apply to the earth.

Aristotle on Empirical Arguments for a Flat Earth

As we have seen, it is difficult to make an empirical argument conclusive. A serious problem with arguments from empirical evidence is that opponents can appeal to them as well. Or perhaps we should say that empirical evidence can be interpreted differently, depending on one's presuppositions. A nice example is Eratosthenes' measurement of the circumference of the earth, which he concluded from differences in shadow lengths. His presupposition was that the earth is spherical. Under the assumption that the earth is flat, the experiment measures the distance from the earth to the sun (see Figs. 11.14a, b).

Aristotle explicitly mentions two arguments that are supposedly in favor of a flat earth, but he has some difficulty in refuting them. Panchenko has shown that Anaxagoras had put forward an ingenious empirical argument, arguing that:

12.11 [The defenders of a flat earth] adduce as evidence the fact that the sun at its setting and rising shows a straight instead of a curved line where it is cut off from view by the horizon, whereas were the earth spherical, the line of section would necessarily be curved[65] (see also Fig. 3.1).

Aristotle, on his turn, argues that these defenders of a flat earth:

12.12 fail to take in consideration either the great distance of the sun from the earth, or the great size of the circumference [of the horizon], and the appearance of straightness which it naturally presents when seen on the surface of an apparently small circle a great distance away.[66]

As for the great distance of the sun, Furley remarks: "it is not clear what this has to do with it," and Heath calls Aristotle's attempt to refute this argument "confused."[67]

[64]Cf. Aristotle, *De Caelo* 287a6–12 and 291b11–23.

[65]Aristotle *De Caelo* 294a1–4; the text between square brackets is my addition. See Panchenko (1997).

[66]Aristotle, *De Caelo* 294a5–7; the text between square brackets is my addition.

[67]Furley (1987, 198), Heath (1913, 235). For a detailed discussion of this argument, see Couprie (2011, 181–188).

Simplicius already wondered: "if the distance and apparent smallness of the size is the cause of the section appearing to be a straight line, why doesn't the same thing happen in the case of solar and lunar eclipses, since the distance and the apparent size are the same?"[68] The great size of the circumference (of the horizon), on the other hand, is perhaps more relevant. Where the text has τὸ τῆς περιφερείας μέγεθος, Guthrie translates "the size of the earth's circumference." I think that what is meant here is not the circumference of the earth, but the great circle of the horizon. Aristotle means to say that, both on a spherical and on a disk-shaped flat earth, if the earth were very small, the section where the horizon intersects the sun would look curved. However, the actual distance between an observer standing on the ground of the spherical earth and the horizon is no more than about 4.6 km and thus the circumference of his horizon roughly 14.5 km. Moreover, as long as our eye is practically in the plane of the horizon, the angular diameter of the sun, which is about 0.5°, will remain 1/720 of the circle of the horizon, which makes it hard to recognize whether or not the line of the horizon that cuts it is curved. Simplicius rephrases the argument, making it is easier to understand:

12.13 Perhaps one should say that if we were outside the earth and saw the sun partially obstructed by the earth, the sections would always appear to us to be curved (Fig. 12.3).[69]

In other words: that we see the sun being cut off by the horizon with a straight line is caused by the fact that we are close to the earth's surface, which enables us to see only a part of the horizon in one glance. As Simplicius says, "a circle which is in the same plane as our eye is seen as a straight line."[70] If we were high enough above the earth, so that our eye was no longer in the plane of the horizon, the section would look curved, both on a spherical and on a disk-shaped earth. Anaxagoras' argument is therefore not decisive. Although his own counter-arguments are not very strong, Aristotle draws the same conclusion:

12.14 This phenomenon gives no cogent ground for disbelieving in the spherical shape of the earth's mass.[71]

But it should be added: "Neither does it provide cogent ground for disbelieving in the flat-disk shape of the earth's mass." In other words, the shape of the horizon does not tell us whether the earth is a sphere or a disk.

Aristotle also mentions another argument, which he ascribes to Anaximenes, Anaxagoras, and Democritus:

12.15 Its immobility necessarily involves [that the earth has] the other shape [sc. flat].[72]

---

[68] Simplicius, *In Aristotelis De Caelo Commentaria* 519.30–33.

[69] Simplicius, *In Aristotelis De Caelo Commentaria* 519.33–520.2.

[70] Simplicius, *In Aristotelis De Caelo Commentaria*, 520.7–8.

[71] Aristotle, *De Caelo* 294a8.

[72] Aristotle, *De Caelo* 294a9–10; texts between square brackets my additions.

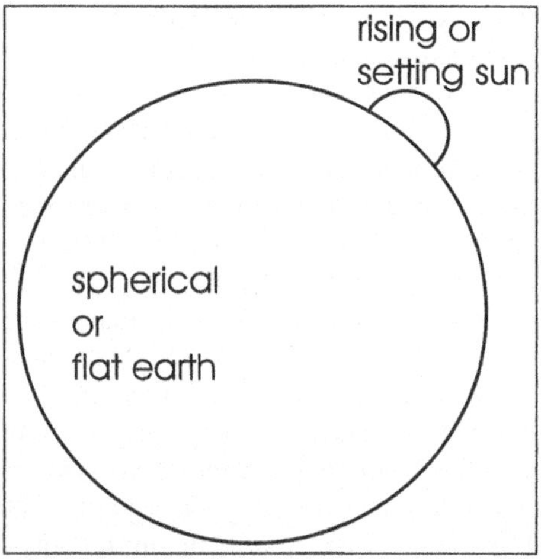

Fig. 12.3 Simplicius' refutation of Anaxagoras' proof that the earth is flat

However, when he discusses this argument, Aristotle formulates it the other way around:

12.16 Anaximenes, Anaxagoras, and Democritus name the flatness of the earth as the cause of its remaining at rest.[73]

The latter was probably the way the argument was used by Anaximenes, Anaxagoras, and Democritus, who were convinced of the flatness of the earth and used it to argue for its immobility. In the context of the arguments concerning the shape of the earth, we do not need to consider the arguments pro and con the immobility of a flat earth. Moreover, Aristotle's argument is begging the question, presupposing what needs to be proven:

12.17 If the earth is not flat, it cannot be owing to its flat shape that it is at rest"[74]

Theoretical Arguments for a Spherical Earth

In order to be able to properly appreciate Aristotle's empirical arguments for the sphericity of the earth, we need to understand the context in which they were used. It is not for nothing, Aristotle placed his empirical arguments as a kind of addendum at the end of his discussion of the earth, its shape, its place in the universe, and why it does not move. According to Aristotle, these are questions that cannot be settled on empirical grounds, but rather in terms of his physical or cosmological theories.[75] His

[73]Aristotle, *De Caelo* 294b14–15.

[74]Aristotle, *De Caelo* 294b23–24.

[75]Cf. Kahn (1994, 118): "In Aristotle's demonstration of the earth's sphericity, general cosmological arguments take precedence over τὰ φαινόμενα κατὰ τὴν αἴσθησιν."

decisive evidence for the sphericity of the earth is of a quite different kind, which has even been called metaphysical, or *a priori*.[76] Aristotle himself puts it this way: "we must consider this question by discussing details but from a universal point of view that takes into account the whole" (ἀλλὰ περὶ ὅλου τινὸς καὶ παντός).[77] In a sense, Aristotle's attitude towards empirical arguments can be seen as a legacy of Plato and even Socrates. Plato did not accept empirical arguments for the sphericity of the earth, but only those that make it clear that it is better and thus inevitable for the earth to have this shape rather than any other.[78] The best empirical arguments can do is to prove *that* something is as it is. Plato wanted a proof that shows *why* something is as it is. Although Aristotle undoubtedly had a more positive attitude towards empirical knowledge, *real* knowledge (ἐπιστήμη) could not be acquired by sense-perception as such, but needed an explanatory syllogism based on indisputable principles, in which it is not only said *that* a fact is so-and-so, but also *why* it is so-and-so.[79] Or, as Aristotle succinctly says: "The study of phenomena is subordinate to astronomy."[80] Elsewhere, Aristotle uses the immobility of the earth as an example. Our empirical knowledge of the immobility of the earth can provide the starting point of the argument, but real knowledge should furnish the reason for its immobility. We can observe that the earth does not move, but the reason why (διὰ τοῦτο) the earth does not move is that it is farthest away from the highest good, the Prime Mover.[81] Aristotle's decisive argument for the sphericity of the earth is not empirical, but an inevitable conclusion from his physics, or more precisely, from his theory of falling:

12.18 That heavy bodies move towards the center of the earth is indicated (σημεῖον ὅτι) by the fact that weights moving towards the earth do not move in parallel lines but always at the same angles to it; therefore, they are moving towards the same center, namely that of the earth.[82]

12.19 [As regards the spherical shape of the earth], there is also the fact (καὶ ὅτι) that all heavy bodies fall at similar angles, not parallel to each other; this naturally means that their fall is towards a body whose nature is spherical.[83]

However, the argument as formulated above is not valid. It is not a *fact* (or: we cannot *observe*) that heavy objects do not fall in parallel lines, so this argument is a *petitio principii*, presupposing what it intends to prove. Moraux has suggested that

---

[76]Dreyer (1953, 109): "In his general conception of the Kosmos Aristotle is guided by purely metaphysical arguments." Dicks (1970, 196): "His arguments are largely *a priori*."

[77]Aristotle, *De Caelo* 294b33.

[78]Plato, *Phaedo* 97D–E.

[79]Cf. Aristotle, *Analitica Posteriora* I.xiii and xxxi.

[80]Aristotle, *Analitica Posteriora* 78b39. In the same sense, Aristotle states in *De Caelo* 297a2–7: "This belief *finds further support* in the assertions of mathematicians on astronomy: that is, *the observed phenomena* (...) *are consistent with* the hypothesis (...) that the earth lies at the centre." (my italics).

[81]Cf. Aristotle, *De Caelo* 292b20.

[82]Aristotle, *De Caelo* 296b18–21.

[83]Aristotle, *De Caelo* 297b20.

Aristotle could have argued that the sun's rays can be considered as being parallel, so that they fall perpendicularly during the summer solstice in a place on the Tropic of Cancer, while they fall at an angle in a place north of the Tropic, but that, in both places, stones fall vertically.[84] This suggestion, however, is also based on a *petitio principii*, since it presupposes a spherical earth, where the sun is so far away that its rays run practically parallel. On a flat earth, where the sun is close and smaller than the earth, the rays of the sun are not parallel (see Figs. 11.1, 11.14b and 13.7). Heavy bodies, on the other hand, fall perpendicularly towards the flat earth's surface (see Fig. 3.8), and this is also what we *observe*, even on a spherical earth, when we look around at fallings things. Perhaps it is better not to take these texts as a separate argument, as Simplicius did,[85] but as another formulation of Aristotle's main argument, based on his physical theory: the spherical shape of the earth is the product of the natural tendency of heavy things to fall towards the center of the spherical cosmos:

12.20 The natural motion of its parts, as well as that of the earth as a whole, is towards the center of the universe.[86]

12.21 It is plain (...) that if particles are moving from all sides alike towards one point, the center, the resulting mass, must be similar on all sides (...). Such a shape is a sphere.[87]

This theory is based on assumptions we would now consider wrong, such as that the heavens are spherical and that heavy and light are opposite qualities of things, causing natural movements downwards and upwards. Still, we can say that these assumptions have an empirical basis: the circular motion of the stars suggests the sphere of the heavens, and we see light things moving up and heavy things falling down. And because it looks as if the earth is in the center of the cosmos, it sounds reasonable to say that "up" means "to the periphery" (centrifugal) and "down" means "to the center" (centripetal):

12.22 The motion of light bodies like fire is contrary to that of the heavy, towards the extremity of the region which surrounds the center.[88]

Aristotle's definite proof of the sphericity of the earth was the conclusion drawn from these fundamental ideas, which were cherished for many centuries to come. Aristotle's physical theories, and especially his theory of falling, solved three problems which Presocratic cosmology was unable to satisfactorily solve: (1) why does the earth not fall although it is not supported by anything, (2) which is the place where the earth dwells within the cosmos, and (3) what is the shape of the earth. The

[84]Moraux (1965, CXXXI, n. 2), see also Jori (2009, 473).

[85]Simplicius, *In Aristotelis De Caelo Commentaria* 545.30.

[86]Aristotle, *De Caelo* 296b7; Guthrie has: "the natural motion of the earth as a whole, like that of its parts."

[87]Aristotle, *De Caelo* 297a22–26.

[88]Aristotle, *De Caelo* 296b13–15.

way he did it was ingenious: he made falling not the problem, but the solution to the problems.[89] In the case of the shape of the earth, knowledge gained from observation not only stood at the beginning of the argument (the heavens look spherical; the earth seems to be in the center; heavy objects fall and light objects rise), but also at the end of the argument (the shape of the earth's shadow during lunar eclipses; stars seen under different angles), where they played the role of confirmation of a result that was built on theoretical considerations. This is why the empirical arguments of the sphericity of the earth appear at the end of book II of *De Caelo*.

Final Remarks

Aristotle was right: the earth is a sphere (or, to be more precise, an oblate spheroid). Critical notes can be made on his empirical arguments, but his successors until Kepler hardly did any better. They all missed the crucial argument about the stars circling around the South Pole. Obviously, they had no reason to look for new arguments, because they were already convinced that the arguments for a flat earth were much weaker. In Aristotle's time, and for many centuries to come, his theoretical arguments seemed undeniable, or at least entirely reasonable. The empirical arguments confirmed the conclusion that was based on the principles of his physics.

One of the most striking features of Aristotle's arguments is that the earth has a shape that seems to be reserved for the noblest bodies, namely the heavenly bodies. The earth, made of the least-noble stuff, is also spherical. For this reason, his arguments for the sphericity of the earth must be different from those for the sphericity of the heavenly bodies.

Copernicus points to the fact that the notion of the earth as one of the heavenly bodies already existed in Aristotle's time in the cosmological image of Philolaus, in which the earth—together with the sun, moon, planets, and stars—revolves around a common center.[90] And Anaxagoras, although he believed in a central flat earth, was already convinced that the moon was "another earth" and that the heavenly bodies were made of stone. Aristotle, however, must reject the idea that the earth was just another heavenly body, or much like the heavenly bodies, because it was excluded by his physical theories. It is precisely at this point that Copernicus attacks Aristotle. If the earth is a sphere, "why should we hesitate any longer to grant to it the movement which accords naturally with its form?" And elsewhere he adds: "So that it can be considered as one of the wandering stars."[91] This also implies another theory of gravity, in which there are many centers, of which the earth is only one: "Since there are many centers (. . .), gravity or heaviness is nothing except a certain natural appetency implanted in the parts (. . .) in order that they (. . .) come together in

[89]See more extensively: Couprie (2011, 213–220).

[90]Copernicus (1543), Preface and Dedication to Pope Paul III.

[91]Copernicus (1543, I.8 and I.9).

the form of a globe. It is believable that this effect is present in the sun, moon, and the other bright planets (. . .)."[92] After almost two millennia, these lines introduced a complete new chapter in astronomy.

References[93]

Bakker, F.A., *Epicurean Meteorology*. Leiden: Brill 2016.

Bigwood, J.M., Aristotle and the Elephant Again. *The American Journal of Philology*, 114 (4) (1993), 537–555

Bowen, A.C. and R.B. Todd, *Cleomedes' Lectures on Astronomy*. Berkeley: University of California Press 2004.

Copernicus, N., *De revolutionibus orbium coelestium*. Nuremberg: Johann Petreius 1543. Available on the internet: http://ads.harvard.edu/books/1543droc.book/.

Couprie, D.L., *Heaven and Earth in Ancient Greek Cosmology*. New York: Springer 2011.

Dicks, D.R., *Early Greek Astronomy to Aristotle*. Ithaca N.Y.: Thames and Hudson 1970.

Diels, H. and W. Kranz, *Die Fragmente der Vorsokratiker*. Zürich/Hildesheim: Weidmann 1951[6] (=DK).

Dreyer, J.L.E., *A History of Astronomy from Thales to Kepler*. Mineola, NY: Dover Publications 1953.

Furley, D., *The Greek Cosmologists. Volume I: The Formation of the Atomic Theory and Its Earliest Critics*. Cambridge: Cambridge University Press 1987.

—, Cosmology. In: K. Algra, J. Barnes, J. Mansfeld and M. Schofield, *The Cambridge History of Hellenistic Philosophy*. Cambridge: Cambridge University Press 2010, 412–451.

Graham, D.W., *The Texts of Early Greek Philosophy*. Cambridge: Cambridge University Press 2010 (=Gr).

—, *Science Before Socrates*. Oxford: University Press 2013.

Hamel, J., *Die Vorstellung von der Kugelgestalt der Erde im europäischen Mittelalter bis zum Ende des 13. Jahrhunderts – dargestellt nach Quellen*. Münster: Lit Verlag 1996.

Hawking, S., *A Brief History of Time*. New York: Bantam Books 1988.

Heath, T., *Aristarchus of Samos. The Ancient Copernicus*. Oxford: Clarendon Press 1913.

Jori, A., *Aristoteles. Über den Himmel*. Berlin: Akademie Verlag 2009.

Kahn, C.H., *Anaximander and the Origins of Greek Cosmology*. Indianapolis/Cambridge Hackettt Publishing Company, Inc. 1994.

Kepler, J., *Epitome Astronomiae Copernicanae*, Francofurti: Schönwetter 1635 (to the best of my knowledge, there is no English translation is available of the quoted Liber primus, Pars prima: *De figura Terrae*).

Kirk, G.S., J.E. Raven and M. Schofield, *The Presocratic Philosophers*. Cambridge: Cambridge University Press 2009 (=KRS).

Laks, A. and G.W. Most, *Early Greek Philosophy VI*, Cambridge, MA 2016.

Liddell, H.G., R. Scott H.S. Jones and R. McKenzie, *Greek-English Dictionary*. Oxford: Clarendon Press 1996. Also available on the internet, e.g.: http://stephanus.tlg.uci.edu/lsj/#eid=1714&context=search (=LSJ).

Mansfeld, J. and D.T. Runia, Aëtiana. *The Method and Intellectual Context of a Doxographer, Volume Two, Part Two*. Leiden: Brill 2009 (=MR).

Moraux, P., *Aristote. Du ciel*. Paris: Collection des universités de France 1965.

[92]Copernicus (1543, I.9).

[93]P = Aëtius in pseudo-Plutarch, *Placita* (numbering according to Dox).
S = Aëtius in Stobaeus, *Anthologium* (numbering according to Wachsmuth and Hense).

Neugebauer, O.E., *A History of Ancient Mathematical Astronomy*. Berlin 1975.
North, J., *Cosmos: An Illustrated History of Astronomy and Cosmology*. Chicago: The University of Chicago Press 2008.
O'Grady, P. *Thales of Miletus: The Beginnings of Western Science and Philosophy*, Aldershot: Ashgate 2002.
Panchenko, D., Anaxagoras' Argument Against the Sphericity of the Earth. *Hyperboreus* 3 (1997), 175–178.
—, The Shape of the Earth in Archelaus, Democritus and Leucippus. *Hyperboreus* 5 (1999), 22–39.
—, Parmenides, the Nile and the Circumnavigation of Africa by the Phoenicians. In: J.M. Candan Morón, F.J. González Ponce and A.L. Chávez Reino, *Libyae lustrare extrema: Realidad y literatura en la visión grecorromana de África. Homenaje al Prof. Jehan Desanges*. Sevilla: Universidad de Sevilla 2008, 189–193.
—, Anaximenean Astronomy in the Light of Chinese Parallels. *Tsinghua Studies in Western Philosophy* 1(2) (2015), 412–426. Text of a paper read at the *Fourth* (not 14th, as mistakenly written in a note on p. 412) *Biennial Conference of the International Association for Presocratic Studies* at the Aristotelian University of Thessaloniki, June 30–July 4, 2014.
Rovelli, C., Aristotle's Physics: A Physicist's Look. *Journal of the American Philosophical Association* (2015), I.1: 23–40.
Tannery, P., *Pour l'histoire de science Hellène : de Thalès à Empédocle*. Paris: Félix Alcan 1887.
Wöhrle, G., *Traditio Praesocratica. Die Milesier: Thales*. Berlin: Walter de Gruyter 2009 (=TP1).
—, *Traditio Praesocratica. Die Milesier: Anaximander und Anaximenes*. Berlin: Walter de Gruyter 2012 (=TP2).

Part II
Ancient China

Chapter 13
An Ancient Chinese Cosmology: Main Features

Contents

The *Gai Tian* Model of a Flat Earth and a Flat Heaven

The ancient Chinese model of the cosmos called *gai tian* ("canopy heaven") that is explained in the *Zhou bi*[1] ("gnomon[2] at Zhou") in the first century BC (but containing much older material)[3] and in additional sources, differs fundamentally

[1]Quotations with an initial # are from Cullen's translation of the *Zhou bi* in Cullen (1996).

[2]In this context, a *bi* or gnomon is just a stick, put perpendicularly on the ground and used to measure its shadow.

[3]For the difficult question of dating the *Zhou bi* as a whole as well as parts of it, see Cullen (1996), 138–145, especially 145. Several items and calculations discussed below seem to belong to the oldest parts, going back to the pre-Qin and early Han periods (about 200 BC). The central idea of a flat circular heaven over a flat square earth is even older (see texts 13.3 and 13.6). In Cullen (2017), 207–212, a concise overview is given of the *gai tian*, as presented in the *Zhou bi*.

D. L. Couprie, *When the Earth Was Flat*, Historical & Cultural Astronomy,
https://doi.org/10.1007/978-3-319-97052-3_13

from the usual archaic conception of a flat earth with a hemispherical celestial vault, like that of the Presocratic Greeks and the Chinese rivaling system called *hun tian*. In the *gai tian* model, the heaven is thought to be flat and parallel to the flat earth. For us, since we are used to think in terms of a spherical earth, it is not so easy to comprehend the implications of an earth conceived of as flat. The conceptual transition to understand the *gai tian* is, as we will see, even "much more difficult to make than the switch from a spherical to a flat earth," as Cullen rightly notes.[4]

Quite apart from its intrinsic interest as a sophisticated attempt to cope with the celestial phenomena under the supposition of a flat earth, the system developed in the *Zhou bi* is worth studying for two specific reasons. The first is that the *gai tian* model provides a solution of a problem that the ancient Greek cosmologists were not able to solve, namely that on a flat earth it is always everywhere the same time. This is so important an effort that it might be called the *raison d'être* of the *gai tian* model. An additional reason to study the *gai tian* model is that recently Panchenko has argued that similar thoughts can be found in the cosmologies of Anaximenes and Xenophanes, implying that they have been introduced from Greece into China. I will argue, on the contrary, that the *gai tian* is an impressively creative, authentic Chinese system.

The conception of the heaven as a plane parallel to that of the flat earth is the main, though implicit, presupposition in the *Zhou bi*, in which it is only remarked that:

13.1. (#A6) The square pertains to earth, and the circle pertains to Heaven. Heaven is a circle and earth is a square.[5]

More explicitly it is stated by Wang Chong in the *Lun heng* (first century A.D.):

13.2. Heaven is level as much as the earth.[6]

As we will see, the measurements of the height of the sun and the pole presuppose a flat earth and result in a heaven that is everywhere 80,000 *li* above the earth. Cullen suggests that this conception of heaven and earth as parallel planes goes back to a model of heaven and earth in a diviner's instrument used to indicate the basic time sequences of the cosmos, like the planisphere or cosmic model (*shi*), dated about 165 BC (Fig. 13.1).

Here already the first important interpretive difficulty rises. The very words *gai tian* mean "canopy heaven" and *gai* is the umbrella-like canopy over an ancient

[4]Cullen (1996), xiii.

[5]Already for this reason the rendition in Needham (1970), 210–216 cannot be right. See, e.g. 210: "The heavens were imagined as a hemispherical cover, and the earth as a bowl turned upside down." Needham (1970, 212, Fig. 87) reproduces Chatley's (1938, 12) misguided drawing. The same holds for several deceitful drawings on the Internet.

[6]Quoted from Forke (1907), 262. See also Cullen (1996), 129, n. 150.

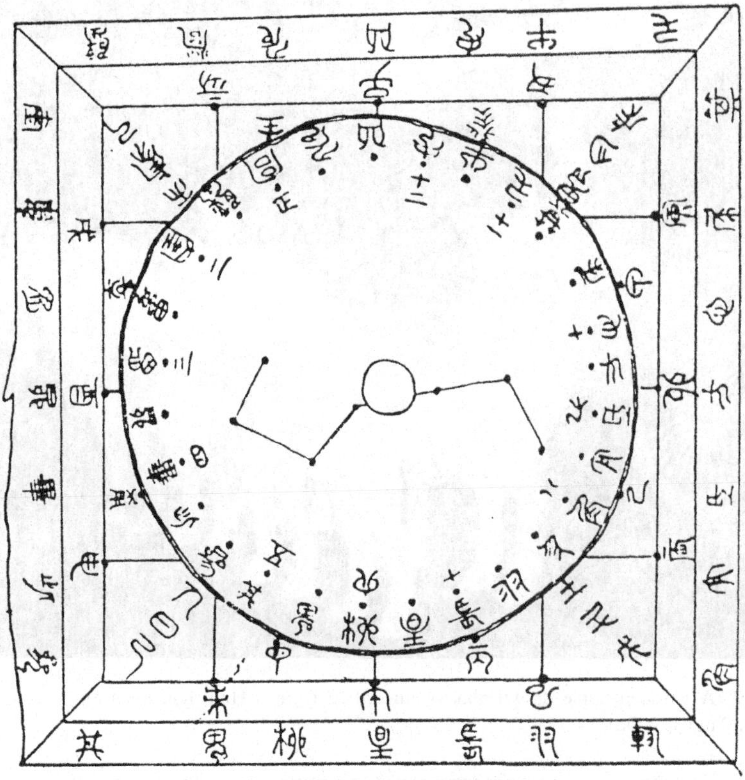

Fig. 13.1 Cosmic model (*shi*), early second century BC (Source: Yin Difei 1978, Figure 3 on 340)

Chinese chariot like in Fig. 13.2.[7] This image is used, for instance, in a poem by Song Yu (c. 300 BC):

13.3. The square earth is my chariot,
the round heaven is my canopy (*gai*).[8]

In the *Huainanzi* no image is used, but there it simply says:

13.4. The way of heaven is called the round
The way of earth is called the square[9]

In the *Zhou bi* the expression *gai tian* is not used, but the rather confusing image of a rain-hat:

13.5. (#A6) One may represent heaven by a rain-hat.

[7]Cf. Cullen (1996), 50.

[8]Quoted from Cullen (2017), 203.

[9]Major (2010), 115.

Fig. 13.2 A wooden horse-drawn chariot unearthed from a Han dynasty tomb (Photo Tomasz Sienicki, August 2006)

Fig. 13.3 A Chinese straw hat (Matt Hahnewald Photography)

The Chinese rain-hat we are acquainted with is a conical straw headgear, like that in Fig. 13.3.

Presumably, the idea behind these images of a canopy or a rain-hat is not to modify the model of a flat heaven into that of a cone shaped heaven, but to stress the circularity of its shape and motion, circling around the vertical polar axis like an umbrella around its stick. This suggestion is strengthened by another document that uses the same metaphor of the canopy of a chariot, in which not the chariot and the canopy themselves, but their roundness and squareness are explicitly said to represent heaven and earth:

13.6. The squareness of the chariot is to represent earth;
The roundness of the canopy is to represent heaven.[10]

However, in a cryptic later development that we will discuss in the next chapter, both heaven and earth are described as somehow sloping:

13.7. (#E6) Heaven resembles a covering rain-hat, while earth is patterned on an inverted pan.

Be this as it may, in the main part of the *Zhou bi* it is clear both from the text and from its detailed calculations that a flat circular heaven, parallel to the flat square earth, is meant. This interpretation is confirmed by Cullen' concise account of the *Zhou bi*: "Parts of the text describe a cosmos which is in effect a *shi* (see Fig. 13.1 above) enlarged to a vast scale, with the *heaven-disc* 80,000 *li* above the flat square earth that lies below it. This view of the cosmos became known as the *gai tian*, chariot-umbrella heaven."[11]

The Movements of the Heavenly Bodies and the Location of Zhou

In the *gai tian* model, the celestial bodies turn around the celestial pole in daily orbits in a plane parallel to the earth's surface. The celestial axis is perpendicular upon the earth and not slanted, as in Presocratic Greek cosmology.[12] This means that in the *gai tian* model there is a subpolar point on earth that we may call "(north) pole," right below the (north) pole of the heavens. In the *Zhou bi*, this point is called "the subpolar point." The Chinese observer is situated eccentric at 103,000 *li* from the pole in a place called Zhou:

13.8. (#B14) (...) the subpolar point is 103,000 *li* to the north of Zhou.
13.9. (#B33) Zhou is 103,000 *li* south of the center of heaven.

Further on we will see how this distance was calculated. That Zhou is not at the other end of the polar axis is a remarkable and revolutionary feature of the model,

---

[10]From the *Kaogong ji* (dated between 475–223 BC), quoted from Lan-ying Tseng (2011), 49; see also Cullen (1996), 50.

[11]Cullen (2017), 208 (my italics). This is more straightforward than his conclusion in Cullen (1996), 115: "The reference to the rain-hat in section #A was clearly unrelated to questions of quantitative cosmography."

[12]See Cullen (1996), 53 ff., and his figure 6 on 65.

because the Chinese were used to consider themselves as living in "The Land of the Middle" (*Zhongguo*). According to Cullen the location of Zhou "is in a latitude close to that of the Yellow River basin, around 35° north."[13] We will discuss his calculation in Chap. 14.

When an observer would be able to go to the subpolar point on earth he would see the heavenly bodies orbiting overhead in circles around the celestial pole. This circular movement, parallel to the plane of the earth, is the essence of the *gai tian* model of the heavens. The model pays special attention to the movements of the sun, making the width of its orbit around the pole vary with the seasons. The implication of the model is that, seen from Zhou, which is the observer's abode, the sun goes also to the opposite part of the heaven, beyond the pole. As will be clarified later on, the *gai tian* system explains why we cannot see the sun when it is at the other side of the pole. Since the pole around which the sun rotates has its counterpart on earth right underneath it in the subpolar point, the circles described by the sun also have their counterparts on earth. This means that the tropics and the equator are not only circles in the flat sky, but also circles on the flat earth. Here I disagree with Cullen, who maintains that "since Chinese astronomers lived in a flat-earth universe the poles and the equator remained solely celestial concepts."[14] On the contrary, I would maintain that it is instructive to speak of a "*Zhou bi* (north) pole" on earth, which is the subpolar point on earth that is mentioned several times in the *Zhou bi*. In an analogous way, we may speak of a "*Zhou bi* equator" on earth, which is the circle on the flat earth, right below the orbit of the sun at the equinoxes, where the sun at the equinoxes stands in the zenith at noon, and similarly of the "*Zhou bi* tropics" and a "*Zhou bi* pole circle on earth. The *Zhou bi* itself does not use these terms, but speaks of "subpolar" and "subsolar" points. To express it shortly: in a *gai tian* world cosmography equals geography. In Chap. 14, we will see that an extrapolation of this conception towards the south will lead to surprising geographical conclusions.

The Shadow Rule and the Fundamental Cosmic Measurements

The ancient Greek cosmologists, who also believed that the earth is flat, were not able to calculate the dimensions of the cosmos. The figures mentioned by Anaximander for the distances of the celestial bodies were purely symbolic. The only real calculation we know of was made by Anaxagoras, who compared the size of the sun with the Peloponnesus, which must have implied some indicative idea of its distance. It is a matter of debate between scholars what kind of calculation he actually used.[15] In the *Zhou bi*, on the other hand, the dimensions of the cosmos were exactly

[13]Cullen (1996), 8.

[14]Cullen (1996), 8.

[15]See Chap. 11.

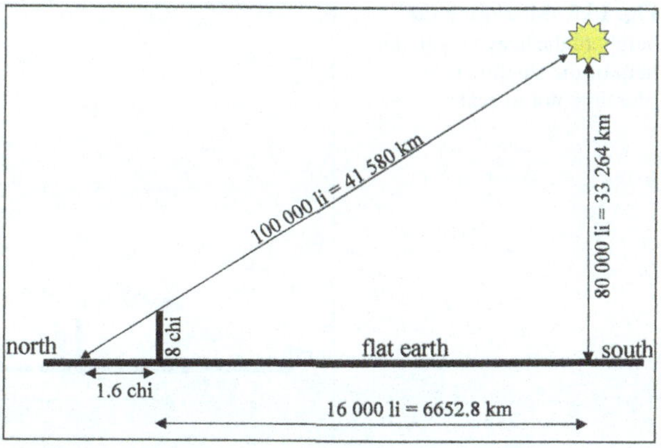

Fig. 13.4 The *Zhou bi* method to measure the height of the sun (drawing not to scale) (For similar pictures, see Cullen 1996, 79, Fig. 7, and Cullen 2017, 209, Fig. 5.9)

calculated. The *Zhou bi* contains many calculations, but the measurement of the height of the sun is the basis of all others.[16]

An essential feature of the measurements is the so called "shadow rule," which can be expressed thus: "one *cun* less shadow for a thousand *li* south." The relevant texts in the *Zhou bi* are:

13.10. (#B10) The *Zhou bi* is eight *chi* in length. On the day of the summer solstice its [noon] shadow is one *chi* and *six* cun. The *bi* is the altitude [of the right-angled triangle], and the exact [noon] shadow is the base. 1000 *li* due south the base is one *chi* and five *cun*, and 1000 *li* due north the base is one *chi* and seven *cun*. The farther south the sun is, the longer the shadow.

13.11. (#B12) Method: the *Zhou bi* is eight *chi* long, and the decrease or increase of the base is one *cun* for a thousand *li*.

13.12. (#D18) The shadow is (…) changing at the rate of one *cun* for every thousand *li*.

To find the height of the heaven the *Zhou bi* argues as follows: in order to get a ratio of 3:4:5 between base, altitude and hypotenuse, choose a shadow of 6 *chi* for a gnomon of 8 *chi* (see Fig. 13.4), and then apply the shadow rule:

13.13. (#B11) Wait until the base is six *chi* in length (…). So start from the base and take the *bi* as altitude. 60,000 *li* from the *bi*, at the subsolar point a *bi* casts no shadow. From this point up to the sun it is 80,000 *li*.

13.14. (#E7) Heaven is 80,000 *li* from earth.

Thus, the number for the distance between earth and heaven is found. The number for the hypotenuse, which is the oblique distance from the observer to the sun, follows from those of the two other sides of the right-angled triangle:

[16]In the Han measurement system one *li* = 1800 *chi*, one *bu* = 6 *chi*, one *zhang* = 10 *chi*, one *chi* = 10 *cun*, one *cun* = 10 *fen*. One *li* equals 415.8 m, one *chi* equals 0.231 m, one *bu* equals 1.386 m.

Fig. 13.5 Measuring the height of the heaven with the help of the shadow rule (drawing not to scale)

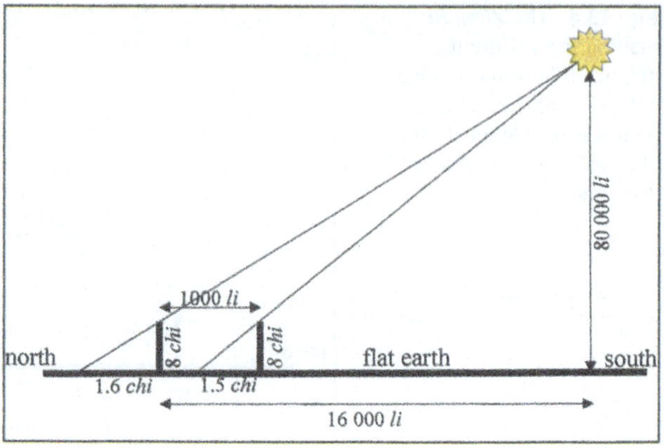

13.15. (#B11) If we require the oblique distance [from our position] to the sun, take [the distance to] the subsolar point as the base, and take the height of the sun as the altitude. Square both base and altitude, add them and take the square root, which gives the distance to the sun. The oblique distance from the position of the *bi* is 100,000 *li*.

The number of 80,000 *li* for the distance between heaven and earth is confirmed by using the shadow rule at the summer solstice. In #B10 (text 13.10) it was said that on the day of the summer solstice the [noon] shadow of an eight-chi gnomon is 1 *chi* and 6 *cun*. The result of the shadow rule is that a gnomon placed 16,000 *li* south of Zhou will cast no shadow at noon at the summer solstice. This is the subsolar point on the northern tropic.

13.16. (#B9) 16,000 *li* to the south at the summer solstice (. . .) if one sets up a post at noon it casts no shadow.
13.17. (#B16) 16,000 *li* south [of Zhou] on the day of the summer solstice (. . .) there is no shadow at noon.
13.18. (#D18) Therefore the solar shadow (. . .) is 1 *chi* 6 *cun* at the summer solstice.
13.19. (#H2, 13) Summer solstice, 1 chi 6 cun.

This result is shown in Fig. 13.5. The height of the heaven is again 80,000 *li*.[17]

[17]The *Huainanzi* has two numbers for the height of the heaven: one calculation with a gnomon of one *zhang* (= 10 *chi*) length, which results, of course, according to the shadow rule, in a distance between earth and heaven of 100,000 *li*. (Major c.s. 2010, 148, ch. 3.45), and another number of 150,000 *li*, without an indication of a method to reach this result (Cf. Major 2010, 117 (3.4). The figure of 150,000 *li* is a correction of the main text, which reads 510,000 *li* (Cf. Major 1993, 68 and note on 294. See also Cullen 1993, 288, and 294, note at 3.IV).

Some More Calculations

Three more calculations are made from the observation of shadow lengths with the help of the shadow rule: the distances from Zhou to the circles of the winter solstice and the equinoxes, and the distance from Zhou to the pole. All other calculations are derived from these. The distance from Zhou to the subsolar point at the winter solstice (on the southern tropic) is calculated in the following way:

13.20. (#B9) (...) and 135,000 *li* to the south at the winter solstice, if one sets up a post at noon it casts no shadow.
13.21. (#B16) (...) and 135,000 *li* south on the day of the winter solstice, there is no shadow at noon.
13.22. (#B34) 135,000 *li* on the day of the winter solstice.[18]
13.23. (#D18) Therefore the solar shadow at the winter solstice is 1 *zhang* 3 *chi* [= 13 *chi*] and 5 *cun*.
13.24. (#H1) The length of the winter solstice shadow is 1 *zhang* 3 *chi* 5 *cun*.
13.25. (#H2,1) Winter solstice, 1 *zhang* 3 *chi* 5 *cun*.

From text 13.23 (#D18), one might infer that the shadow length was found from the distance from Zhou to the subsolar point south of Zhou during the winter solstice. Actually, of course, the procedure must have been the other way around: the distance from Zhou to the subsolar point at the winter solstice (texts 13.20, 13.21, and 13.22) is found by applying the shadow rule to the noon shadow of the gnomon at Zhou at that time (texts 13.23, 13.24, and 13.24).

For the shadow length and the subsolar point at the equinoxes (on the equator), another procedure was followed:

13.26. (#F9) At the spring equinox, and at the autumn equinox, the sun is on the middle *heng*. (...) The middle *heng* is 75,500 *li* from Zhou.
13.27. (#H2, 7) Spring equinox 7 *chi* 5 *cun* 5 *fen*.
13.28. (#H2, 19) Autumn equinox 7 *chi* 5 *cun* 5 *fen*.

The shadow length at noon at the equinoxes was not calculated by observing shadow lengths but by simply (and wrongly) dividing by 2 the sum of the shadow lengths of the solstices: (1 *chi* 6 *cun* + 13 *chi* 5 *cun*) ÷ 2 = 7 *chi* 5 *cun* 5 *fen*. Similarly, the distance from Zhou to the circle of the equinox was calculated by dividing by 2 the difference between the distances from Zhou to the circles of the solstices: (135,000 − 16,000) ÷ 2 = 59,500 *li*. However, since the equinox sunray from the top of a gnomon is the bisector of the sunrays of the solstices, the length of the equinox shadow does not end halfway the shadows of the solstices (see Fig. 14.1). The Chinese astronomers abusively thought it was better to calculate the length of the equinox shadow than to observe it on the ground. The origin of their mistake was that what they thought to be distances on the flat plane of the heavens

[18]The text of #B34 in Cullen (1996), 181 has "on the day of the summer solstice", but this is a misprint, Christopher Cullen assured me, because the Chinese text has 冬至日.

which could be translated into analogous measurements of shadow lengths on the ground were in fact angular distances between the sun at the solstices and the sun at the equinoxes. In the first section of Chap. 14, we will elaborate on the shadow measurement during the year. As a result of the use of the shadow rule all calculations result in a height of the sun, and thus of the heaven, of 80,000 *li*, which is in accordance with the idea of the heaven as a plane parallel to the flat earth.

The third measurement, the distance from Zhou to the subpolar point, is needed to find the sun's orbit, of which the pole is the center. This measurement, too, is done with the help of the shadow rule, although the measurement is done by night, so that we will have to use what Cullen calls the "polar shadow,"[19] which is found by sighting towards the pole along a cord tied to the top of the gnomon:

13.29. (#B14) Now set up a gnomon eight *chi* tall, and sight on the pole: the base is one *zhang* three *cun* [= 10.3 *chi*]. Thus it can be seen that the subpolar point is 103,000 *li* to the north of Zhou.

13.30. (#F2) To fix the pivot of the north pole, the center of the *xuan ji*, to fix the center of the heaven (. . .): At the winter solstice, at the time when the sun is at *you*,[20] set up an eight-*chi* gnomon, tie a cord to its top and sight [along the cord] on the large star in the middle of the north pole [constellation]. Lead the cord down to the ground and note [its position].

13.31. (#F4) [The positions] where it is noted that the cords reach the ground are 1 *zhang* 3 *cun* from the gnomon, and therefore the center of heaven is 103,000 *li* from Zhou.

The *xuan ji* is a fictitious star[21] circling around the pole at a distance of 11,500 *li* that we will discuss later. The "large star in the middle of the north pole" mentioned in #F2 is probably β Umi[22] (Kochab), which was in the Han period at an angular distance of about 8° from the real pole.

When we add the number of the distance from Zhou to the subpolar point and that of the distance from Zhou to the subsolar point at the summer solstice circle (the northern tropic), we get 103,000 + 16,000 = 119,000, which is the radius of the sun's orbit at the summer solstice. Similarly, the orbits of the sun at the winter solstice and the equinoxes can easily be calculated. The relevant texts of the *Zhou bi* are:

13.32. (#B16) 16,000 *li* south [of Zhou] on the day of the summer solstice, and 135,000 *li* south on the day of the winter solstice, there is no shadow at noon. From this we can see that from the pole south to noon at the summer solstice is 119,000 *li*, and it is the same distance north to midnight. The diameter overall is 238,000 *li*, and this is the diameter of the solar path at the summer solstice.

---

[19]See Cullen (1996), 105–106.

[20]Due east of the pole, see Cullen (1996), 191 n. 213. The obvious meaning is that the sun is beyond the range of visibility of an observer in Zhou, because it is night.

[21]Cf. Cullen (1996), 190.

[22]See Cullen (1996), 191 n. 214.

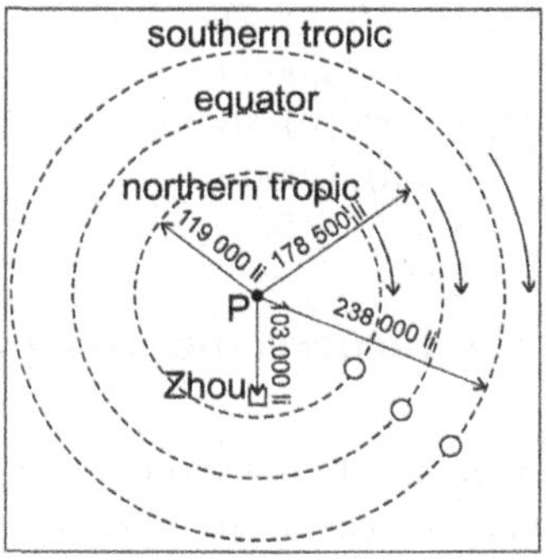

Fig. 13.6 The *gai tian* model of the heaven according to the *Zhou bi* (For similar pictures, see Cullen 1996, 130, Fig. 12, and Cullen 2017, 211, Fig. 5.11)

13.33. (#B17) From the summer solstice noon to the winter solstice noon is 119,000 *li*, and it is the same distance north to the subpolar point.

13.34. (#B18) From the pole south to the winter solstice noon is 238,000 *li*, and it is the same distance north to midnight. The diameter overall is 476,000 *li*, and this is the diameter of the solar path at the winter solstice.

13.35. (#B19) From the equinoctial noon north to the subpolar point is 187,000 *li*. The diameter overall is 357,000 *li*, [and this is the diameter of the solar path at the equinoxes].

13.36. (#B21) [If one measures] south to the summer solstice noon and north to the winter solstice midnight (in Fig. 8.6 from the northern tropic through the pole to the southern tropic) or south to the winter solstice noon and north to the summer solstice midnight (in Fig. 8.6 from the southern tropic through the pole to the northern tropic) in both cases the diameter is 357,000 *li*.

13.37. (#B34) The diameter of the solar path at the winter solstice is 476,000 *li*.[23]

It should be noted that the orbits of the sun make full circles above and parallel to the earth and not around and under the earth, as the ancient Greeks who believed that the earth is flat were used to say. In Fig. 13.6, a plan view of the heaven is drawn according to the measurements of the *Zhou bi*. In Figs. 13.17–13.20 several measurements are rendered in elevation view.

The *gai tian* model has some other extraordinary features. As is clear from Fig. 13.6, one of them is that the sun's orbit in summer is much smaller than in winter. This entails that the sun in summer moves much slower (two times) than in winter:

13.38. (#B16) The circumference [of the solar path at the summer solstice] is 714,000 *li*.[24]

[23]The same numbers for the first, fourth, and seventh *heng* in #D8 (diameter 238,000 *li*), #D11 (diameter 357,000 *li*), and #D14 (diameter 476,000 *li*). The concept of the seven *heng*, imaginary circles around the pole, will be discussed in Chap. 14. Here it suffices to say that the first, fourth, and seventh *heng* coincide with the northern tropic, the equator, and the southern tropic of Fig. 13.6.

[24]In the calculations of the circumference of a circle the *Zhou bi* uses $\pi = 3$. The same numbers also in #D8, #D11, and #D14.

13.39. (#B18) The circumference [of the solar path at the winter solstice] is 1428,000 *li*.

13.40. (#B19) Its circumference (viz. of the solar path at the equinoxes) is 1071,000 *li*.

The Incorrectness of the Shadow Rule

At first sight the shadow rule looks like based upon observation, until one realizes that this cannot be right, for 16,000 *li* equals 6652.8 km, whereas the Tropic of Cancer is only about 1200 km south of Zhou. The reason for the shadow rule's being mistaken can hardly lie in the difficulty to measure distances over the surface of the earth, because for that the discrepancies such as the above-noted are too big. Cullen remarks: "As to the origins of the shadow rule, I do not think there is any evidence on which to base worthwhile conclusions. The most striking fact about the rule is how completely wrong it is."[25] As a consequence of the shadow rule being wrong, all measurements that depend on it are mistaken as well. However, Cullen's verdict needs to be differentiated. The shadow rule as such is right on a flat earth in so far as the noon shadow shortens with equal steps for gnomons placed at successive equal distances on a north-south line (a meridian). This can be explained by the drawing in Fig. 13.7.

AN, BO, CP, etc., up to LZ, are gnomons, placed at equal distances from one another. The ends of the shadows of the gnomons are indicated as A′, B′, C′, etc., up to L′. The noon sun S stands right above gnomon AN, so that this gnomon does not throw a shadow and point A′ coincides with point N. Triangle BOB′ is similar to triangle SNB′, because ∠BB′O = ∠SB′N, while ∠BOB′ and ∠SNB′ are 90°; CPC′ is similar to SNC′; DQD′ similar to SND′, etc., up to triangle LZL′, which is similar to triangle SNL′. Because the gnomons are placed at equal distances, the distances NO, NP, NQ etc., up to NZ are growing continuously, and because the small triangles and the big triangles are similar, the distances OB′, PC′, QD′ etc., up to ZL′ must also grow continuously.

The main problem with the shadow rule is that the Chinese astronomers calculated the shortening of the shadow about 5.5 times too small. Panchenko has put forward the intriguing suggestion that "the shadow rule was established somewhere outside of China and that, in the process of the transmission, the Chinese *li* was substituted for a foreign measure."[26] With his words "somewhere outside of China," Panchenko means Greece. I think there is an easier and more natural explanation: The ancient Chinese astronomers were convinced of the enormous size of both earth and heaven. They started with the standard triangle of a (6, 8 and 10 *cun*) ratio between base, altitude and hypotenuse because that gave a ratio of 3:4:5 with which

[25]Cullen (1996), 113.

[26]Panchenko (2002b), 252.

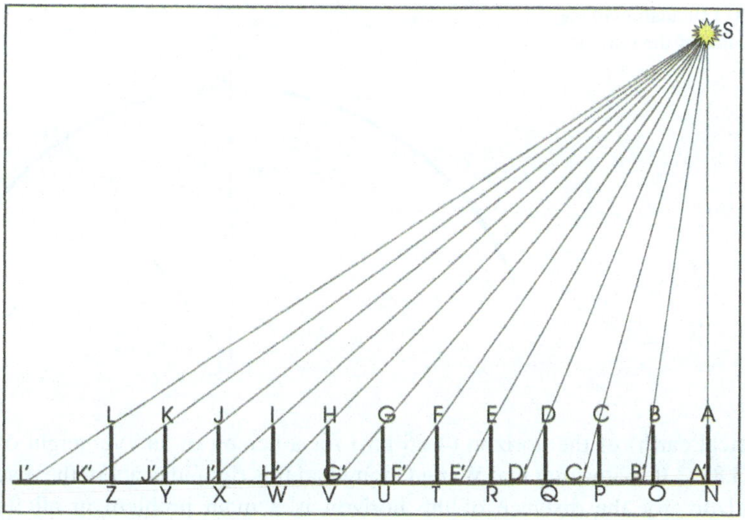

Fig. 13.7 The shadow rule explained (drawing not to scale)

it was easy to calculate. Then they deliberately chose the numbers of the shadow rule so as to produce a huge triangle of similar numbers that were also easy to calculate with: 60,000, 80,000, and 100,000 *li*. And the other calculations were made according to the "rule" thus found. In other words, the shadow rule was not based on observation but construed for the sake of easy calculating. The ancient Chinese were not so much interested in the actual distances, but all the more in the functioning of the calculating system.

The Horizon and the Rising and Setting Sun as Optical Illusions

Another important element of the model also not explicitly mentioned in the *Zhou bi* but necessary for its understanding, is that the horizon is an optical illusion. Wang Chong expresses it thus:

13.41. To men heaven and earth seem to unite at a distance of no more than ten *li*. That is the effect of the distance, for they do not come together in fact.[27]

That heaven and earth seem to touch each other at the horizon everyone can observe, but the point here is that they are thought of as two parallel planes, the joining of which at the horizon is an optical illusion. Curiously, what Wang Chong calls the apparent distance of the horizon, ten *li*, is just about the real distance (i.e. on

[27]Forke (1907), 261.

Fig. 13.8 Calculation of the real distance of the horizon

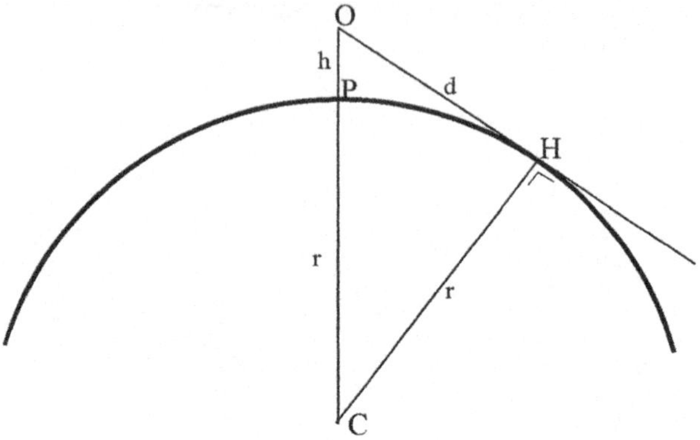

a spherical earth) of the horizon (4.66 km) for a person of an eye height of 1.7 m (Fig. 13.8).[28] It is amazing that Wang Chong had this right idea about the distance of the horizon, for the distance of the horizon is a main problem in all flat earth cosmologies.

The *Zhou bi* places the horizon at the limit of men's range of visibility (which in the below quoted text of Zhao Shuang is called "the farthest extent of the human eye's gaze"), much farther away than Wang Chong's apparent distance of the horizon of ten *li*, namely at 167,000 *li*, as we shall see. The same explanation of the phenomenon of the optical illusion of the horizon as the limit of our range of visibility points is made by Zhao Shuang, a third century commentator on the *Zhou bi*:

13.42. It is not that they really join, rather that the human eye is at the farthest extent of its gaze, and so heaven and earth join.[29]

Accordingly, the phenomena of sunrise and sunset (and the risings and settings of the other heavenly bodies) are optical illusions as well. This can be illustrated by the flight of an airplane (although of course these machines did not exist in the time of the *Zhou bi*). When an airplane is right above our head and not too far away, e.g. when it has just started from the airport, we can see it as a rather big object in the air. But when it flies away, it seems to approach the horizon before it becomes invisible. In the *Lun heng*, Wang Chong uses another image, a torch disappearing out of sight:

[28]The derivation of the formula for the distance d to the horizon of a spherical earth is found by Pythagoras' theorem:

In the right-angled Δ CHO, where C the center of the earth, r is the earth's radius, O the observer's eye, OP = h the distance from the observer's eye to the earth, and H the horizon. Then $d = \sqrt{(r+h)^2 - r^2} \rightarrow d = \sqrt{r^2 + 2rh + h^2 - r^2} \rightarrow d = \sqrt{2hr + h^2} \rightarrow d = \sqrt{h(2r+h)}$.
When we insert the radius of the earth (6378 km) for r, and the height of the observer (0.0017 km) for h, then $d \approx 4.66$ km.

[29]Quotation from Cullen (1996), 221–222.

13.43. When we behold the sun setting, he does not set either, it is also the distance. (...) Let a man take a big torch, and walk at night on a level road, where there are no gaps. He will not have walked to a distance of one *li* from us, before the light of the fire is gone out. It does not go out, it is the distance. In the same manner the sun revolving westward and disappearing does not set.[30]

The distance from which an object is still visible depends on its size and on its brightness, as the torch in the example of Wang Chong. The brightest object of all is the sun, and even this disappears out of our sight because of the distance, a phenomenon that we are used to call the setting of the sun at the horizon. The same holds for the other heavenly bodies, even though they are not by far as bright as the sun. The *Zhou bi* does not explain how, if the visibility of an object partly depends on its brightness, it is possible that we see stars at the horizon, which are as far away as the sun but by far not as bright as the sun.

An obvious objection against the interpretation of the phenomena of the rising and setting sun as an optical illusion is that the sun, just before it disappears at the horizon, precisely because of an optical illusion, looks much bigger than at noon, whereas the airplane gets smaller and smaller before disappearing at the horizon. I do not know of a serious Chinese answer to this objection. Wang Chong's remark[31] that some savants say that the rising and setting sun are near and the sun at noon far away cannot hold for the flat heaven of the *gai tian*, because there the opposite is the case. The oblique distance from the observer to the sun varies with the season; in the situation of Fig. 13.4 it is 100,000 *li* (the hypotenuse of the big triangle), while in Fig. 13.5 the oblique distance between Zhou and the sun at the summer solstice is $\sqrt{16,000^2 + 80,000^2} \approx 81{,}584$ *li*. The distance to the horizon, on the other hand, is always 167,000 *li* according to the *gai tian*, as will be discussed below. Wang Chong's own solution, though ingenious, can hardly be taken seriously:

13.44. When the sun is culminating, the brightness of daylight makes him appear small, and when the sun is rising or setting, daylight is fading, and he looks larger in consequence. In the same manner, a fire looks small at day-time, but big at night. (...) When the sun approaches the horizon, and is about to set, his light fades, and he appears bigger.[32]

[30]Forke (1907), 261–262. Wang Chong's explanation holds only for the setting sun, because according to him, "when the sun rises, he is near, when he sets, he is far and becomes invisible. Hence the term setting (...)" (Forke 1907, 261).

[31]Cf. Forke (1907), 263.

[32]Forke (1907), 263 and 264–265.

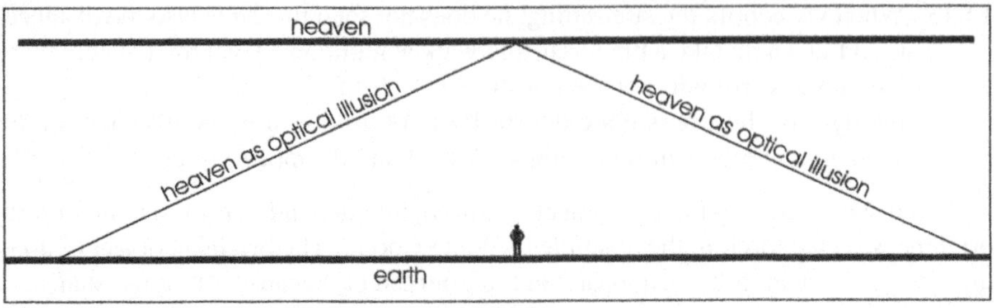

Fig. 13.9 First suggested interpretation of the heaven as optical illusion

Questionable Interpretations of the Heavens as an Optical Illusion

We may conclude that the optical illusion is conceived of as twofold: (1) heaven and earth seem to touch one another at the horizon, although in the *gai tian* reality they are two parallel planes, and (2) the sun seems to set at the horizon, but in the *gai tian* conception the sun does not set but simply disappears out of our sight (and *mutatis mutandis* for the rising sun). How precisely must we visualize this optical illusion? It cannot be that the sun at the horizon is the only optical illusion, whereas the sun at all other times of the day could be seen at its right place. There must be some gradual decline by which the sun is seen lowering and attaining the horizon, although in the *gai tian* reality it is always at the same 80,000 *li* from the surface of the earth. The first possibility that comes to mind is that heaven as an optical illusion must behave as in Fig. 13.9, a cross-section of heaven and earth. In this option, the resulting three-dimensional heaven as optical illusion would have a conical shape.

Perhaps a picture like this inspired the simile of the heaven as a Chinese hat. When we imagine the picture of Fig. 13.9 in three dimensions, the heaven as an optical illusion will appear as a huge cone-shaped hat all around the observer. However, there is something uncomfortable with this picture: the heaven would seem to make a sudden angle in the zenith, and all celestial bodies would be seen as optical illusions, attached to the heaven that would appear as a whole as an optical illusion, except for one point in the zenith. Another possibility would be to let the slope of the heaven be more gently, in a huge curvature, as in Fig. 13.10. The three-dimensional shape of the heaven as optical illusion would be a kind of huge inverted wok, as worded by Wang Chong.

13.45. Heaven appears to us in the shape of a bowl turned upside down.[33]

This looks somewhat better, but here, too, all celestial bodies and the heaven as such would be seen as optical illusions. Moreover, Wang Chong's image of an inverted bowl is somewhat confusing, just like that of Fig. 13.9, because they both

[33]Forke (1907), 261.

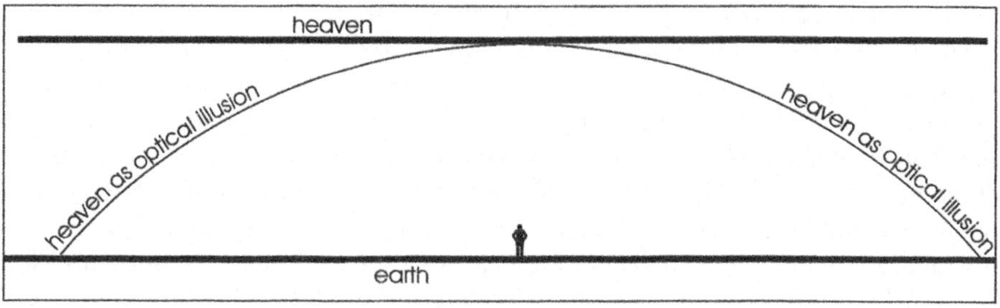

Fig. 13.10 Second suggested interpretation of the heaven as optical illusion

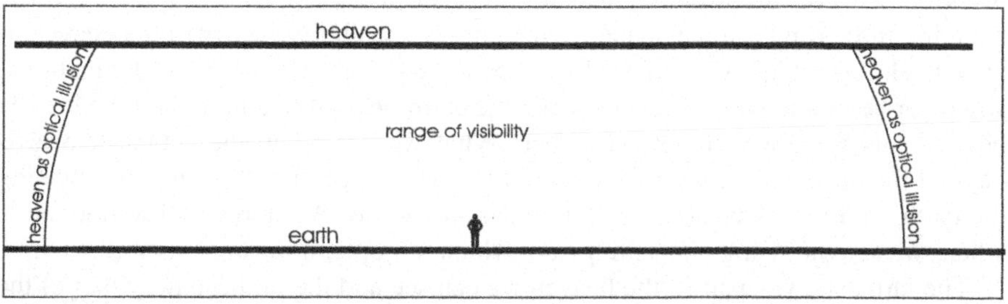

Fig. 13.11 My interpretation of the heaven as an optical illusion, in line with the ideas of the *Zhou bi*

look like another version of the image of a rain hat or canopy. These images, however, are not used in the *Zhou bi* to indicate the heaven as an optical illusion, but to emphasize that the heaven is a circular plane, as explained at the beginning of this chapter. The main reason, however, why I think another interpretation of the heaven as an optical illusion is needed is the applicability of the shadow rule. As we have seen, the measurement of the height of the heavens by means of the measurement of the height of the sun or the pole resulted always in 80,000 *li*. It is inconceivable how this result could be received with a heaven that would be seen as bent or curved as in Figs. 13.9 or 13.10. Therefore, I will argue that in the *Zhou bi* the conception of the heaven as an optical illusion looks like Fig. 13.11, in which it is linked with the idea of the limited range of visibility.

The Heaven as an Optical Illusion and the Range of Visibility

In order to understand the concept of range of visibility, which in the *Zhou bi* is called "the distance to which human vision extends," and which Zhao Shuang expresses as "the farthest extent of the gaze of the human eye" (see text 13.42), I will make use of an image. When I am in a room without windows (and with the light on) my vision is limited physically by the four walls around me, the ceiling above my head, and the floor under my feet. In the *gai tian* model, my vision is limited again, by the ground under my feet and by the heaven above my head. But around me my vision is also limited, even though there are no physical walls. My power of vision does not extend infinitely, but has an absolute limit beyond which I cannot see. This limit shows itself where the two planes of earth and heaven come together at the horizon. The horizon defines the limit of my range of visibility. If there were not the flat earth underneath me and the flat heaven above me, the complete shape of my range of visibility would display the shape of a sphere around me. Since my view is limited above me by the heaven and below me by the earth, we get the actual extent of my range of visibility when we draw between the earth and the heaven a part of a circle with the observer in the center and with a radius that equals the distance between the observer and his horizon, as is shown in Fig. 13.11. The resulting three-dimensional shape of this range of visibility, which is the heaven as an optical illusion, is a slice of a sphere. Each of us bears such a slice with him or her wherever he or she goes. We cannot see the earth beyond this limit, and the heaven beyond this limit we cannot see either. What we see when the actual heaven is beyond this limit is the optical illusion of the heaven bending toward and touching the earth, and this is what we call the phenomenon of the horizon. The horizon is where we see the rising and setting sun, which are optical illusions as well.

The farthest away part of the heaven we can see and the farthest away part of the earth we can see touch one another at the horizon. Therefore, the limits of the range of visibility must be identical with the optical illusion of the heaven. The idea of the heaven as an optical illusion was not always well understood, as a critical objection by Huan Tan—in his *Xin Lun* around the beginning of the Christian era—shows. He relates how he and Yang Ziyun were sitting on a veranda. This meeting probably took place at daybreak.[34] Huan Tan tells:

13.46. Because of the cold we turned our backs to the sun, which warmed them for a while, and then the sun's rays moved away, and our backs were no longer warmed by them. I used this as an illustration for Ziyun, saying 'If heaven really turned like a cover, carrying the sun towards the west, its rays should still be shining under this veranda, but just have moved a little towards the east. As this is not the case, on the contrary [the facts] correspond to the methods of the *hun tian* school. Ziyun thereupon destroyed [the device] he had made.[35]

As far as I can see, there was no compelling need for Yang Ziyun to destroy his device, for he could have answered that his opponent did not take into account the optical illusion which makes heaven and earth look like touching each other at the horizon and makes the sun seem to rise and set. Not only the direction in which we

[34]Cf. Cullen (1996), 60, n. 60: "It will help us to understand the references to the movement of the sun if we recall that court business normally began at dawn."

[35]Quoted from Cullen (1996), 60.

see the rising and setting sun is an optical illusion, the direction from where its warmth is a felt to come when the sun rises or sets is an illusion as well.

The Interrelation of the Range of Visibility and the Area of Sunlight

In order to know when we see the sun at its real (*gai tian*) place on the heaven and when we see it as an illusion on the inclined heaven, we must know the size of our range of visibility or, in other words, we must know how far away the horizon is. For this we will have to consider another concept, that of the area of sunlight, because according to the *Zhou bi* the range of visibility is identical with the area of sunlight, which is called in the *Zhou bi* "the extent of solar illumination" or "the area illuminated by the sun:"

13.47. (#B25) (. . .) the illumination of the sun extends 167,000 *li* to all sides. The distance to which human vision extends must be the same as the extent of solar illumination.

The formulation "illumination of the sun" lays bare a certain ambiguity in the concept of the area of sunlight. In the *Zhou bi*, its meaning is always the two-dimensional circular area that is thrown by the sun on the surface of the earth and not the three-dimensional shape of sunlight between the sun and the earth. In the literature, the area of sunlight is usually explained by supposing that the sun functions as a flashlight or a searchlight, or like the light cast as a conical beam by a lampshade, throwing a circle of light on the surface of the earth.[36] Within the circle of light, which is the area of sunlight, it is day, and outside this circle it is night. Interpreted like this, the three-dimensional shape of sunlight of the *gai tian* model would look like a Chinese rain-hat, as illustrated in Fig. 13.12.

That the range of visibility is equal to the area of sunlight can be explained with the help of Fig. 13.12. An observer A, standing outside of the area of sunlight will not see the sun; for him it is night. When he approaches the rim of the area of sunlight (or when the area of sunlight approaches him), as in B, he will see the sun (albeit as an optical illusion) at the horizon. This means that the radius of his range of visibility equals the radius of the area of sunlight, as was said in #B25 (text 13.47). When he would enter the area of sunlight further, he would stand in broad daylight, as in C, and if he would go further southwards, he could even come at a subsolar place at the northern tropic, where the sun once a year, at the summer solstice, is right above his head. Figure 13.13 is a three-dimensional picture of this *gai tian* model with the

[36]See Needham (1970), 211: "as if by a kind of searchlight-beam." Nakayama (1969), 30: "like that cast by a lampshade." The same image also in Lan-ying Tseng (2011), 113 fig. 5.11. Cullen avoids any suggestion of its shape.

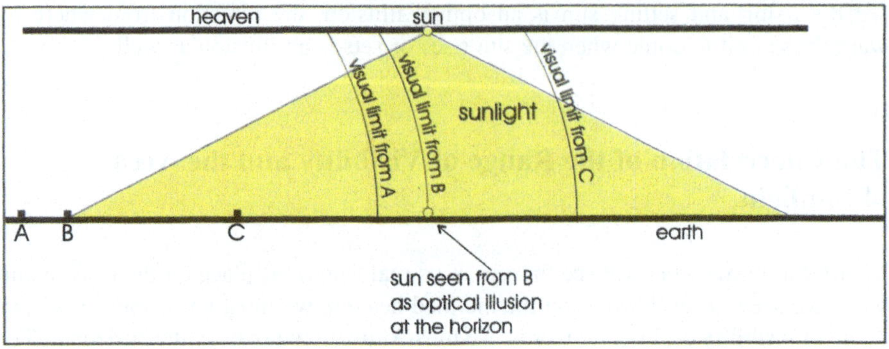

Fig. 13.12 Entering the allegedly conical shape of sunlight in the usual interpretation

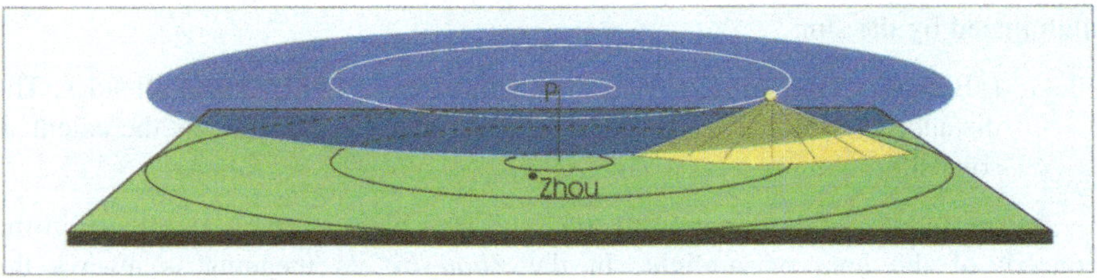

Fig. 13.13 Three-dimensional impression of the usual interpretation of the *gai tian* conception of heaven and earth with an allegedly conical shape of sunlight; it is night in Zhou

circle of the heaven over the square of the earth, and the conical shape of sunlight in between.

Another Interpretation of the Three-Dimensional Shape of Sunlight

The text of the *Zhou bi* does not mention any shape of the sunlight, but on second thought, the image of the light of the sun as a conical beam, a kind of searchlight circling over the surface of the earth cannot be right. One wonders whether the ancient Chinese would not have thought that the light of the sun shines towards all sides, just like a candle, or like the torch that becomes invisible because of the distance in the story told by Wang Chong. It is easy to read this story as meaning that the light of the torch has a certain limit beyond which it cannot shine, and this could hold as well for all other lights, including the sun. Moreover, a curious consequence of the representation as in Figs. 13.12 and 13.13 is that if someone (or a bird) could fly high up in the sky near the sun, he would nevertheless be in the dark. For these reasons, and as a kind of counterpart of our range of visibility, which I described above as a virtual sphere around any observer, I think better imagine the light of the

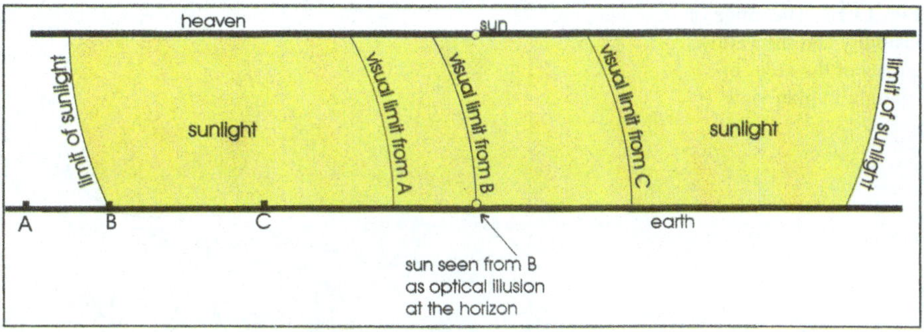

Fig. 13.14 Entering the slice-of-a-sphere shaped sunlight in my interpretation

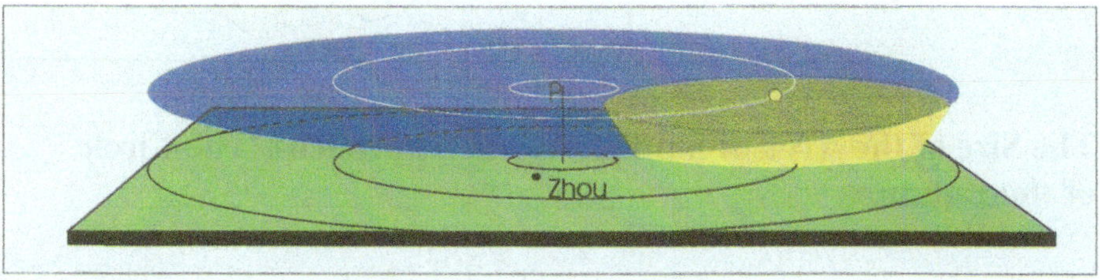

Fig. 13.15 My interpretation of the *gai tian* conception of heaven and earth; it is night in Zhou

sun with a spherical limit, beyond which there is no sunlight. The sphere of sunlight is also cut of above by the plane of the heaven and below by the earth's surface, which results in a slice-of-a-sphere shape of sunlight between heaven and earth. I surmise that this is more like what the author of the *Zhou bi* had in mind. It is illustrated in Fig. 13.14. An observer at A is outside the area of sunlight; for him it is night. An observer at B sees the rising (or setting) sun—as an optical illusion—at the horizon. An observer at C stands in broad daylight.

In Fig. 13.15, the same idea is rendered in a three-dimensional view.

Fortunately, for all calculations of the *Zhou bi* discussed here it makes no difference which version one prefers, because both options result in the same circular area of sunlight on the flat earth. In the plan view of Fig. 13.16 I have rendered the same situations as in the cross-sections of Figs. 13.12 and 13.14: in the yellow circle on top (A), representing the area of sunlight, the sun is too far away to be seen by an observer at Zhou (the little square to the south of the pole P), so that for him it is night; in the right yellow circle below (B) he just sees the rising sun, and in the left one (C) he stands in broad daylight. In a plan view like this, the sun's orbit and the sun itself at different positions on the same day (the little circles in the center of the three big circles of the areas of sunlight) are projected upon the earth's surface. This picture, by the way, also makes clear that an observer at Zhou cannot see the sun when it is at the other side of the pole: then the sun it beyond his range of visibility.

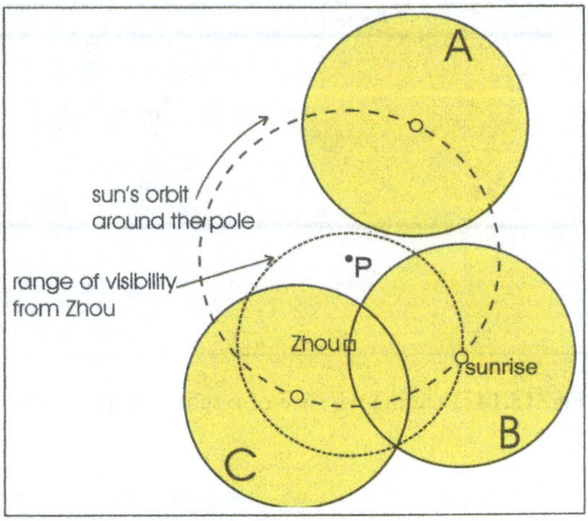

Fig. 13.16 The range of visibility and the yellow circles of the areas of sunlight in plan view

The Size of the Area of Sunlight (First Approach); The Circle of the Equinox

The *Zhou bi* contains an absolute measure for the area of sunlight (167,000 *li*), and thus also for the range of visibility, but we have not yet seen how this number was calculated. The radius of the area of sunlight (and thus of the range of visibility) is also said to be equal to the radius of the circle of the equinox:

13.48. (#B22) From the division of day and night at the autumn equinox to the division of day and night at the spring equinox, there is never sunlight at the subpolar point. Therefore, at the time of division of day and night at the spring and autumn equinox, the area illuminated by the sun reaches just up to the pole. This is the equal division of *Yin* and *Yang*.

13.49. (#B24) Therefore the illumination of the noon sun at the spring and autumn equinoxes reaches north to the subpolar point, and the illumination of the midnight sun likewise reaches south to the pole. This is the time of division between day and night.

Cf. in the *Lü shi chun qiu* (239 BC)[37]:

13.50. At the summer solstice the sun moves along the closest track and reaches the highest point. Beneath the pivot there is [then] no [alteration of] day and night.

This is what an observer at the north pole of a spherical earth would see during the equinoxes: the sun, orbiting on the equator revolves around the horizon. But how did the Chinese come to the same conclusion on their flat earth? Obviously, they had noticed that the line from an observer at Zhou to the celestial pole and the line from the same observer to the sun due south during the equinox make an angle of 90°.

[37] Quoted from Cullen (1996), 51.

They concluded that the line to the celestial pole from an observer standing at the (north) pole and the line from the same observer to the sun at noon at the equinox must also make an angle of 90°. This entails that an imaginary observer at the subpolar point would see the equinox sun on the horizon. Since the radius of the circle of the equator equals 178,500 *li*, as we have already calculated, the radius of the area of sunlight and the radius of the range of visibility must also be 178,500 *li*. Cullen mentions the same identification of the three circles: "This accurate statement of polar conditions implies that the range of the sun's rays should be equal to the radius of the equinoctial orbit of the sun round the pole (...) which is 178,500 *li*," and: "the range of the sun's rays, which is the same as the extent of human vision."[38]

The Size of the Area of Sunlight (Second Approach); The *xuan ji*

Here we are confronted with a strange complication. Notwithstanding these clear utterances in #B22 (text 13.48) and #B24 (text 13.49), the *Zhou bi* gives in #B25 (text 13.47) the somewhat smaller number of 167,000 *li* for the radius of the area of sunlight instead of the 178,500 *li* that we would expect. We must therefore assume that the range of visibility, which is the same size as the area of sunlight, also extends 167,000 *li* in all directions, while the radius of the circle of the equinox is 178,500 *li*. Cullen suggests that the deviant number of 167,000 *li* was due to the fact that the Chinese astronomers wanted to get at the perfect number of 810,000 *li* ($81 = 3^4$) for the outer limit of the region ever illuminated by the sun. They reached this number by adding twice the radius of the circle of the winter solstice (the southern tropic) to twice the radius of the area of sunlight, which should therefore be 167,000 *li*, because then we get: $2 \times 238{,}000$ li $+ 2 \times 167{,}000$ *li* $= 810{,}000$ *li*.[39]

I would like to suggest another origin of this difference. Perhaps, the ancient Chinese astronomers discovered that when they considered the distance between Zhou and the subpolar point (103,000 *li*) on the one hand,[40] and the distance between Zhou and the subsolar point at noon at the equinox (75,500 *li*) on the other hand,[41] the angle between the line from Zhou to the pole and the line from Zhou to the sun at noon at the equinox was more than 90° (about 95.5°, see Fig. 13.17).

This angle, however, should be 90° to allow the sun to stand exactly on the horizon for someone at the subpolar point during the equinox, as required in #B22 (text 13.48), where it was said that "at the spring and autumn equinox, the area illuminated by the sun reaches just up to the pole." The ancient Chinese astronomers, I imagine, noticed that when they used as reference point a star that orbited around the pole at a

[38]Cullen (1996), 124.

[39]See ibidem.

[40]See Fig. 13.6 and cf. texts 13.31(#F4), 13.8 (#B14), 13.9 and 14.28 (#B33).

[41]See Fig. 13.6 and texts 13.26(#F9) and the shadow lengths in 13.27 (#H3, 7) and 13.28 (#H2, 19).

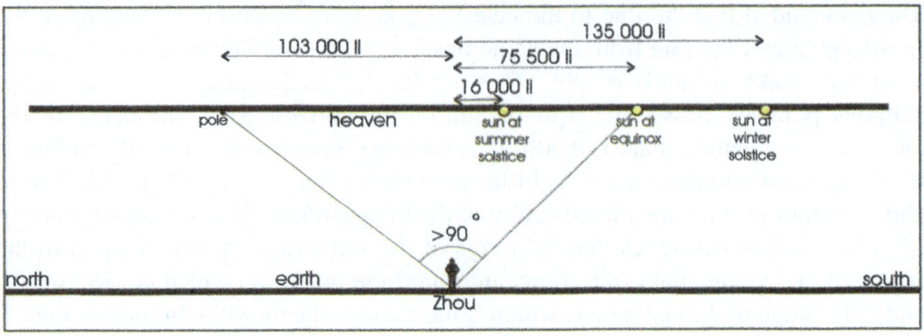

Fig. 13.17 With the *Zhou bi* numbers, the angle between the sun at the equinox and the pole is more than 90° (For a picture of the heavenly distances, similar to this, see Cullen 2017, 210, Fig. 5.10)

distance of 11,500 *li* and which they called the *xuan ji*, and then measured the angle between this star at the southernmost point of its orbit and the sun at noon at the equinox, the angle had the required value of 90°. Actually, the angle still was slightly too large, about 92°, but we must take into account that the measurement of the altitude of the *xuan ji* had to be done at night which would have hampered its accuracy (the distance of the sun during the equinox was not the result of observation but a purely mathematical interpolation, as we have seen). For clarity's sake, I repeat the above-quoted statement (text 13.30) of the *Zhou bi*, in which this star is mentioned, and add two others.

13.51. (#F2) To fix the pivot of the north pole, the center of the *xuan ji*, to fix the center of the heaven (...): At the winter solstice, at the time when the sun is at *you*,[42] set up an eight-*chi* gnomon, tie a cord to its top and sight [along the cord] on the large star in the middle of the north pole [constellation]. Lead the cord down to the ground and note [its position].

13.52. (#F6) The diameter of the *xuan ji* [circle] is 23,000 *li* and its circumference is 69,000 *li*. This [means that] the *Yang* is cut off and the *Yin* manifests itself [within this region], so that it does not give birth to the myriad [living] things.

13.53. [To speak of the sunlight] reaching the pole [at the equinoxes] means [reaching] the limit of the *xuan ji* where the *Yang* is cut off and the *Yin* manifests itself (...).[43]

The commentary of Zhao Shuang is clear: "reaching the pole" means "reaching the limit [of the orbit] of the *xuan ji*." Perhaps the best way to understand what the ancient Chinese astronomers had in mind is not to consider of the celestial pole as a point but as a relatively small circular area with a radius of 11,5000 *li*. This is what #F2 (text 13.51) seems to imply with the words "the pivot of the north pole, the

[42]Due east of the pole (as seen from Zhou), see Cullen (1996), 191 n. 213.

[43]From Zhao Shuang's commentary on the *Zhou bi*, quoted from Cullen (1996), 126.

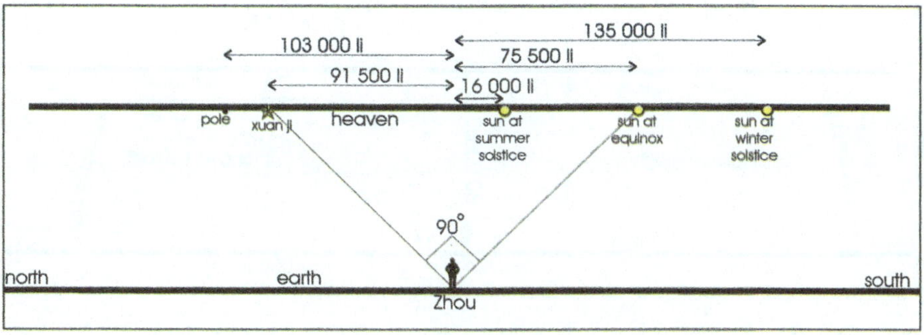

Fig. 13.18 The angle between the sun at the equinox and the *xuan ji*

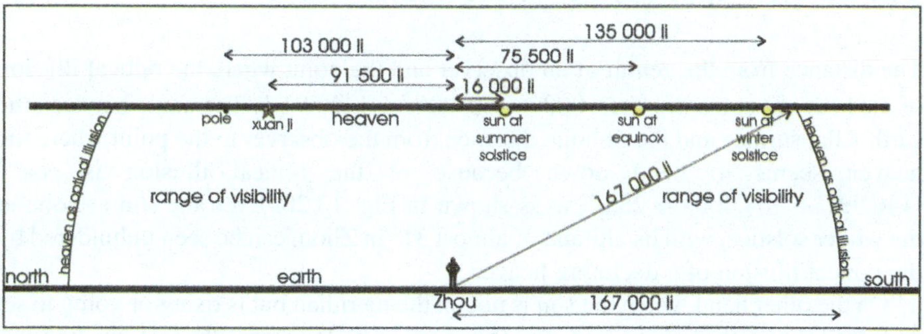

Fig. 13.19 The heavenly bodies and the range of visibility

center of the *xuan ji*" and "the middle of the north pole." Cullen adds the word "constellation" between square brackets, but I think the word "region" expresses the meaning better. If the Chinese astronomers conceived of the pole as a region, they could maintain that the sunlight during the equinox was "reaching the pole." The result of this operation was that the area of sunlight (and thus also the range of visibility) was somewhat smaller than the distance from the pole (the center of the *xuan ji* region) to the equinox, respectively 167,000 and 178,500 *li* (respectively 91,500 + 75,500, and 103,000 + 75,000 in Fig. 13.18).

How We See the Sun; The Shadow Rule Once Again

When we fill in the radius of 167,000 *li* for the range of visibility and plot the sun on the southern heaven at noon at the respective distances in the seasons and the pole and the *xuan ji* star at their distances on the northern heaven, we may conclude that all these objects can be observed at their real (*gai tian*) places on the plane of the heaven. Figure 13.19 shows that the calculations of their distances from Zhou do not interfere with the heaven as an optical illusion.

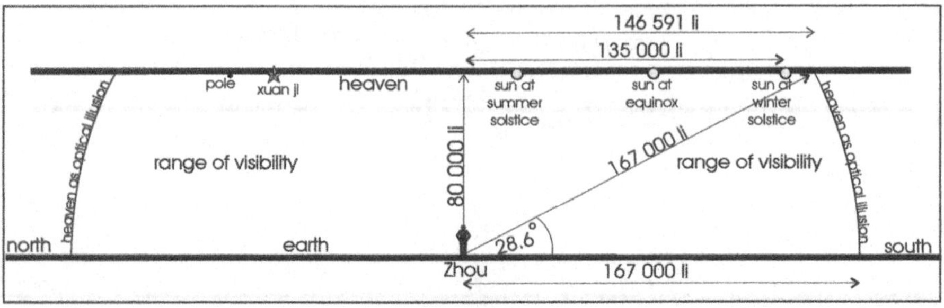

Fig. 13.20 How the point from where the heaven seems to decline is determined

The Limited Applicability of the Shadow Rule

The distance from the zenith of an observer and the point where the optical illusion of the heaven begins is $\sqrt{167,000^2 - 80,000^2} = 146{,}591$ *li*. The angle between the earth's flat surface and the oblique distance from the observer to the point where the heaven seems to bend down because of the optical illusion is $\cos^{-1}$ (146,591 ÷ 167,000) ≈ 28.6°, as is shown in Fig. 13.20. Even the sun at noon at the winter solstice, with its altitude of almost 31° in Zhou, can be seen unhindered by the optical illusion of a declining heaven.

On the other hand, when the sun is not on the meridian but is rising or going to set and stands lower than about 28.6° (in Zhou), we do no longer see the real sun, because in that case its oblique distance is beyond our range of visibility and further away than 146,591 *li* from the zenith of the observer. However, we can still see the sun as an optical illusion, until the sun appears to be at the horizon, at a distance of 167,000 *li*, although the sun in (*gai tian*) reality is still 80,000 *li* above the earth's surface.

The strange consequence of this is that we see the rising or setting sun as an optical illusion at the horizon, although the actual sun is at that time beyond our range of visibility. In Fig. 13.21, I drew the rising and setting sun not only as illusions at the eastern and western horizon, but also at their real (*gai tian*) places on the heaven. In this picture, the situation during the equinox is rendered, when the sun rises due east and sets due west. When oblique distance to the sun is farther away than $\sqrt{80,000^2 + 167,000^2}$ =185,173 *li*, we are no longer able to see it, even as an optical illusion. Then it is night.

The calculations with the shadow rule, as we saw at the beginning of this chapter, always resulted in a heaven 80,000 *li* above the earth. We must now conclude that these calculations concerned the distances of the pole and the sun during the solstices and equinoxes, all of which lie within the area where the heaven is not yet seen as an optical illusion. Heavenly bodies outside that area, more precisely lower than 28.6° above the horizon and between the distances of 146,591 *li* and 167,000 *li* from the zenith of an observer, are seen as optical illusions and therefore cannot be measured with the shadow rule. Not only the rising and setting sun, but also a bright star like Sirius, for instance, is most time of its visibility in the winter season only visible as

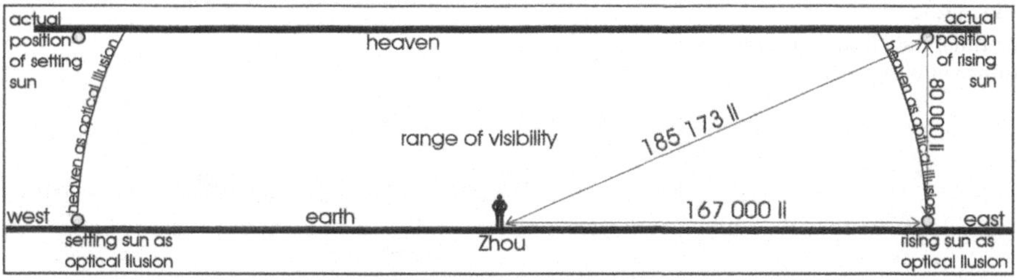

Fig. 13.21 The range of visibility and the rising and setting sun

an optical illusion. The distance of 146,591 *li* from the zenith of an observer is the limit of the applicability of the shadow rule. Heavenly objects further away than 167,000 *li* from the zenith of an observer are invisible.

The Cardinal Directions

The combined concepts of the area of sunlight and range of visibility yield the most brilliant feature of the *gai tian* model: it can account for the differences in time at different places on earth, which is impossible in the Presocratic flat earth system with a (hemi)spherical heaven, where the sun rises all over the earth at the same time. This feature can be explained by noticing that in the *gai tian* model, for an observer anywhere on earth the direction of "north" is always from the observer to the pole, and "south" is the opposite direction. As a result, the four cardinal directions are not absolute, but depend on the position of the observer. Actually, each place on earth has its own north-south and east-west coordinates, as is shown in Fig. 13.22 for three at random placed observers, A (blue), B (red) and C (green). This means, as the picture shows, that all directions are relative to the observer. The "east" and "west" directions are especially ambiguous. For an observer standing still, east and west are perpendicular to his north-south line, but going east or west would mean moving in a circle, always keeping the same distance to the pole. This movement eastward or westward is indicated for observer A with the blue dotted circle.

As Zhao Zhuang says:

13.54. What people call east, west, south and north are no fixed places. Everyone calls the place where the sun rises "east," where the sun is centered "south," where the sun goes in "west" and where the sun is absent "north."[44]

And in the *Zhou bi*:

13.55. (#E5) Now in [each of] these four regions heaven and earth have their four poles.

---

[44]Quoted from Cullen (1996), 222.

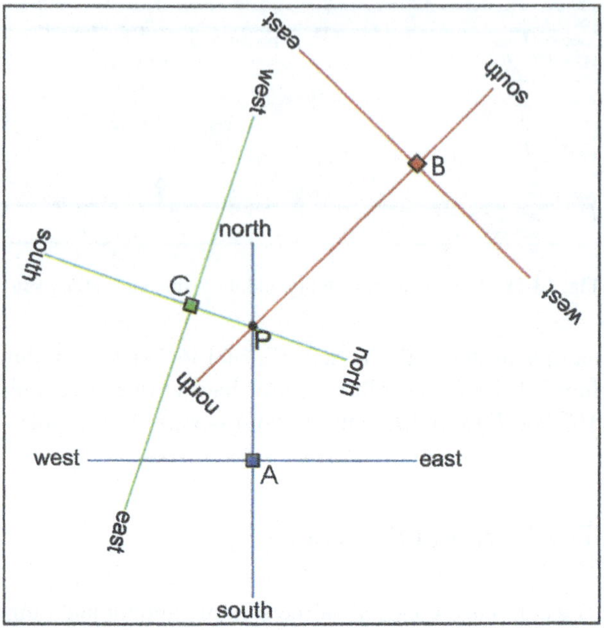

Fig. 13.22 The four cardinal directions for three different observers

The meaning of this somewhat cryptic line is obviously: each region (and in the end, every human being) has its own cardinal directions: north, south, east, end west. Whereas the text of the *Zhou bi* only applies it to the four directions of an observer at Zhou, Zhao Zhuang correctly formulates this crucial point in more general terms. The consequence of Fig. 13.21 is that each observer sees the sun rise and set not only elsewhere than the other observers, but also at a different time. The same point is made in the *Zhou bi*:

13.56. (#E4) Therefore when the sun's rotation has brought it to a position north of the pole, it is noon in the northern region and midnight in the southern region. When the sun is east of the pole, it is noon in the eastern region and midnight in the western region. When the sun is south of the pole, it is noon in the southern region and midnight in the northern region. When the sun is west of the pole, it is noon in the western region and midnight in the eastern region.

With regard to the observation that one man's sunset is another man's sunrise, and that our noon is the midnight of an observer on the other side of the pole, Cullen remarks: "I can think of no likely way in which this view could be the result of actual geographical information."[45] Yet, one gets the impression that this view cannot only be based on theoretical considerations about the implications of the model, but must also be the result of information from travelers translated in terms of the *gai tian* model. However, it is not easy to imagine how those travelers, without a clock or

[45]Cullen (1996), 131.

modern communication devices, could conclude that elsewhere the sun rises earlier or later than at home. Perhaps travelers to the north and south reported that in summer the days in the north last longer than in the south, which entails that the sun rises at different times for people situated at different latitudes. Likewise, people at different degrees of longitude could have reported that a solar or lunar eclipse started, for example, in one place when the sun or moon rose at the horizon, while at another, distant longitude in the east the same eclipse started when the sun or moon was already relatively high in the sky. In this sense, Panchenko provides data on solar eclipses that could have been observed in Greece and in Persia. His examples, however, remain speculative, as no confrontations of different observations have been recorded.[46] Similarly in ancient China: although we have records of observations of eclipses, we have no evidence of such observations and discussions of the same eclipse seen at different longitudes.

Figures 13.13, 13.15, 13.20, and 13.21 illustrate the revolutionary character of the *gai tian* system. Unlike the ancient Greek systems with their oblique celestial axis and the heavenly bodies that pass under the earth, the *gai tian* system is able to account for time differences on a flat earth. For example, it may happen that the sun rises for an observer in A in Fig. 13.21, while it is noon for an observer at the subsolar point due west of A (Cf. also Fig. 14.13 in the next chapter). Figure 13.21 can also clarify why the shadow rule is only applicable on the north-south line, which is the meridian of the observer. The north-south line is the only one on which the sun can be seen at the same time (noon), while it is a different time of the day at any other point outside the north-west line.[47]

References

Chatley, H., "The Heavenly Cover." A Study in Chinese Astronomy. *Observatory* 61 (1938), 10–21.

Cullen, C., A Chinese Eratosthenes of the Flat Earth. Appendix A in: J.S. Major, *Heaven and Earth in Early Han Thought*. Albany: State University of New York Press 1993.

—, *Astronomy and Mathematics in Ancient China: the Zhou bi suan jing*. Cambridge: Cambridge University Press 1996.

—, *Heavenly Numbers: Astronomy in Early Imperial China*. Oxford: Oxford University Press 2017.

Forke, A., *Lun-Hêng. Philosophical Essays of Wang Chung*. London: Luzac & Co 1907. Available on the Internet: https://archive.org/details/lunhng00forkgoog.

Lan-ying Tseng, L., *Picturing Heaven in Early China*. Cambridge (Mass.) and London: Harvard University Asia Center 2011.

Major, J.S., *Heaven and Earth in Early Han Thought*. Albany: State University of New York Press 1993.

[46] See Panchenko (1999), 37–38.

[47] Cullen (1996), 41, insists that Chinese astronomers "took special note of what we call meridian transits." And more specifically: "all the solar shadows mentioned in the *Zhou bi* are noon values" (id., 103).

—, *The Huainanzi*. New York: Columbia University Press 2010.

Nakayama, S., *A History of Japanese Astronomy. Chinese Background and Western Impact.* Cambridge, MA: Harvard University Press 1969.

Needham, J., *Science and Civilisation in China, Vol. 3. Mathematics and the Sciences of the Heavens and the Earth*. Cambridge: Cambridge University Press 1970.

Panchenko, D., The Shape of the Earth in Archelaus, Democritus and Leucippus. *Hyperboreus* 5 (1999), 22–39.

—, The City of the Branchidae and the Question of the Greek Contribution to the Intellectual History of India and China. *Hyperboreus* 8 (2002), 244–255.

Yin Difei, 西汉汝阴侯墓出土的占盘和天文仪器 / Xi Han Ru Yin hou mu chu tu de zhan pan he tian wen yi qi (The divining board and the astronomical instrument excavated from the Western Han tomb of the Marquis of Ru Yin). *Kaogu* 5 (1978), 338–343.

Chapter 14
An Ancient Chinese Flat Earth Cosmology: Details and Calculations

Contents

Based on the introductory representation in the previous chapter, this chapter provides more details, calculations, and an extrapolation.

The Location of Zhou

According to the *Zhou bi*, the distance between Zhou and the pole, or more precisely, the subpolar point, is 103,000 *li*,[1] which equals 42,827.4 km. This absurd distance is the consequence of the shadow rule which says that 1 cun shadow more or less equals 1000 *li* north- or southward.[2] Cullen has tried to determine the actual latitude of Zhou. He obtains a value of 35.33°, which is, as he says, easy to calculate.[3] Cullen's easy calculation presupposes that the shadow measurements at the summer solstice (1.6 *chi*, cf. texts 13.18 and 13.19) and at the winter solstice (13.5 *chi*,

[1]Cf. texts 13.8 and 13.29 (#B14), 13.9 and 14.28 (#B33).
[2]Cf. texts 13.10 (#B10), 13.11 (#B12), and 13.12 (#D18).
[3]Cullen (1996), 104–105.

D. L. Couprie, *When the Earth Was Flat*, Historical & Cultural Astronomy,
https://doi.org/10.1007/978-3-319-97052-3_14

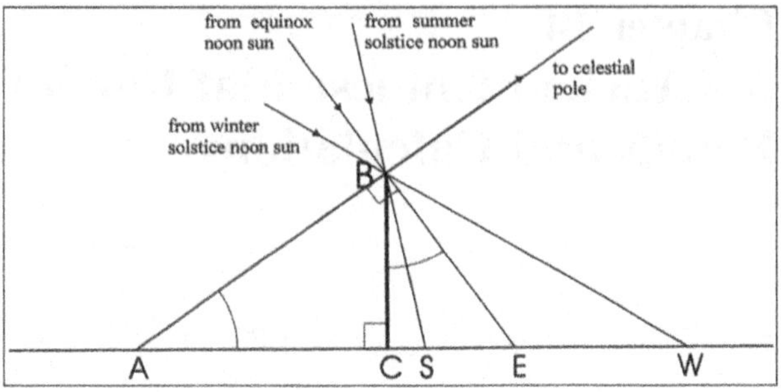

Fig. 14.1 Calculation of the latitude of Zhou (Adapted after Cullen 1996, 105, Fig. 10)

cf. texts 13.23–13.25) with a gnomon of 8 *chi*, were based on actual observations.[4] The calculation goes as follows (see Fig. 14.1): the angle between the gnomon BC and the sunray at the summer solstice (∠CBS) = $\tan^{-1}$ (1.6/8) ≈ 11.31°. The angle between the gnomon and the sun ray at the winter solstice (∠CBW) = $\tan^{-1}$ (13.5 ÷ 8) ≈ 59.35°. The angle between the sunray at the summer solstice and the sunray at the winter solstice (∠SBW) = ∠CBW – ∠CBS = 59.35° – 11.31° = 48.04°. The equinox sunray is the bisector of ∠SBW, so ∠SBE = 48.04° ÷ 2 = 24.02°. ∠CBE, which is the sum of ∠CBS and ∠SBE, = 11.31° + 24.02° = 35.33°. Since Δ BCE and Δ ABE are similar, ∠CBE = ∠BAE = 35.33°, which is the latitude of Zhou.

To an angle of 35.33° belongs a shadow CE of 8 ÷ tan (90 – 35.33) = 5 *chi* 6 *cun* 7 *fen*, which differs from the 7 *chi* 5 *cun* 5 *fen* in the *Zhou bi* (cf. texts 13.26–9.128). Trying to calculate the latitude of Zhou with the help of the length of the shadow at the equinoxes according to the text of the *Zhou bi* results in a latitude for Zhou of $\tan^{-1}$ (7.55 ÷ 8) ≈ 43.34°. The cause of this mistake is that out of the 24 shadow lengths in the table of #H2 only those of the summer and winter solstices were (more or less) based on observation, while the other values were not calculated but interpolated are equal distances between the measured shadow lengths of the gnomon at noon at the winter and summer solstices.[5] We already saw this in the previous chapter, Sect. *The Shadow Rule and the Fundamental Cosmic Measurements*, for the shadow length at noon during the solstices. In the table of #H2, the shadow rule is not expressed as "one *chi* for a thousand *li*", but by the assumption that the shadow length of the gnomon over the year from one winter solstice (when the shadow equals 13.5 *cun*) until the next summer solstice (when the shadow equals 13.5 *cun*) shortens 12 times with the same amount of

[4]The differences between the actual figures for the shadow lengths at the solstices and those of the *Zhou bi* are only slight and could be due to the inaccuracy of the measurements. At the summer solstice, the maximum altitude of the sun at 35.33° N was 78.2° and the shadow length 8 ÷ tan 78.2 = 1 *chi* 6 *cun* 7 *fen* (according to the *Zhou bi*: 1 *chi* 6 *cun*); at the winter solstice, the minimum altitude of the sun at 35.33° N was 59.3° and the shadow length 8 ÷ tan 59.3 = 1 *zhang* 3 *chi* 4 *cun* 7 *fen* (according to the *Zhou bi*: 1 *zhang* 3 *chi* 5 *cun*).

[5]Cf. Cullen (1996), 196: "[the] sole purpose [of this section #H] is to provide a list of noon gnomon shadows for each of the 24 *qi* throughout the year. This is done by linear interpolation between the values for the summer and winter solstices, which are the only data to bear any close relation to observation."

9 *cun*, 9 *fen* and 1/6 *fen*, and then grows again 12 times with an equal amount until the next winter solstice. This version of the shadow rule is formulated in the following lines of section H of the *Zhou bi*:

14.1. (#H1) For the 24 *qi* (...), the decrease or increase [of the shadow] for one *qi* is 9 *cun* 9 *fen* and 1/6 *fen*. The length of the winter solstice shadow is 1 *zhang* 3 *chi* 5 *cun*, and the length of the summer solstice shadow is 1 *chi* 6 *cun*.
14.2. (#H3) For the 24 *qi* (...), the decrease or increase [of the shadow] for one *qi* is 9 *cun* 9 *fen* and 1/6 *fen*. The winter and summer solstices are the beginnings of decrease and increase.
14.3. (#H4) Method:
Set up the winter solstice shadow, subtract the summer solstice shadow, and the difference is made the dividend.
Take 12 as the divisor.
The integral quotient gives the cun.
Multiply the remainder by ten and divide again to obtain the fen.
Make the [final] remainder the numerator over the determinator.

The text of 14.3 (#H4) is somewhat cryptic. First of all, "dividend" and "numerator" are interchangeable terms, and so are "divisor" and "determinator." The first line, expressed in *cun*, results in $135 - 16 = 119$ *cun*. Divided by 12 this makes about 9.917 *cun* in our decimal notation. The term "integral quotient" refers to the integer of the quotient (which is, 9 *cun*). "The remainder" should be taken to mean the quotient minus the integer. This remainder multiplied by ten gives 9.17, of which the integer is 9 *fen*. The expression "divide again" is somewhat obscure and should be understood as something like "take the integer again to obtain the *fen*." "The final remainder" should be taken to mean the number 17, and "making it the numerator over the determinator" something like "divide it by 100", which makes 17/100 or 1/6 *fen*.

When we try to measure the latitude of Zhou by means of the "polar shadow" of 10.3 *chi* (see #B14 in text 13.29), the result is $\tan^{-1}(80{,}000 \div 103{,}000) \approx 37.84°$, whereas to a latitude of 35° belongs a length of the "polar shadow" of 11.4 *chi* ($8 \div \tan 35 \approx 11.4$), as Cullen points out.[6] The distance from the pole to the point in the zenith of Zhou should therefore be 114,000 *li* instead of the 103,000 *li* given by the *Zhou bi*. Cullen's conclusion is that the Chinese somewhat cheated to get "a neatly arranged universe in which the summer and winter paths were related in size by a factor of two," even taken into account that "the celestial pole cannot in any case be observed directly."[7]

[6]See Cullen (1996), 106.
[7]Ibidem.

Measuring the Sun's Diameter

In the *Zhou bi*, not only the distance to the sun is measured, as indicated in the previous chapter,[8] but also the sun's diameter. Two essential distinctions must be kept in mind to ensure a proper understanding of the following. First, the angular diameter versus the actual diameter of the sun. The angular diameter is the angle of an object as seen by an observer. The angular diameter of the sun is about 0.5°. The real diameter of the sun is 1392 million km, but for the ancient Chinese, who believed that the earth is flat and therefore the sun is not at an enormous distance, it must have been much smaller. Second, the oblique distance versus the vertical distance of the sun. The oblique distance of the sun is the distance from an observer to the sun. The vertical distance is the distance from the flat earth to the flat heaven, which according to the *Zhou bi* is always 80,000 *li*. The measurement of the size of the sun in the *Zhou bi* sun is otherwise than that discussed in Chaps. 3 and 11. Since in the *gai tian* system, the sun does not turn around the earth but parallel to the earth's flat surface, the Chinese could not use the distance from the observer to the sun at the summer solstice as the radius of the sun's orbit. So, they used this distance in another, ingenious way, by using a sighting tube:

14.4. (#B11) Wait until the base is six *chi* in length, then take a bamboo [tube] of diameter one *cun*, and of length eight *chi*. Catch the light [down the tube] and observe it: the bore exactly covers the sun, and the sun fits into the bore. Thus it can be seen that an amount of eighty *cun* [=8 *chi*] gives one cun of diameter. (...) The oblique distance to the sun from the position of the *bi* is 100,000 *li*. Working things out in proportion, eighty *li* gives one *li* of diameter, thus 100,000 *li* gives 1250 *li* of diameter. So, we can state that the diameter of the sun is 1250 *li* (Fig. 14.2).

The oblique distance of 100,000 *li* from the observer to the sun was calculated using a gnomon of 8 *chi* (80 *cun*) with a shadow length of 6 *chi*, as shown in Fig. 13.4. The calculation of the diameter of the sun is a simple equation: $80 : 1 = 100{,}000 : x \rightarrow x = 1250$. Converted in kilometers, the diameter of the sun is: $1250 \times 0.4158 = 519.75$ km.[9] With regard to this calculation of the sun's diameter, Cullen notes: "The figures given here predict an apparent solar diameter of 43′, which is about 10′ greater than the value actually observed."[10] The calculation behind this remark is that when we take 100,000 *li* as the radius of a circle with the observer as its center, the circumference of this circle $= 2\pi \times 100{,}000 = 600{,}000$ *li*.[11] The apparent diameter of the sun is about

[8]See Figs. 13.4 and 13.5 and the accompanying text.

[9]The calculation given in Couprie (2011), 198–199 and Fig. 16.6 is needlessly complicated and partly makes use of wrong numbers, e.g. for the oblique distance of the sun, although the right number for the diameter of the sun is mentioned on p. 198.

[10]Cullen (1996), 128.

[11]Taking $\pi = 3$.

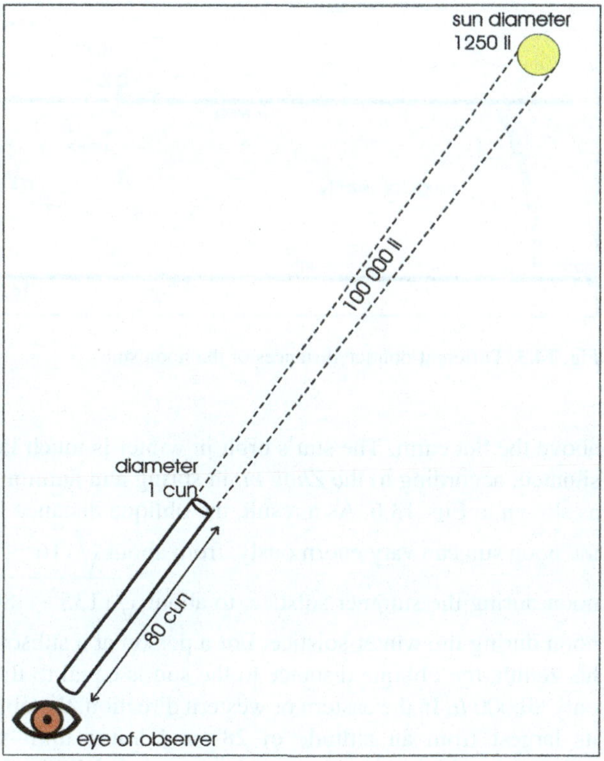

Fig. 14.2 The sighting tube and the diameter of the sun (not to scale) (In Couprie 2011, 199, Fig. 16.6, I needlessly complicated this picture, calculating with similar triangles)

0.5°, which would result in a diameter of the sun of 600,000 ÷ 720 ≈ 833 *li*. By the measurements with the sighting tube in the *Zhou bi*, however, the diameter is 1250 *li*. The apparent diameter of the sun belonging to that number is (1250 ÷ 873) × 0.5 ≈ 0.716°; or expressed in minutes, as Cullen does: 0.716 × 60 ≈ 43′. According to Cullen, this difference must be due to the deliberately sized sighting tube, whose length is exactly equal to the length of the standard gnomon and its bore exactly one *cun*.[12] This means that the Chinese astronomers consciously chose the dimensions of the sighting tube in order to bring them into line with their calculations of the celestial dimensions. This is clear from the fact that the distance of 100,000 *li* belongs to a gnomon of 8 *chi* with a shadow length of 6 *chi* (see again Fig. 13.4). Since the calculation of the distance of the sun was wrong, as explained in the previous chapter, the calculation of the sun's diameter, which depends on it, must also be wrong.

However, there is something more fundamentally wrong with this measurement. In the *gai tian* system, the sun on the flat heaven orbits at a distance of 80,000 *li*

[12]Cf. Cullen (1996), 128.

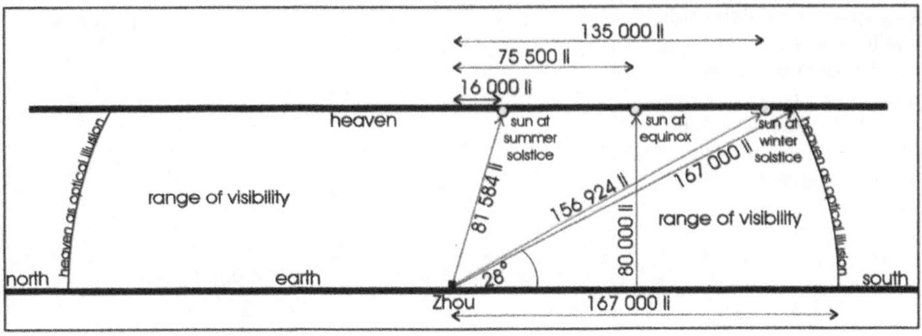

Fig. 14.3 Different oblique distances of the noon sun

above the flat earth. The sun's orbit in winter is much larger than in summer, and is situated, according to the *Zhou bi*, in spring and autumn exactly between these two, as shown in Fig. 13.6. As a result, the oblique distance from an observer in Zhou to the noon sun can vary enormously, from about $\sqrt{(16^2 + 80^2)} \times 1000 = 81{,}584$ *li* at noon during the summer solstice, to about $\sqrt{(135^2 + 80^2)} \times 1000 = 156{,}924$ *li* at noon during the winter solstice. For a person at a subsolar point, i.e. with the sun in his zenith, the oblique distance to the sun is equal to the vertical distance, which is only 80,000 *li*. In the eastern or western direction, the oblique distance of the sun is at its largest from an altitude of 28° and lower until sunrise and sunset, namely 167,000 *li*, which is the limit of the range of visibility (see Fig. 14.3).

We can learn from this analysis that the astronomers of the *Zhou bi* used their sighting tube of 80 cun with a bore of 1 cun in the situation of Fig. 13.4 (when the oblique distance to the sun equals 100,000 *li*) and not at other times of the day or year, because otherwise they would have discovered that they had to choose between two options, both of which would have undermined their ideas of a flat heaven over a flat earth. If they would have believed that the actual size of the sun is always the same, which they would probably have done, then they would have concluded that the sun was always at the same oblique distance from the observer, because in whatever direction and at whatever time of the day or year they would have looked at the sun, they would have been able to use exactly the same sighting tube with its dimension of 80 chi length and 1 cun diameter. This would have been contrary to the presuppositions of their *gai tian* system, according to which not the oblique distance from the observer, but the vertical distance from the earth is always and everywhere assumed to be the same.

On the other hand, if they were to adhere to the presuppositions of the *gai tian* system, they would have been confronted with the strange consequence that the size of the sun varied considerably with time, both during the day and throughout the year, and also with place. If the sun should always be seen from Zhou as having the same angular diameter, then the actual diameter of the sun should vary between 81,584 ÷ 80 = 1019.8 *li* at noon during the summer solstice and 167,000 ÷ 80 = 2087.5 *li* at sunrise or sunset on the horizon. In other words, the actual diameter of the sun on the horizon should be about twice as large as the sun in the zenith. This is purely the result of the system's conceptions and has nothing to do with the optical illusion that the sun seems to be bigger at the horizon, because if we use a sighting tube, that illusion disappears. Moreover, the consequence would have been that for an observer with the sun in his zenith the actual size of the sun would have been much smaller than the same sun seen from Zhou.

The Extension of the Solar Illumination

The *Zhou bi* provides a number of calculations that can be visualized in plan view drawings. Of course, the distances in these calculations and drawings do not have any relation to reality, because they are all ultimately based on the measurement of the sun's height, which was, in its turn, based on the wrong shadow rule. Whether or not the authors of the *Zhou bi* were aware of this, we do not know. Yet, these sections of the *Zhou bi* give the impression that they made their calculations not only to provide a better insight into the movements of the sun and the relation of its area of light to the range of visibility and the seasonal circles, but also just for fun. Anyway, I enjoyed drawing the pictures that visualize the calculations in the *Zhou bi*. We may readily surmise that the ancient Chinese astronomers made similar drawings to illustrate their calculations.

In the Figs. 14.4, 14.5, 14.6, 14.7, and 14.8, we must imagine the flat heavens projected on the surface of the flat earth. The distance from Zhou to the pole is therefore not the oblique distance, but the distance to the subpolar point on earth, and so on for all celestial points and circles. Zhou is indicated as a black square. The dotted circle is the range of visibility from Zhou. The yellow circle is the area of sunlight with the sun always in its center and rotating around the pole. I have added as black dots the points of the summer solstice noon sun, the spring/autumn equinox noon sun, and the winter solstice noon sun, all due south of Zhou. Characteristically, the *Zhou bi* also speaks of "the extent of solar illumination *at midnight* on the winter solstice," and "the extent of solar illumination *at midnight* on the summer solstice" (see texts 14.9 and 14.11 (#B28.1 and 2), quoted below). Accordingly, I added the points of the midnight winter sun, midnight spring/autumn sun, and midnight summer sun, all due north of Zhou. In Zhou, these midnight suns are invisible, because they are beyond the range of visibility of an observer in Zhou. The pole is also marked as a black dot, surrounded by the orbit of the *xuan ji* star. To understand the calculations, readers are requested to consult Figs. 13.6 and 13.18 in the previous chapter. All numbers in the next drawings must be multiplied by 1000. In the explanation of the drawings, I will use the following abbreviations for the distances we already became acquainted with in the previous pages:

AS = 167,000 *li* = the radius of the area of sunlight
ES = 59,500 *li* = the distance between the equator and the southern tropic
NE = 59,500 *li* = the distance between the northern tropic and the equator
NS = 119,000 *li* = the distance between the northern tropic and the southern tropic
PE = 178,500 *li* = the radius of the equator
PN = 119,000 *li* = the radius of the northern tropic
PS = 238,000 *li* = the radius of the southern tropic
PX = 11,500 *li* = the radius of the orbit of the *xuan ji* star
RV = 167,000 *li* = the radius of the range of visibility from Zhou
ZE = 75,500 *li* = the distance from Zhou to the equator
ZN = 16,000 *li* = the distance from Zhou to the northern tropic
ZP = 103,000 *li* = the distance from Zhou to the pole
ZS = 135,000 *li* = the distance from Zhou to the southern tropic
ZX = 91,500 *li* = the distance from Zhou to the orbit of the *xuan ji* star

Some calculations do not need extra drawings, for instance:

14.5. (#B21) [If one measures] south to the summer solstice noon and north to the winter solstice midnight, or south to the winter solstice noon and north to the summer solstice midnight, in both cases the diameter is 357,000 *li* and the circumference is 1,071,000 *li*.

The measurement is from the pole, so that we get: PN + PS or PS + PN; the circumference is calculated with $\pi = 3$.

The ancient Chinese astronomers were well aware of the fact that, as Zhao Shuang expresses it:

14.6. Beneath the north pole, the sun is in sight for 6 months, and out of sight for 6 months. For the 6 months from the spring to the autumn equinox[es] the sun is always in sight, while for the 6 months from the autumn to the spring equinoxes the sun is always out of sight. (. . .) What is called a year is a day and a night below the pole."[13]

The Figs. 14.4–14.8 not only show that they were acquainted with this fact, but also that they were able to account for it on a flat earth. In the previous chapter, we already saw that, quite surprisingly, the radius of the area of sunlight (AS) and the range of visibility (RV) was not given as 178,500 *li* and equal to the radius of the equator (PE), but as 167,000 *li*. The difference is due to a star, *xuan ji*, which allegedly describes a small circle with a radius of 11,500 *li* (PX) around the pole. In Fig. 14.4 is illustrated how the difference between the radius of the area of sunlight and the radius of the equator originates: PE − AS = PX.

[13]Quoted from Cullen (1996), 222.

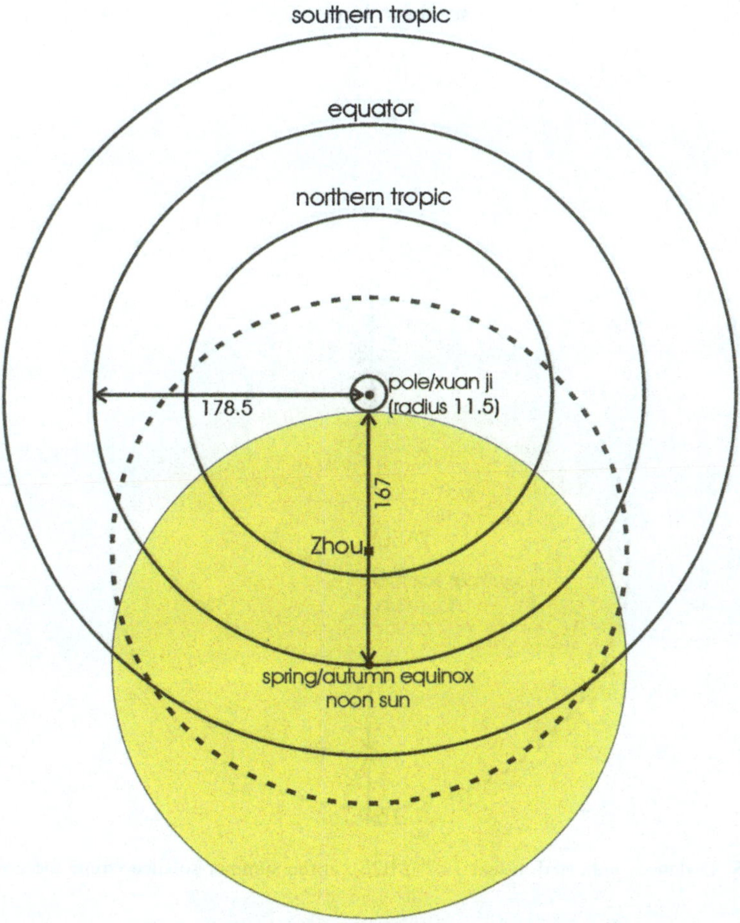

Fig. 14.4 The relation of the area of sunlight to the pole at the equinoxes. For a similar indication of limit the range of visibility from Zhou (the dotted line), see Cullen 2017, 211, Fig. 5.12 (there called: "range of sight" and "limit of sight")

During the summer solstice, the sun circles on the northern tropics and shines all day at the pole. The following relationships are calculated in the *Zhou bi* and shown in Fig. 14.5 as a result of Zhou's being situated 16,000 *li* north from the nearest point of the northern tropic (the distance ZN).

14.7. (#B26) At noon on the summer solstice the solar illumination extends 48,000 *li* south beyond the winter solstice noon. It extends 16,000 *li* south beyond the limit of human vision, 151,000 *li* north beyond Zhou and 48,000 *li* north beyond the pole.

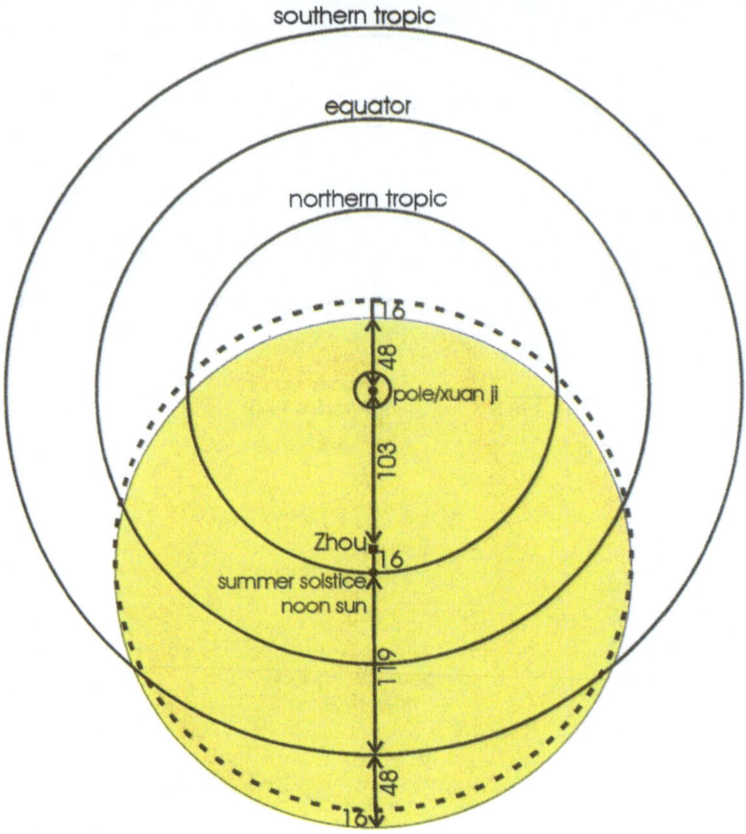

Fig. 14.5 Distances indicated in text 14.7 (#B26), at the summer solstice (some more measures added)

The calculations are: AS = NS + ZN + x → x = 48,000 *li*, AS = ZN + ZP + y → y = 48,000 *li*, and ZP + y = 151,000 *li*. Figure 14.5 also visualizes another calculation:

14.8. (#B25) The extent of vision from Zhou reaches 64,000 *li* north beyond the pole, and 32,000 li south beyond the winter solstice noon point.

Respectively: (48 + 16) × 1000, and (48 − 16) × 1000.

Figure 14.6 shows why we, when we are at Zhou, we can never see the sun when it is on the other side of the pole: even in summer, when the sun is at its nearest, the "midnight sun" will remain beyond our range of visibility and the area of sunlight will not reach Zhou. The number of 96,000 *li* is twice the value for y, found above.

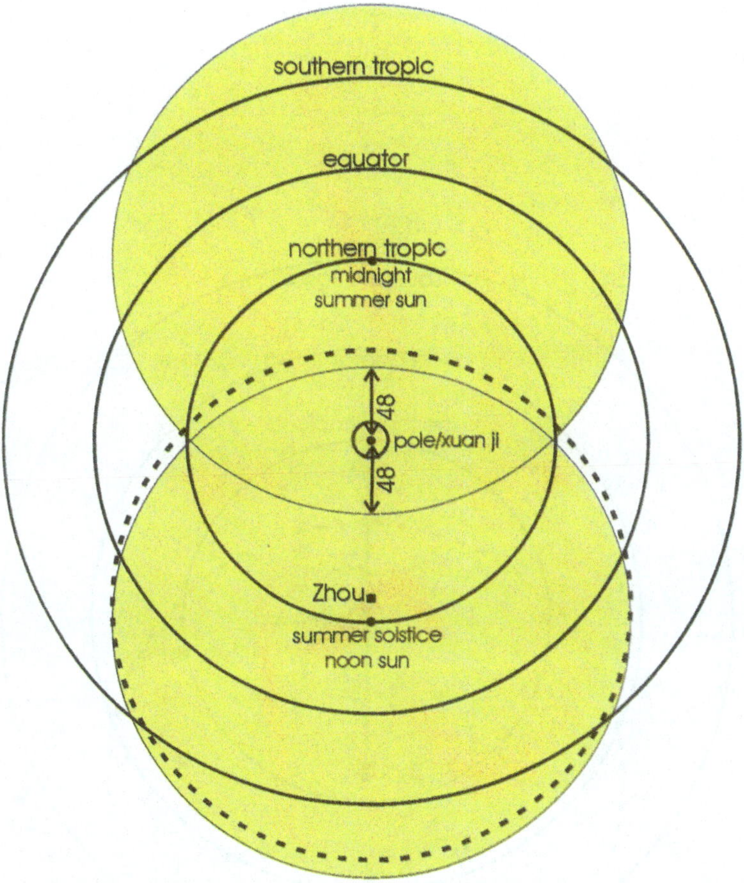

Fig. 14.6 Distances indicated in #B28.1 (text 14.9); the sun at noon and the "midnight sun" at the summer solstice

14.9. (#B28.1) At the summer solstice the illumination of the sun at noon and the illumination of the sun at midnight overlap by 96,000 *li* across the pole.

For the winter solstice, when it is night on the pole all day, the *Zhou bi* first gives calculations for the "midnight sun:"

14.10. (#B27) At midnight on the winter solstice the extent of solar illumination southwards falls short of the limit of vision of the human eye by 7000 *li*, and falls 71,000 *li* short of the subpolar point.

The number 7 in Fig. 14.7 we found already in the two previous drawings. The calculation is: PN − 48,000 *li* = 71,000 *li*; ZP + 71,000 *li* = 174,000 *li*, which is 7000 *li* more than RV.

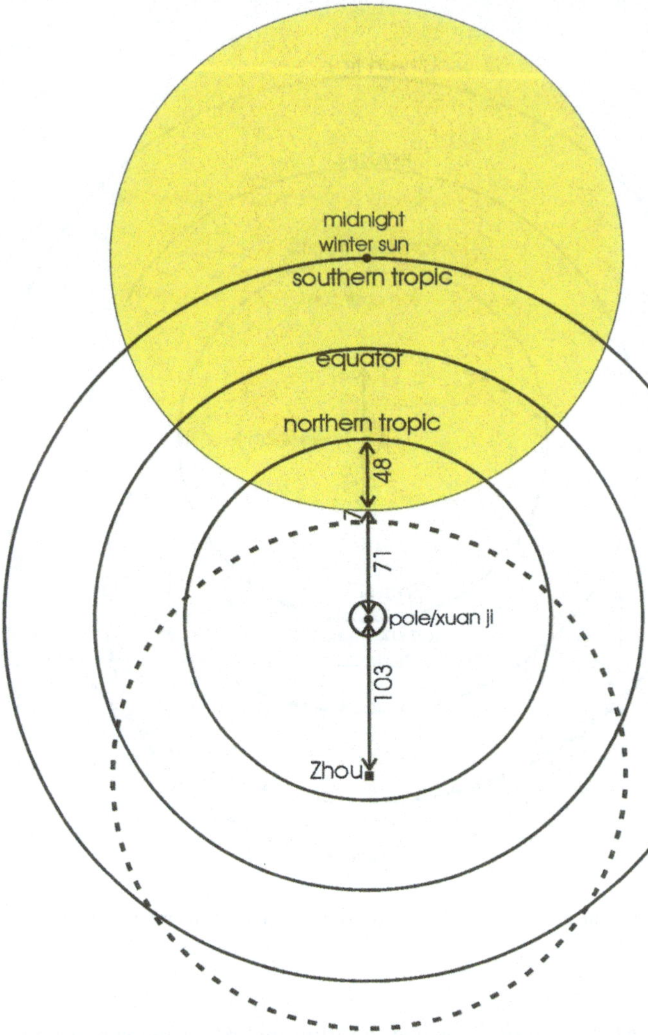

Fig. 14.7 Distances indicated in #B27 (text 14.10); the "midnight sun" at the winter solstice

And finally, the calculations for the sun at noon and the "midnight sun" at the winter solstice follow easily from the previous drawing, as shown in Fig. 14.8.

14.11. (#B28.2) At the winter solstice the illumination of the sun at noon falls 142,000 *li* short of meeting the illumination of the sun at midnight, and falls 71,000 *li* short of the subpolar point.

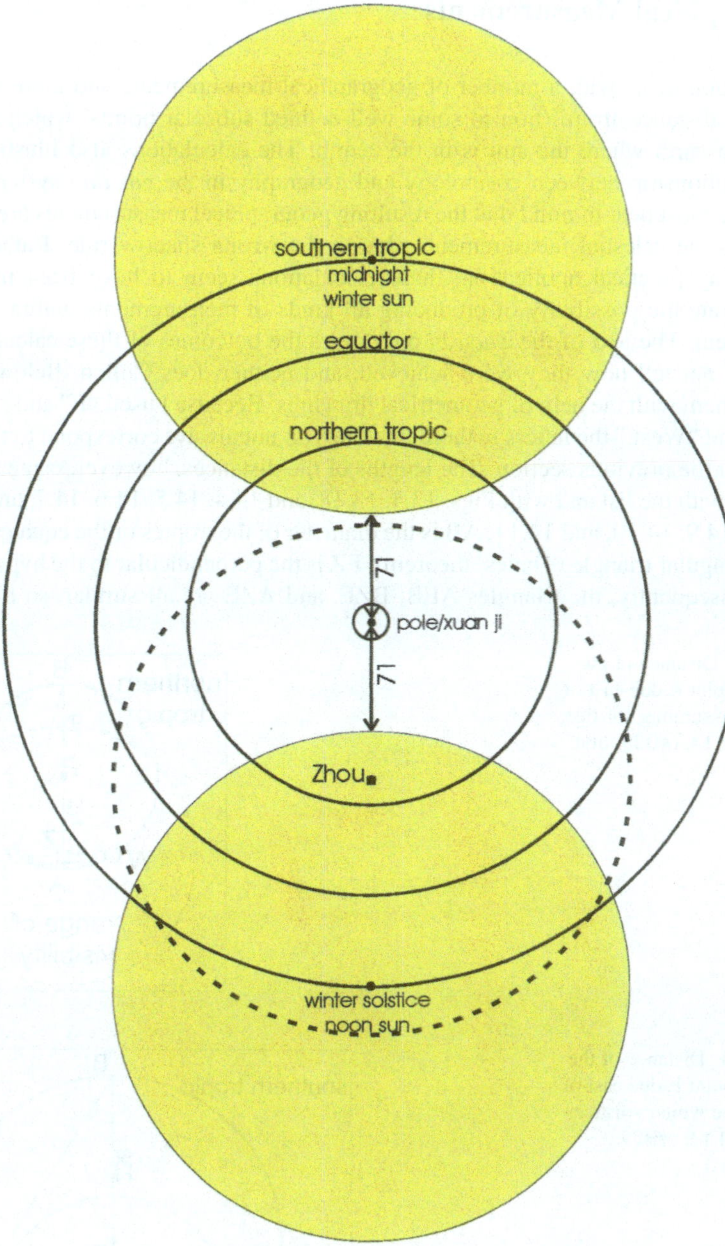

Fig. 14.8 Distances indicated in #B28/2 (text 14.11); the sun at noon and the "midnight sun" at the winter solstice

Geographical Measurements

This section deals with a number of geographical measurements and more specifically the distance from Zhou to some well-defined subsolar points, which are the points on earth where the sun is in the zenith. The calculations also illustrate the close relationship between cosmology and geography in the *gai tian* system. And again, we must bear in mind that the resulting geographical measurements are just as wrong as the celestial measurements, due to the wrong shadow rule. Rather than having any practical application, these calculations seem to have been made to demonstrate the possibility of producing all kinds of measurements within the *gai tian* system. The text of the *Zhou bi* only gives the outcomes of these calculations, but does not tell how they were achieved, and neither does Cullen. Below I will present them with the help of geometrical drawings. Because I used "E" and "W" for "East" and "West," the letters in these drawings do not always correspond to those of the list in the previous section. The lengths of the distances, however, can easily be checked with the list and with Figs. 13.4, 13.18, and 14.4, 14.5, 14.6, 14.7, and 14.8. In Figs. 14.9, 14.10, and 14.11, AB is the diameter of the tropics or the equator, AEB is a rectangular triangle (Thales' theorem), EZ is the perpendicular to the hypotenuse AB. Consequently, the triangles AEB, BZE, and AZE are all similar, so that AZ:

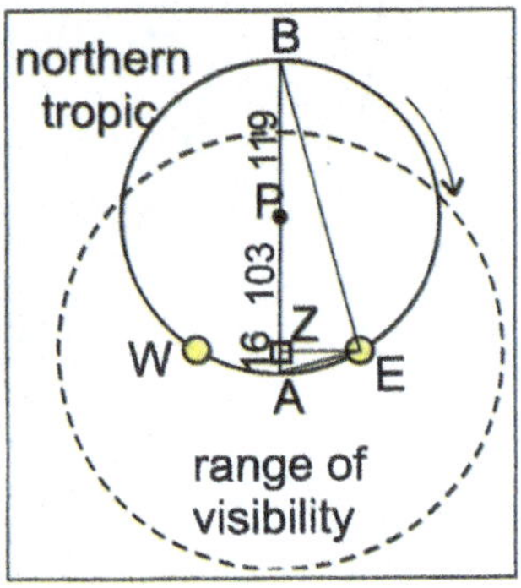

Fig. 14.9 Distance of the subsolar point E due east of Zhou at the summer solstice; see text 14.13. (#B29, first part)

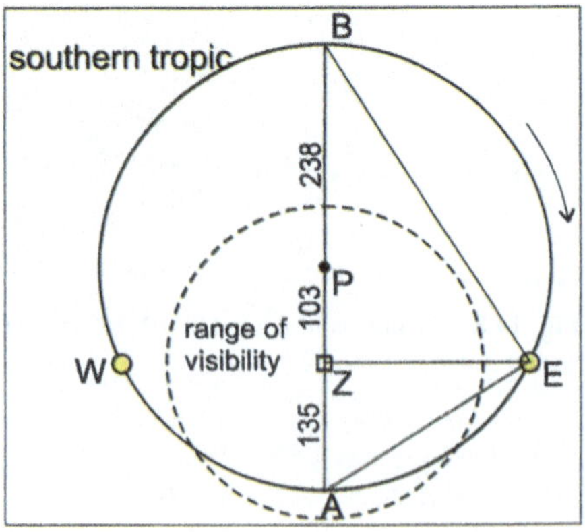

Fig. 14.10 Distance of the subsolar point E due east of Zhou at the winter solstice; see text 14.14. (#B29, second part)

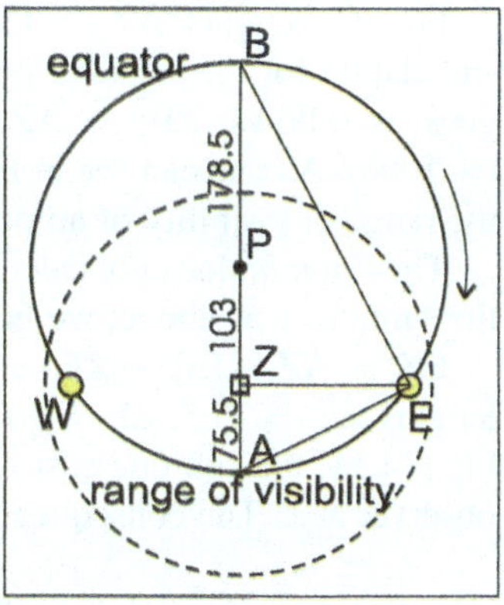

Fig. 14.11 Distance of the subsolar point E due east of Zhou at the equinoxes; see text on p. 308, at the end of this section

EZ = EZ: BZ → EZ^2 = AZ × BZ. Again, all numbers must be multiplied by 1000; the square Z is Zhou, the black dot P is the pole (i.e. the subpolar point).

Wang Chong remarks,

14.12. At the time when the sun sets in the west, the people living there will perhaps say that he is culminating, and looking from the point where the sun is setting, eastward to our world, heaven and earth may appear to the beholder joined together.[14]

The *Zhou bi* gives the calculations for the summer and winter solstices that belong to this remark.

14.13. (#B29, first part) On the day of the summer solstice, if one sights due east and west of Zhou then from the subsolar points directly due east and west of Zhou it is 59,595½ *li* to Zhou.

The calculation for the subsolar points due east and west of Zhou at the summer solstice goes as follows:

EZ^2 = AZ × BZ → ZE^2 = 3552 → ZE = 59.598,657,7 (×1000). The lengths AZ = 16, ZP = 103, and PB = 119 can be easily deduced from Fig. 13.6. As one can see in Fig. 14.9, the sun due east or west of Zhou is within the range of visibility of an observer at Z, as the *Zhou bi* says. Both this as well as the next two calculations indicate, by the way, that the Chinese astronomers recognized that there should not only exist places on the meridian where the sun could be observed right overhead (at a sub-solar point due south), but also elsewhere.

14.14. (#B29, second part) On the day of the winter solstices the sun is not visible in the regions due east and west, [however] by calculation we find that from the subsolar points it is 214,557½ *li* to Zhou.

---

[14]Forke (1907), 263–264.

For the lengths AZ = 135, ZP = 103, and PB = 238, see Fig. 13.6. The calculation for the subsolar points due east and west of Zhou at the winter solstice goes as follows: $ZE^2 = AZ \times BZ \rightarrow ZE^2 = 46{,}035 \rightarrow ZE = 214.557{,}684{,}6$ ($\times$ 1000). As one can see in Fig. 14.10, the sun due east or west of Zhou is beyond the range of visibility of an observer at Zhou, as the *Zhou bi* says.

The *Zhou bi* does not calculate the figures at the equinoxes, but when we do it in the same way as above, we get:

$EZ^2 = AZ \times BZ \rightarrow ZE^2 = 21{,}253.25 \rightarrow ZE = 145.784{,}944{,}4$ ($\times$ 1000). For the lengths AZ = 75.5, ZP = 103, and PB = 178.5, see Fig. 13.6. As one can see in Fig. 14.11, the sun due east or west of Zhou is within the range of visibility of an observer at Z. The consequence of this feature will be discussed in the next section.

Sunrise and Sunset Seen from Zhou

The teachings of the *Zhou bi* as explained in the previous section are not without problems. In Fig. 14.11, at the time of the equinox, the points W and E, representing the sun due east and west of Zhou, lie within the range of visibility of an observer at Zhou. Therefore, the sun must have risen further north-east of Zhou and set further north-west of Zhou, at the intersections of the circles of the equator and the range of visibility. This means that, according to the *gai tian* system, the day at the equinoxes in Zhou is shorter than the night, as can be seen in Fig. 14.12, where red is the part of the sun's orbit during the day and blue the part of its orbit at night. In reality, however, an observer at Zhou at the time of the equinoxes sees the sun rise due east and set due west, and the length of the day is equal to the length of the night. The wrong effect is due to the location of Zhou south of the subpolar point. This anomaly must have been the reason why the *Zhou bi* does not give a calculation for the

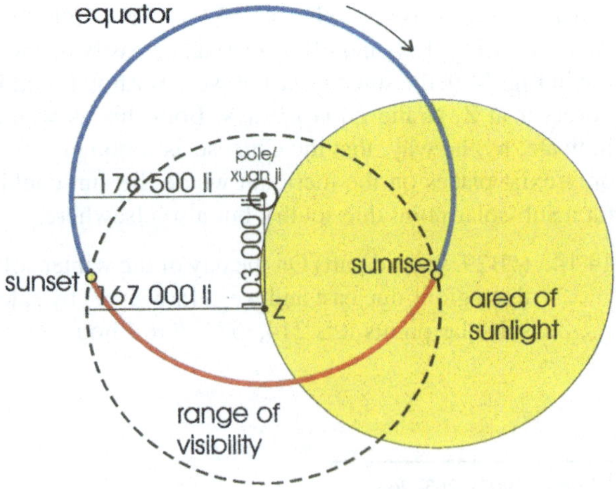

Fig. 14.12 Sunrise (and sunset) for an observer at Zhou at the equinox

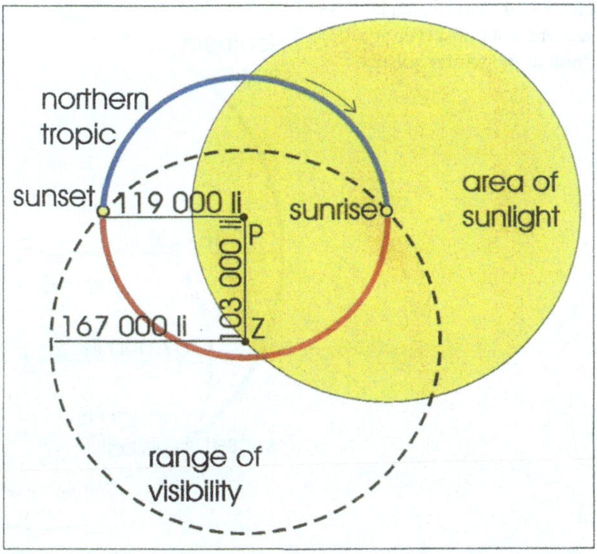

Fig. 14.13 Sunrise and sunset for an observer at Zhou at the summer solstice

equinoxes.[15] Perhaps it was also one of the reasons why the number of 167,000 *li* for the extension of the illumination of the sun, and thus also of the range of visibility was smuggled in: it weakens the wrong effect and makes the sun rise more to the east and set more to the west. If the range of visibility was much smaller than 167,000 *li*, namely 145,785 *li* (the distance ZE in Fig. 14.11), then, during the equinoxes, the sun would rise in Zhou due east and set due west. In that case, however, the difference in length between night and day would be greater.

In a fragment of his lost book *Xin Lun*, Huan Tan already made a similar point: the diagram drawn by his opponent Yang Ziyun could not be right, because the observer is south of the pole, and thus during the equinoxes his east-west line can never cut the circles of the orbit of the sun around the pole in two equal halves:

14.15. [Since heaven] is turning like a cover, that means that the northern [part of the sun's] track is distant from us, and the southern part is close. So how could the length of day and night be equal? Ziyun had no explanation.[16]

An analogous inconvenience appears with the summer solstice. As can be seen in Fig. 14.13, in the *gai tian* system at the summer solstice, when the sun is within the range of visibility of an observer at Zhou (the red half of the circle), the day appears to be about as long as the night, when the sun is beyond his range of visibility (the blue half of the circle). In reality, however, for an observer at Zhou during the summer solstice the day lasts much longer than the night, and he sees the sun rise in the north-east and set in the north-west.

---

[15]Cullen (1996), 131–132 makes the same point, although from his text it is not immediately clear that the wrong effect is due to the location of Zhou.

[16]Quoted from Cullen (1996), 60. See also Cullen (2017), 229, with Figure 5.15.

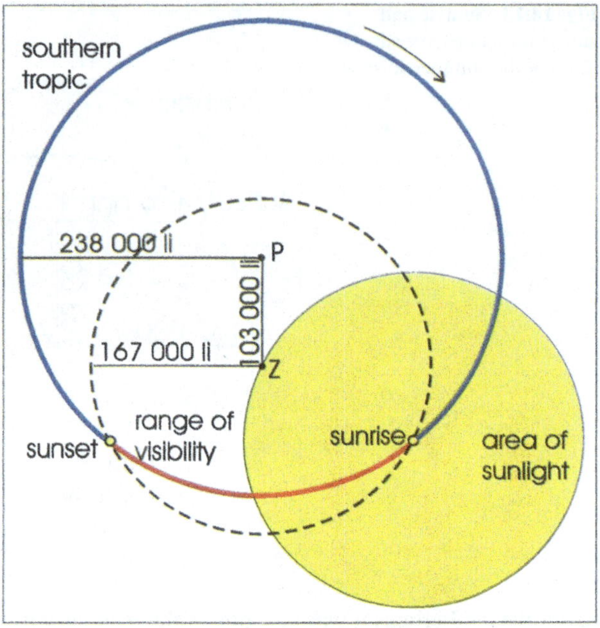

Fig. 14.14 Sunrise and sunset for an observer at Zhou at the winter solstice

At the winter solstice, an observer at Zhou will see the sun rise and set to the south of his east-west line, and the day is shorter than the night. This is also the case in the picture of the *gai tian* system, although the difference between the duration of day and night is clearly too great, and the day almost 3½ times as long as the night (see Fig. 14.14).

Of course, the anomalies shown in Figs. 14.12, 14.13, and 14.14 increase the further one goes south. Cullen also makes this remark with regard to the division between night and day at the equinoxes, without mentioning, however, the problems at the solstices.[17]

The Seven *Heng* and the Limit of the Cosmos

We have already seen that the circle nearest to the pole is that of the *xuan ji* star, with a diameter of 23,000 *li*. In the *Zhou bi* seven more concentric and equidistant circles are drawn around the pole, called *heng*, some of which we have already met: the first *heng* is the circle of the summer solstice (the northern tropic), the fourth *heng* is the circle of the equinoxes (the equator), and the seventh *heng* is the circle of the winter solstice (the southern tropic).

[17]See Cullen (1996), 132 and n. 153 in which he refers to a later Chinese critic, Wei Cheng.

14.16. (#D8) The first and innermost heng: diameter: 238,000 *li* (. . .).
14.17. (#D9) Next is the second *heng*: diameter: 277,666 *li* 200 *bu* (. . .).[18]
14.18. (#D10) Next is the third *heng*: diameter: 317,333 *li* 100 *bu* (. . .).
14.19. (#D11) Next is the fourth *heng*: diameter: 357,000 *li* (. . .).
14.20. (#D12) Next is the fifth *heng*: diameter: 396,666 *li* 200 *bu* (. . .).
14.21. (#D13) Next is the sixth *heng*: diameter: 436,333 *li* 100 *bu* (. . .).
14.22. (#D14) Next is the seventh *heng*: diameter: 476,000 *li* (. . .).

After these, one more circle is mentioned, the farthest extent of the area of sunlight. This circle is the ultimate boundary of the *gai tian* cosmos:

14.23. (#D15) Next comes the limit of solar illumination at the winter solstice. This goes 167,000 *li* beyond the outermost *heng*. This gives a diameter of 810,000 *li* (. . .).
14.24. (#D19) The diameter of the four poles is 810,000 *li* (. . .).
14.25. (#E3) Therefore the diameter of the outward extent of the sun's rays is 810,000 *li* (. . .).
14.26. (#B32) From the extent of the differences of the figures and the limit of solar illumination, the diameter of the four poles is 810,000 *li* (. . .).
14.27. (#D16) Nobody knows what is beyond this.

Since there is no sunlight beyond this ultimate limit, we might say that this is where the eternal night begins. A picture of these nine circles looks like Fig. 14.15;

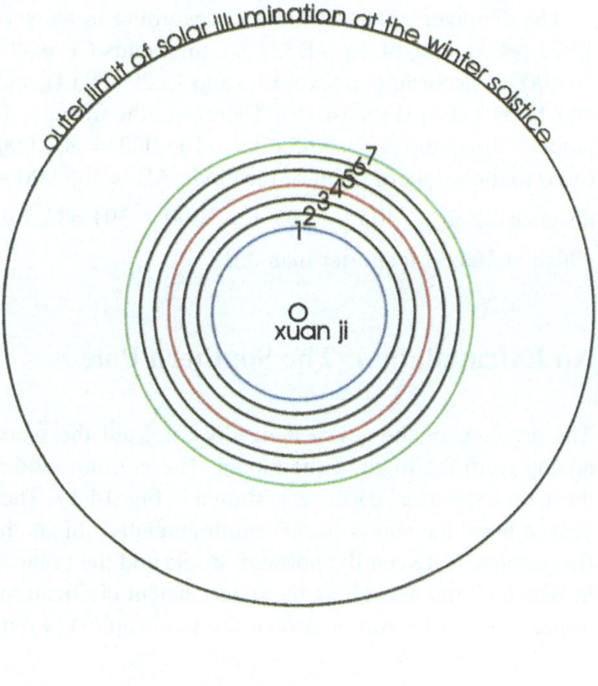

Fig. 14.15 The seven *heng*, the *xuan ji* circle, and the limit of solar illumination (For a similar picture, see Cullen 2017, 211, Figure 5.12)

[18]One *bu* = 6 *chi*, and 300 *bu* = one *li*, so one *bu* = 1.386 m (see also note 841).

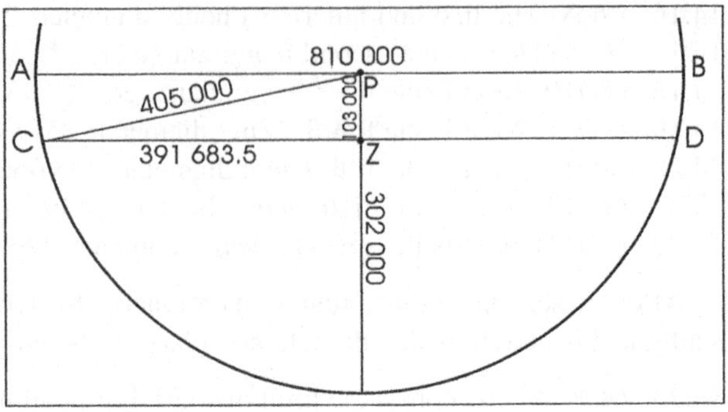

Fig. 14.16 The distance from Zhou to the outer limit of solar illumination

the blue circle (first *heng*) is the northern tropic, the red circle (fourth *heng*) is the equator, and the green (seventh *heng*) circle is the southern tropic.

The distance from Zhou to the outer limit of solar illumination can be calculated:

14.28. (#B33) From Zhou southwards to the [furthest]place illuminated by the sun is 302,000 *li*, and northwards to the [furthest] place illuminated is 508,000 *li* from Zhou. The distances east and west [from Zhou to the furthest points illuminated] are each 391,683½ *li*. Zhou is 103,000 *li* south of the center of heaven, and therefore the east-west measurement is shorter than the central diameter by just over 26,632 *li*.

The diameter AB is 810,000 *li*, according to texts 14.23 (#D15), 14.24 (#D19), 14.25 (#E3), and 14.26 (#B32); so, the radius CP is 405,000 *li*. The distance PZ is 103,000 *li*, according to texts 13.8 and 13.29 (#B14), and 14.28 (#B33), 13.30 (#F2), and 13.31 (#F4) (Fig. 14.16). Therefore, the distance from Z(hou) to the southern point of the circle AB is 405,000 − 103,000 = 302,000 *li*, and the distance from Z (hou) to the northern point of the circle AB is 405,000 + 103,000 = 508,000 *li*. The distance CZ is $\sqrt{(405^2 - 103^2)} \times 1000 = 391{,}683.5$ *li*. The distance CD is 2x CZ, which is 26,633 *li* shorter than AB.

An Extrapolation: The Southern Pole

The drawing of the seven *heng* does not tell the whole story of the implications arising from the logic of the model. The *gai tian* model displays regions similar to those on a spherical earth, as is shown in Fig. 14.17. There is a region where 2 days a year at noon the sun is in the zenith annually; this is the tropical zone between the two tropics. Between the northern tropic and the polar zone lies the temperate zone, in which China as well as the other ancient civilizations are situated. Note that the tropical zone (the zone between the two tropics) is twice as wide as the temperate

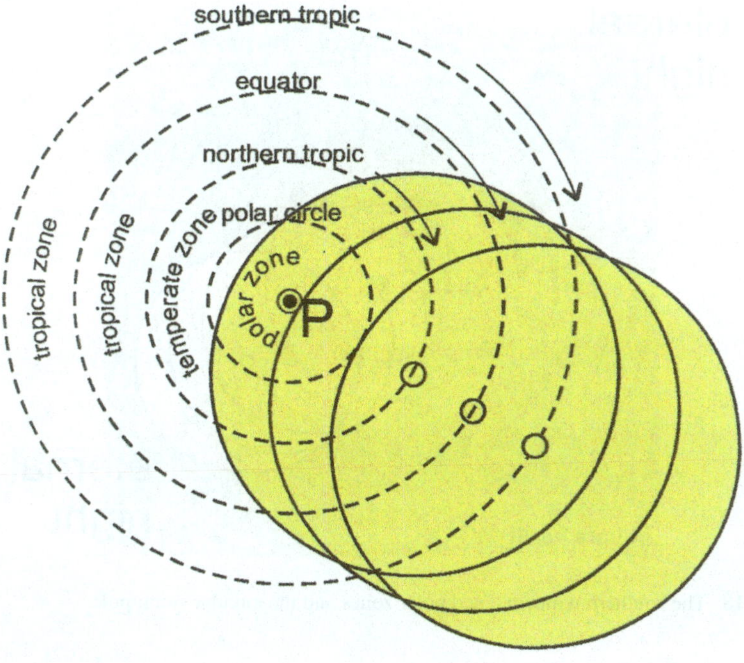

Fig. 14.17 The (northern) polar, temperate, and tropical zones

zone. The previous sentences presuppose that, next to the equator and the tropics, in the *gai tian* model there must also be a circle that corresponds to the arctic circle on a spherical earth. The area within the arctic circle is the polar zone. As can be seen in Fig. 14.17, the polar zone does not coincide with the area around the pole that corresponds with the orbit of the *xuan ji*. On a spherical earth, the arctic circle is defined as the periphery of an area around the north pole which would theoretically experience annually at least one 24-h period in which the sun is continuously above the horizon and at least one 24-h period in which the sun is continuously below the horizon.[19] However, we must not forget that the radius of the area of sunlight is assumed to be slightly smaller than that of the equator (167,000 *li* vs. 178,500 *li*). The definition of the polar circle is therefore not exactly applicable to the *gai tian* model. In Fig. 14.17, I have chosen to let the polar circle border the area of sunlight during the winter solstice.

Perhaps the strangest consequence of the model appears when we realize that there must be another temperate zone south of the southern tropic, bounded by another polar circle, beyond which there is another polar zone, and finally another,

[19]See *Wikipedia*, article "Polar circle."

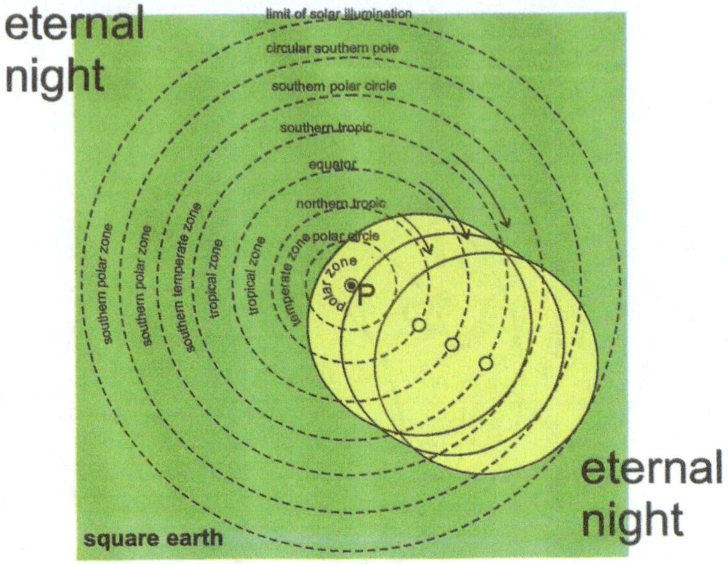

Fig. 14.18 The southern temperate and polar zones and the circular south pole

circular pole that can be called the south pole of the *gai tian* system. However, since in this system the days become shorter as one goes more southwards and the climate colder, the name "temperate zone" seems less suitable here. In this system, the south pole is not a point but a large circle at the edges of the earth. Moreover, at the outer limit of this circular south pole, which is also the limit of solar illumination, there is daylight only once a year, while it is night for the rest of the year.

As mentioned in the previous section, the *Zhou bi* uses the expression "the four poles" as an indication of the outer limit of solar light. Obviously, the idea is that not only the one central pole exists, but also four peripheral poles, which should be understood as the farthest places east, west, north and south at the outer limit of sunlight. Taking into account that for every observer, wherever on earth, the direction to the central pole is "north" and the opposite direction "south," it is less confusing to speak of one circular south pole instead of "the four poles."

The idea of southern circles and zones is rendered in Fig. 14.18 on the contours of a square earth. This picture also shows, incidentally, how small, as compared to the surface of the earth as a whole, the Chinese astronomers who adhered to the *gai tian* must have imagined their own country, somewhere in the northern temperate zone. The outermost circle is the limit of the area that can lit by the sun, albeit only once a year. Beyond this circle it is eternally night.

Major argues that the image of the canopy of a chariot defines a square within a circle. The concept of a circle within a square, as in Fig. 14.18, is however confirmed

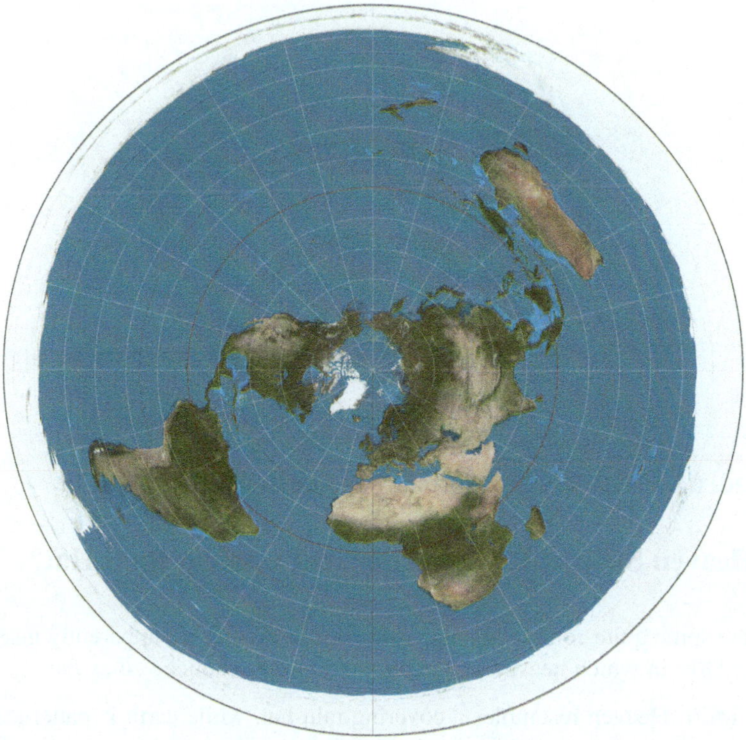

Fig. 14.19 A polar azimuthal equidistant projection of the earth (Courtesy Daniel R. Strebe, date listed for the image upload 15 August 2011)

by Shan Juli's critical remark to his teacher Zeng Shen (505–435 BC), that if heaven were round and earth square, then the four angles of the earth would not be well covered.[20]

The central north pole and the circular south pole are features that can be compared to those on a map of the spherical earth in a polar azimuthal equidistant projection of our spherical earth (see Fig. 14.19), although there it is an effect of mapmaking, whereas in the *gai tian* it is the result of the conception of heaven and earth.

How great was my surprise when I discovered that the same kind of projection is used in the emblem of the United Nations. The *gai tian* astronomers would have loved it (Fig. 14.20).

[20]Major (1993), 35 and figure 2.4 on 36. Shan Juli's remark quoted from Lan-ying Tseng (2011), 50.

Fig. 14.20 The emblem of the United Nations on the fence of the Geneva office

The Heaven Shaped Like a Truncated Conical Rain Hat?

There are some quite confusing phrases in section #E that are apparently inserted by a later editor, in which heaven and earth are no longer conceived as flat.

14.29. (#E6) Heaven resembles a covering rain-hat, while earth is patterned on an inverted pan.

14.30. (#E2) As for the subpolar point, it is 60,000 *li* higher than where humans live, and the pouring waters run down on all sides. Likewise the center of heaven is 60,000 *li* higher than its edges.

14.31. (#E8) Heaven is 18,000 *li* from earth. Even though the winter solstice sun is on the outer *heng*, it is still 20,000 *li* above the land below the pole.

These lines look like a kind of compromise between the *gai tian* and the *hun tian* systems, which fall outside the scope of this book, so I do not have to go into detail. As far as I can see it is impossible to keep the overall distance of 80,000 *li* intact when the earth is square and the heaven conical. This also remains a problem in Cullen's interpretation of these texts, in which he gives both heaven and earth the shape of a Japanese rain hat as a truncated cone.[21] Moreover, when the earth is shaped like an inverted pan, it obviously is no longer regarded as a square. And finally, "in a universe where heaven and earth are not flat and parallel, the shadow rule cannot rationally be applied," as Cullen remarks.[22]

---

[21]See Cullen (1996), 135ff., and especially figure 13 on 136.

[22]Cullen (1996), 135. In Chap. 16, however, we will examine Cosmas Indicopleustes' limited version of the shadow rule, which is independent of the shape of the heaven, but only depends on Cosmas' assumption that the sun orbits parallel to the surface of the flat earth.

A Short Evaluation of the *Gai Tian* System in the *Zhou Bi*

At first glance, it may be tempting to judge that the *gai tian* system as a whole is of no value because it is fundamentally mistaken. Yet, the *gai tian* system must be considered as an impressive and unique conceptual construction. Let us enumerate once more its fundamental ideas: not only the earth but also the heaven is flat; the heavenly bodies describe circles overhead around the pole; rising and setting of the heavenly bodies are optical illusions; human sight is not infinite but limited; the sun throws a limited circle of light on the surface of the earth. It is amazing that with these innovations, all of which we would call erroneous, the early Chinese astronomers were able to achieve an acceptable picture of the universe, at least as seen from Zhou. The system makes it possible to measure not only the height of the heaven above the flat earth, but also many more distances in the universe.

Of course, based on suppositions we now know to be wrong, the system entails several serious problems. The picture of the sun orbiting overhead around the pole makes the sun's orbit in summer much smaller than in winter, and thus the summer sun's orbital velocity correspondingly slower. As a result of the introduction of the fictitious *xuan ji* star, the radius of the area of sunlight, and consequently that of the range of visibility is reduced to 167,000 *li* instead of 178,500 *li*. For an observer at Zhou, the calculations in the *Zhou bi* result an equinoctial day that is shorter than the night, while at the summer solstice the day appears to be about as long as the night. The more an observer goes in a southerly direction, the bigger this anomaly grows. An extrapolation of the *gai tian* model as a kind of azimuthal equidistant projection of the earth leads to unrealistic geographical dimensions, culminating in a circular south pole. The shadow rule used to measure the cosmological distances is obviously wrong and not based on observation. Because of a well-known optical illusion, the rising or setting sun appears larger than the sun high in the sky, but in the *Zhou bi* the measurement of the sun's diameter leads to an angular diameter of the sun as seen from a subsolar point that is bigger than from a point that is not directly below the sun. Moreover, more distant objects usually look smaller, which means that for an observer at Zhou the sun (and the moon) should seem smaller when they are farther away in their orbit around the pole. The *gai tian* also cannot explain why the setting sun is cut off at the horizon: the sun should become smaller, but not cut off. It remains unexplained that we can see stars at the horizon, at the limit of our visual field, although they are by far not as bright as the sun. It turned out that all heavenly bodies lower than 28° should be considered as optical illusions. Not only the rising and setting sun are illusions, but also the direction where their warmth comes from, as the story of Huan Tan and Yang Ziyun sitting on a porch shows. Even more important is that it remains strange that the sun and a part of the heaven, being beyond the range of visibility, are yet visible because of an optical illusion (a kind of fata morgana). The model cannot explain why from Zhou, being 103,000 *li* south of the pole, the orbits of the celestial bodies are seen as circles instead of ellipses. Perhaps one of Yang Xiong's objections points in the same direction: objection 8 points out that in the *gai tian* system, two stars should appear closer together when they move north of the pole than when they are in the south.[23] A serious problem is also that the system may be able to explain the occurrence of solar eclipses, it has difficulty explaining lunar eclipses.

In spite of all this, the *gai tian* system of heaven and earth that is presented in the *Zhou bi* is a highly sophisticated whole. It is not easy to fathom its basic concepts: a flat heaven over a flat earth, the range of visibility, the area of sunlight, the heaven and the rising and setting of heavenly bodies as optical illusions, the shadow rule, and the way they are all intertwined. The most extraordinary feature of the *gai tian* model is that it is able to account for the different times of the day on different parts of the earth. The authors of the *Zhou bi* were quite aware of this feature of their system, as evidenced by the way the movements of the sun and the limitation of the area if sunlight were described. When they spoke about the "midnight sun," these words were their way of expressing the idea of time differences all over the earth. It is my conviction that this feature was the very reason why the whole system was invented. As such it was an impressive intellectual achievement.

References

Couprie, D.L., *Heaven and Earth in Ancient Greek Cosmology*. New York: Springer 2011.

Cullen, C., *Astronomy and Mathematics in Ancient China: the Zhou bi suan jing*. Cambridge: Cambridge University Press 1996.

—, *Heavenly Numbers: Astronomy in Early Imperial China*. Oxford: Oxford University Press 2017.

Forke, A., *Lun-Hêng. Philosophical Essays of Wang Chung*. London: Luzac & Co 1907. Available on the Internet: https://archive.org/details/lunhng00forkgoog.

Lan-ying Tseng, L., *Picturing Heaven in Early China*. Cambridge (Mass.) and London: Harvard University Asia Center 2011.

Major, J.S., *Heaven and Earth in Early Han Thought*. Albany: State University of New York Press 1993.

Wikipedia, article "Polar circle."

[23]Quoted in *Sui shu* 19:4a, compiled by Wei Cheng and others, 636-656 BC Kind information by Christopher Cullen.

Chapter 15
Ancient Chinese Versus Greek Flat Earth Cosmology

Contents

Two Kinds of Flat Earth Cosmology Compared

In the two parts of this book, I discussed two concepts of flat earth cosmology which tried in different ways to solve the problems that arise from the premise that the earth is flat. There is little point in trying to decide which system succeeded best, but it is worthwhile to recapitulate the differences. The main difference is that the ancient Greeks, with the possible exception of Anaximander, conceived of the heavens as a hemispherical dome over or as a full sphere around the earth, while the Chinese *gai tian* imagined a flat circular heaven parallel to a flat square earth. Another important difference, which was immediately related to the difference in conception of the shape of the heaven, was that the Greeks considered Delphi as the center of the earth and thus the celestial axis, around which the heavenly bodies orbit, as tilted, while the Chinese thought that the celestial axis was perpendicular between heaven and earth and their land at a certain distance from the center of the earth. A major drawback of the Greek conception was that they were unable to solve the problem of time differences on a flat earth. As far as we know, they even seem to have almost completely ignored it. The great achievement of the Chinese *gai tian* system was that it made time differences on a flat earth understandable. Their flat heaven and perpendicular celestial axis made it possible to accomplish this, even though they needed a set of auxiliary constructions, the most important of which were the limitation of the area of sunlight and the limitation of the range of visibility. However, these were notions of common sense. The example of the receding torch that becomes invisible after a while, illustrated both concepts at the same time. The horizon was considered as the visual the boundary beyond which nothing can be seen.

D. L. Couprie, *When the Earth Was Flat*, Historical & Cultural Astronomy,
https://doi.org/10.1007/978-3-319-97052-3_15

In the Greek view, the heavenly bodies, with the exception of the circumpolar stars, rise on the eastern horizon, follow a curved path along the heaven, and set on the western horizon. Here, common sense was on the side of the ancient Greeks. In the Greek conception with a hemispherical vault, however, the question of where the celestial bodies were located after they had set was given two unsatisfactory answers, either by imagining that they were traveling along the horizon behind mountains, or that they were new every day. In the Greek conception with a spherical heaven, on the other hand, the problem of where the heavenly bodies were located when they had set, as well as the problem of how they managed to rise at the expected point, were elegantly solved. Moreover, this insight paved the way for the transition to the conception of a spherical earth. In the Chinese *gai tian* system, the heavenly bodies always moved above the earth and thus never set. To explain the evidence of the rising and setting sun (as well as the other rising and setting heavenly bodies), the Chinese used the sophisticated auxiliary construction of optical illusion. The *gai tian* system did not provide any basis for an eventual transition to the conception of a spherical earth.

In both the Greek and the Chinese systems, the heaven and the celestial bodies are relatively nearby (with, in the case of Greek cosmology, the possible exception of Anaximander). The Chinese calculated the distance between heaven and earth, but because of an incomprehensible error, they made it too big. Precise calculations of the height of the heaven made by Greek flat earth cosmologist have not been handed down, although it can be argued that Anaxagoras' estimate was based on a calculation very similar to that of the Chinese astronomers. Based on the measurement of the distance between heaven and earth, the Chinese calculated several distances both in heaven and on earth, although these calculations suffered from the same systematic error. The Presocratic Greeks, even though familiar with the measurement procedure since Thales, did not provide any such measurements.

The tilted celestial axis was another serious problem of the Greek system, because the counterpart on earth of the celestial pole was an arbitrarily chosen center of the flat earth, which was traditionally thought to be Delphi. Therefore, the amount of the tilt of the celestial axis and thus the tilt of the heaven, was also arbitrary and related to the height of the celestial pole in whichever place the center of the earth was deemed to be. The Chinese choice of the subpolar point (the north pole) as the other end of the celestial axis, was the only one logically possible, given the idea that the heavenly bodies circle on a flat plane parallel to the earth.

Finally, the geography resulting from the two systems was completely different. The result of the tilt of the celestial axis was that the equatorial plane was to be imagined as cutting the flat earth in an east-west line, the "Ionian equator", and the tropics as defined by the points of sunrise and sunset in summer and winter as seen from Delphi. Accordingly, the northern part of the earth was conceived of as the colder and the southern part as the warmer. The geography of the Chinese system was completely different, with a central north pole and a circular south pole, and between these two poles a circular equator and circular tropics, right below their counterparts in the heaven.

Greek Influence on the *Gai Tian* Flat Earth Cosmology?[1]

Panchenko has recently argued that the idea of the *gai tian* was derived from ancient Greek flat earth cosmology and that, in retrospect, the cosmology of Anaximenes, but perhaps also that of Xenophanes, can be better understood from this perspective. His book "On the Eastern Slope of Olympus: The Impact of Greek Ideas on Chinese Cosmology" has so far been published in Russian only, so until an English translation will see the light of day, we will have to resort to his articles, conference papers, and personal information. The main article on this subject, *Anaximenean Astronomy in the Light of Chinese parallels*, which contains the text of a conference paper, has appeared in a Chinese journal. Panchenko's interpretation of Anaximenes is inspired by Hippolytus' text, from which, as he says, the main evidence comes,[2] and which was already quoted at the beginning of Chap. 7:

15.1 (Anaximenes) denies that the heavenly bodies move under (ὑπὸ) the earth, as others suppose, but he says they turn around (περὶ) the earth, like a felt cap (πιλίον) turns around our head (περὶ τὴν ἡμετέραν κεφαλὴν στρέφεται). The sun is hidden not by going under the earth, but by being covered by the higher parts of the earth and by being a greater distance away from us.[3]

Panchenko refers to Kirk's interpretation of the cap simile, namely that "the cap-image must illustrate the hemispherical shape of the sky, not its obliquity; it is difficult to see, indeed, why the cap should be imagined as being tilted on the head," and he adds that Kirk "is right on this point."[4] This is a mistake, however, because in the sentence following the one just quoted, Kirk clearly says that he means to say that "Anaximenes appears to have accepted the broad structure of the naïve world-picture."[5] And on the previous page he explained this "naïve world-picture" as "the pre-philosophical world-picture, where the sun, at least, floats round river Okeanos to the north."[6] By these words he can only mean the archaic world-picture as visualized in Fig. 7.13, with the double bend of the solar path, which is certainly not how Panchenko imagines Anaximenes' world-picture.

Be that as it may, Panchenko writes: "the slant of heaven in this system [sc. that of Anaximenes] is illusory. It is our off-centre position that accounts for the apparent downwards tilt of the celestial pole. This is crucial for our understanding of Anaximenes' cap or hat simile."[7] In Panchenko's interpretation, it is not only the

[1]Several of the critical remarks in this section also apply to Kočandrle (2018), which is not yet available in English.

[2]Panchenko (2015, 413).

[3]Hippolytus, *Refutatio Omnium Haeresium* 1.7.6 = DK 13A7(6) = LM ANAXIMEN. D3(6) = Gr Axs12(6) = TP2 As56[7.6] = KRS 156.

[4]Panchenko (2015, 416), quotation from KRS, p. 157.

[5]KRS, p. 157.

[6]KRS, p. 156.

[7]Panchenko (2015, 416).

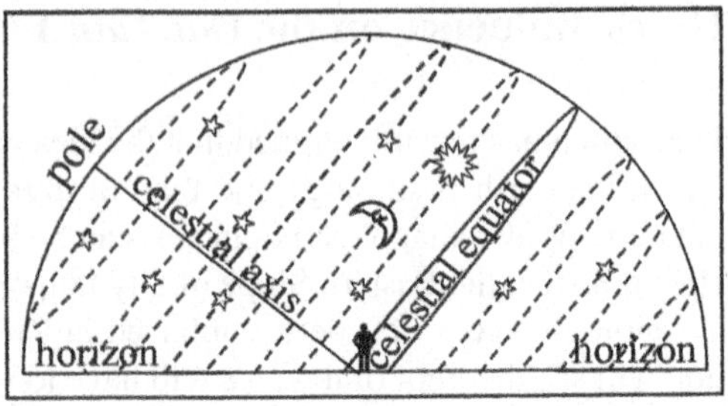

Fig. 15.1 The cap simile as an optical illusion; summer in Athens (tentatively after Panchenko)

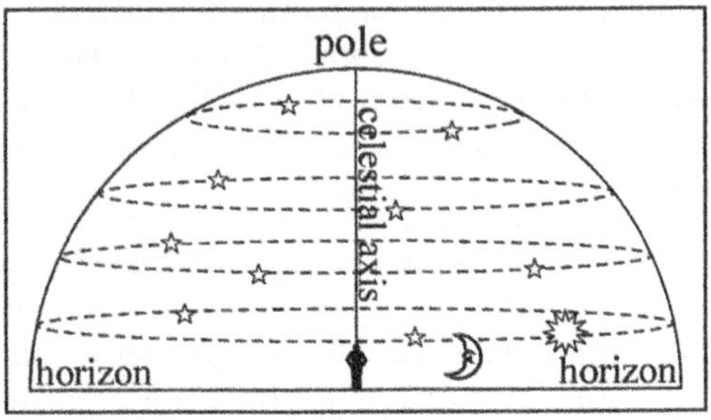

Fig. 15.2 The cap simile as an optical illusion; summer right under the celestial pole (tentativey after Panchenko)

tilting of the heaven which is illusory, but also its curvature. According to Panchenko, Anaximenes' cap simile of the heaven should not be understood as an image of the real hemispherical shape of the heaven, but as an image that depicts the heaven as an optical illusion. When he compares Anaximenes' cosmology with that of the *gai tian*, Panchenko quotes with approval Wang Chong, who uses an image reminiscent of Fig. 13.10:

15.2 Heaven *appears to us* in the shape of a bowl turned upside down[8]

This optical illusion of a hemispherical heaven changes with the place of the observer, who takes it with him, as it were, wherever he goes. The celestial axis will seem to rise when an observer approaches the subpolar regions. If Fig. 15.1, for example, were to depict the heaven as it appears to an observer in Athens, one would expect that, to an observer right below the pole, the heaven would look like Fig. 15.2, in which the sun would seem to orbit for half a year above the horizon and for half a year below the horizon.

It should be noted that all this is much more than Hippolytus' text says. The problem with this interpretation is that the cap simile is not presented by Hippolytus as an image of the optical illusion of the heaven, but as an image of the real heaven and the movements of the heavenly bodies. The point of Hippolytus' rendition is not that the heaven *looks curved* like a cap, but that the heaven *is curved* like a cap. The very notion of optical illusion is not present in Hippolytus' cap simile.

[8]Panchenko (2015, 417). Quotation from Forke (1907, 261, my italics).

If, however, the curved heaven is supposed to be an optical illusion, then the next question that arises is what its real shape should be. For this, Panchenko refers again to Hippolytus' text (text 15.1), in which it is said that, according to Anaximenes, the heavenly bodies do not move under the earth, but turn around the earth. Panchenko does not pay attention to the strange use of the preposition περὶ, the meaning of which meaning is "around", not "above," as it should be because it contrasts with ὑπὸ ("under"). He does also not discuss Pseudo-Plutarch's version of Aëtius, in which it is said that, according to Anaximenes, the heavenly bodies go under the earth, just as Anaximander already taught (See Chap. 7, section *A Fresh Look at the Doxography*). Panchenko's interpretation of the phrase "the heavenly bodies do not move under the earth, but turn around the earth" is that Anaximenes' conception of the heaven is similar to that of the *gai tian*: "in reality, it lies in a plane parallel to the surface of the earth. The celestial pole is the centre of heaven, and the heavenly bodies, including the sun, describe their orbits round it."[9] We should note again that this is much more than Hippolytus' text says.

However, Panchenko's interpretation is even more complicated than this, for he also suggests that Anaximenes' heaven not only *looks* like a cap, due to an optical illusion, but, at least partly, *is* also a kind of cap in reality. I quote: "I propose that not only the track of the sun is alternatively expanding and contracting, but also that the sun goes half a year down and half a year up; from solstice to solstice, it describes a truncated cone."[10] Consequently, the sun is thought not to move in the plane of the heaven, as was said in the last quotation, but under it. For his idea of the orbit of the sun moving up and down during the year, Panchenko refers to "some support in Chinese material,"[11] apparently alluding to some confusing passages in the *Zhou bi* (see texts 14.29–14.31), of which I have argued in Chap. 14, section *The Heaven Shaped Like a Truncated Conical Rain Hat?* that these are later inserts to reconcile the essential characteristic of the *gai tian*, the flat heaven, with the alternative system of *hun tian*. Apart from this, Panchenko's interpretation of the sun's orbit moving up and down during the year seems to have been derived from Cosmas Indicopleustes, who, as he says, "employs some ideas that can be traced back to Ionian science"[12] (cf. Fig. 5.3), rather than being inspired by the *gai tian* with its flat heaven. In a recent publication on Cosmas, this relationship is questioned: "The problem (...) is that (...) there was little direct access to the Pre-Socratics from the Hellenistic era onward."[13]

Apparently, Panchenko's intention is that we must combine somehow the idea of the heaven as a plane parallel to the surface of the flat earth with the idea that the sun's annual up and down movement describes a truncated cone (and perhaps also the moon in its monthly movement, and the planets?). The meaning is, probably, that

---

[9]Panchenko (2015, 416–417).

[10]Panchenko (2015, 423).

[11]Panchenko (2015, 423).

[12]Panchenko (2015, 420).

[13]Anderson (2013, 44).

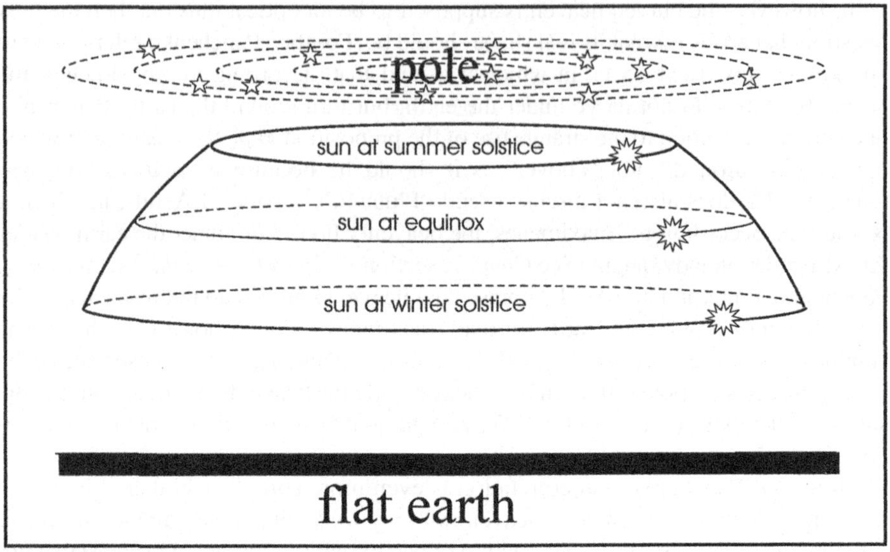

Fig. 15.3 The plane of the starry heaven and the truncated cone of the up and down movement of the sun during the seasons in Panchenko's interpretation of Anaximenes

the stars are supposed to circle in a plane parallel to the surface of the earth, and that the sun in its orbit moves up and down under the starry heaven. I have tried to drawn this proposal as I understand it in Fig. 15.3. The thick lines indicate the "truncated cone." This picture shows the heaven as it is supposed *to be* and at the same time as it would *be seen* by an observer on the subpolar point. I wonder, however, if anyone could seriously maintain that this can be read in Hippolytus' text on Anaximenes' cosmology.

The words "The sun is hidden not by going under the earth, but (...) a greater distance away from us" in Hippolytus' text, Panchenko interprets as referring to the limited extension of both our range of visibility and the area of sunlight, just as in the *gai tian*: "If days are shorter in winter than in summer, then this is because the sun for a shorter time remains within the range of visibility or one can say that solar illumination for a shorter time extends as far as to our location (...) because the winter sun track is farther from us than the summer sun track."[14]

Both some of the Chinese material, to which Panchenko refers in his description of the *gai tian*, and the main Greek source he uses for his interpretation of Anaximenes, are rather dubious. He extensively quotes Wang Chong, which he calls "the most important source,"[15] but of whom Cullen wrote: "his low status and general unfashionableness in

[14]Panchenko (2015, 414).

[15]Panchenko (2015, 417). In an earlier article (Panchenko 2002, 251) he rightly called the *Zhou bi* "the most important text."

his day make his testimony of rather marginal value."[16] The most important source is, without a doubt, the *Zhou bi*, and Wang Chong should only be quoted as an additional source that should be treated with caution. Panchenko also explicitly refers to lines that were inserted later in the *Zhou bi* and have nothing to do with the main system. As far as the interpretation of Anaximenes is concerned, Hippolytus is not only a late source but also a disputable one, who, for example in the case of Anaxagoras, has not shown much understanding of Presocratic flat earth cosmology. It is indicative that the cap simile and the idea that the sun is hidden because of the distance are not found in other sources on Anaximenes. Panchenko rightly emphasizes that the *gai tian* system offers a solution to the problem of time differences on a flat earth. However, there is not a scratch of evidence in the Greek sources for his claim[17] that Anaximenes should have used a similar system to solve this problem which Anaximander's cosmology left unsolved.

In passing, Panchenko also mentions "Xenophanes, who like Anaximenes assumed that the celestial bodies move only above the earth."[18] He refers to a text from Aëtius, which he translates, with Kirk, as:

15.3 The sun goes onwards in infinitum, but seems to move in a circle.[19]

Panchenko's explanation, however, is rather cryptic. That the sun seems to move in a circle, "refers to the arcs described by the sun through the points of sunrise, culmination and sunset." This, however, is "illusory", because "in reality, Xenophanes meant, during each the day (...) the sun remains at the same height."[20] Although he does not say it in so many words, the *real* movement of Xenophanes' sun is, according to Panchenko, not in a straight line but in a circle above the earth, parallel to the earth's surface. This is clear from a picture on the same page, on which the path of the sun is rendered as a circle,[21] just like in several pictures in the two previous chapters, e.g., Figs. 13.16, 14.14, and 14.17. It is also expressed in the puzzling words, "The contrast here is hardly between movement along a straight line and movement in a circle."[22] To me it is not clear how it can be read at all in Aëtius words (text 15.3, see also text 7.28), that that the curved path of the sun from sunset to sunrise is an optical illusion, but that the sun in reality moves in a circle parallel to the earth's flat surface.

Finally, Panchenko not only states that Anaximenes' cosmology (as well as that of Xenopanes) should be understood as a system similar to the *gai tian*, but even that this version of Ionian cosmology stood at its cradle: "in fact the *gai tian* and the main contents of the *Zhou bi* are of Greek origin, and they preserve for us an otherwise

16Cullen (1996, 61).

17Cf. Panchenko (2015, 423–426).

18Panchenko (2015, 414–415).

19Panchenko (2015, 415). P 2.24.9 ≈ S 1.25.3 = DK 21A41a = LM XEN. D35 = Gr Xrs66 = MR 563 = KRS 179.

20Panchenko (2015, 415)

21See Panchenko (2015, 415, Fig. 1).

22Panchenko (2015, 415).

completely lost chapter in the history of early Ionian science."[23] Here, the evidence is, at best, circumstantial. Panchenko's suggestion is that "the shadow rule was established somewhere outside of China and that, in the process of the transmission, the Chinese *li* was substituted for a foreign measure."[24] With his words "somewhere outside of China," Panchenko means Greece. However, this will remain a gratuitous remark, unless it is clearly demonstrated which Greek measures could have been involved and how they were converted into Chinese ones. In the same article, Panchenko suggests that the transfer from Greece to China could have taken place via the city of the Branchidae, a temporary Milesian colony practically on the future Silk Road in present-day Uzbekistan, the inhabitants of which were massacred by Alexander's army in 329 BC. The people of this city, Panchenko claims, were "the most likely link between the worlds of Greek and Eastern thought."[25] Intriguing though this suggestion may be, it can hardly be used as evidence.

The mere existence of reports that according to some Presocratics the heavenly bodies do not go under the earth, the use of the image of a cap in Hippolytus' report on Anaximenes, the idea that the sun becomes invisible "because of the distance" in the doxography on Anaximenes and Xenophanes, the existence of a Milesian colony north of India, and the strange mistake in the shadow rule, do not justify the idea that the Chinese *gai tian* system had Greek roots, nor that these ancient Greek flat earth cosmologies can better be understood, in retrospect, by meanss of the *gai tian* system. If anything like the *gai tian* would have been taught by some of the Presocratics, it would have left clear traces in the works of Aristotle, Theophrastus and Aëtius, who wrote about Presocratic flat earth cosmologies. My critical remarks are based on Panchenko's articles in English. If an English edition of his book *On the Eastern Slope of the Olympus*, which is in Russian, becomes available, it might lead me to reconsider my judgement.

References[26]

Anderson, J.C., *The Christian Topography of Cosmas Indikopleustes*. Roma: Edizioni di Storia e Letteratura 2013.

Cullen, C., *Astronomy and Mathematics in Ancient China: the Zhou bi suan jing*. Cambridge: Cambridge University Press 1996.

Diels, H. and W. Kranz, *Die Fragmente der Vorsokratiker*. Zürich/Hildesheim: Weidmann 1951[6] (=DK).

Forke, A., *Lun-Hêng. Philosophical Essays of Wang Chung*. London: Luzac & Co 1907. Available on the Internet: https://archive.org/details/lunhng00forkgoog.

[23]Panchenko (2015, 419).

[24]Panchenko (2002, 252).

[25]Panchenko (2002, 249).

[26]P = Aëtius in pseudo-Plutarch, *Placita* (numbering according to Dox).
S = Aëtius in Stobaeus, *Anthologium* (numbering according to Wachsmuth and Hense).

Graham, D.W., *The Texts of Early Greek Philosophy*. Cambridge: Cambridge University Press 2010 (=Gr).

Kirk, G.S., J.E. Raven and M. Schofield, *The Presocratic Philosophers*. Cambridge: Cambridge University Press 2009 (=KRS).

Kočandrle, R., Charakteristické aspekty Anaximenovy kosmologie [Characteristic Aspects of Anaximenes' Cosmology]. *Reflexe* 54 (2018), 19–42.

Laks, A. and G.W. Most, *Early Greek Philosophy VI*, Cambridge MA 2016.

Mansfeld, J. and D.T. Runia, Aëtiana. *The Method and Intellectual Context of a Doxographer, Volume Two, Part Two*. Leiden: Brill 2009 (=MR).

Panchenko, D., The City of the Branchidae and the Question of the Greek Contribution to the Intellectual History of India and China. *Hyperboreus* 8 (2002), 244–255.

—, Anaximenean Astronomy in the Light of Chinese Parallels. *Tsinghua Studies in Western Philosophy* 1(2) (2015), 412–426. Text of a paper read at the *Fourth* (not 14th, as mistakenly written in a note on p. 412) *Biennial Conference of the International Association for Presocratic Studies* at the Aristotelian University of Thessaloniki, June 30–July 4, 2014.

—, *On the Eastern Slope of Olympus: The Impact of Greek Ideas on Chinese Cosmology*. St. Petersburg: Academy of Sciences [Nauka], 2016 (in Russian).

Wöhrle, G., *Traditio Praesocratica. Die Milesier: Anaximander und Anaximenes*. Berlin: Walter de Gruyter 2012 (=TP2).

Chapter 16
Two Appendices: Cosmas Indicopleustes and Samuel Birley Rowbotham

Contents

Cosmas Indicopleustes and the Shadow Rule

In Book 6 of his *Christian Topography*, Cosmas Indicopleustes, the famous defender of the flatness of the earth who lived in the sixth century A.D., intended to prove that the sun is much smaller than the earth. Cosmas applied to his flat earth the ancient Greek geographical idea of κλίματα (climates) on a spherical earth.[1] This is not the place to elaborate on this concept, so I quote the most essential characteristics from a recent article: "The term κλίμα originates from the verb κλίνω and thus was intended to note the angle of inclination of the celestial sphere and the terrestrial latitude characterized by this angle (...) and in most cases, it can be safely be translated as 'latitude.'"[2] "There were "two ways two express the latitude numerically: (1) by the ratio of the gnomon to the shadow of its equinox, and (2) by the length of the longest or shortest day (or night) at the solstice, or by their ratio."[3] Sometimes, the term is used as a synonym for 'zone,' to divide the spherical earth from north to south into parallel zones of equal width. In this sense, it was used in Chap. 8, Fig. 8.2c.[4]

Cosmas measured the shadow lengths at the summer solstice. He saw with his own eyes that in Meroë (16.15 N, 34.30; approximately 200 km north-east of

[1]I used McCrindle's translation (McCrindle 1897), but refer to the Greek text of Cosmas' chapter six (ζ), as indicated in the Bibliography.

[2]Shcheglov (2003–2007), 160.

[3]Shcheglov (2003–2007), 160–161.

[4]Cf. Shcheglov (2003–2007), 162, n. 8.

D. L. Couprie, *When the Earth Was Flat*, Historical & Cultural Astronomy,
https://doi.org/10.1007/978-3-319-97052-3_16

present-day Khartoum), at noon on the date of the summer solstice, his shadow pointed to the south and that it measured more than one foot. He tells that, again at noon on the day of the summer solstice, in Alexandria (31.12 N, 29.55 E, the place where Eratosthenes had also placed a gnomon), his own shadow was one foot long. He was informed by a certain Abott Stephanus that in Antioch (near present-day Antakya), the shadow was one and a half feet long on that day and time, and in Byzantium (present-day Istanbul) two feet.[5] These measurements were, of course, very rough. If we assume that the angle of the shadow line at the summer solstice in Alexandria was 7° (as in Eratosthenes' measurement), then either Cosmas must have been a giant of about ($1 \div \tan 7 \approx$) 8.1 feet (which is about 2.50 m, the ancient Greek foot set at 30.82 cm),[6] or he must have had unusually small feet (assuming that Cosmas' actual length was 1.70 m, we obtain: $170 \div \tan 7 \approx 21$ cm).

The picture in Fig. 16.1 shows the climates, from south to north, indicated by the names of places that were supposed to be at approximately equal distances from each other. In the drawing, nine of these places are indicated by gnomons, with the names of the places (below) and the shadow-lengths (above) attached to them, from south (left) to north (right). The name of the southernmost climate is written to the left of the leftmost gnomon. The name of the northernmost climate, for which no gnomon has been drawn, is written to the right of rightmost gnomon.[7]

This picture is both helpful and confusing. Most confusingly, as Cosmas' text says, there is no shadow in Meroë, which is indicated by the word ἄσκιον to the right of the gnomon of Meroë. The perpendicular ray of sunlight, however, is directed at the gnomon of Syene, where Cosmas says it should cast a shadow of half a foot to the north. In reality, however, a gnomon in Syene casts no shadow at noon at the summer solstice, while a gnomon in Meroë will cast a shadow to the south. So we are faced with a double error here: Cosmas wrongly believes that the Tropic of Cancer passes over Meroë instead of Syene, and the drawing misrepresents Cosmas' intention (perhaps the latter is a copyist's error). The shadow at Axomis (second from left) should be one foot, as results from the drawing, while the text elsewhere says that it is "more than one foot to the south."[8] This is all the more strange because it concerns one of the two measurements he claims to have carried out himself (the other being that in Alexandria). Of the ten climates shown in Fig. 16.1, only four were actually measured, either by Cosmas himself, or by Abott Stephanus, the others being extrapolated. See Table 16.1, last column.

The climates from Meroë to the Borysthenes River are the seven main climates of the ancient Greek world. In the second column, the latitudes are added which,

---

[5]Cosmas ζ.4.

[6]A similar remark in Kominko (2013b), 193.

[7]See also Kominko (2013b), 191: "The text mentions ten lines marking *klimata*, but only nine appear in the drawing." McCrindle's translation (McCrindle 1897) also has a drawing with nine gnomons. The Greek text I used (see Bibliography) has a picture with ten gnomons.

[8]Cosmas ζ.6.

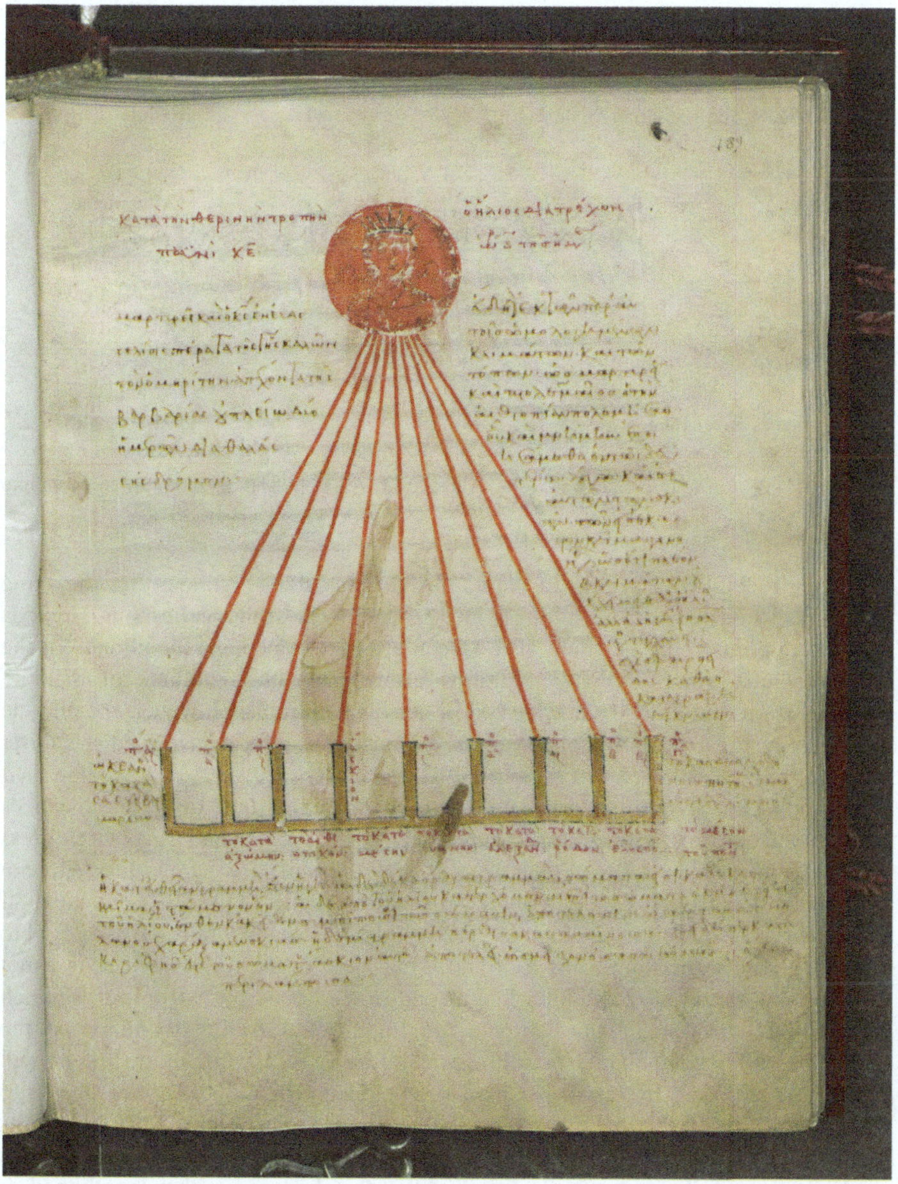

Fig. 16.1 The κλίματα and the shadow rule according to Cosmas Indicopleustes (Florence, The Biblioteca Medicea Laurenziana, ms. Plut. 9.28, f. 189r. Reproduced with permission of MiBACT. Further reproduction by any means is prohibited)

Table 16.1 The κλίματα according to Cosmas Indicopleustes

Climate	Latitude (Ptolemy)	Shadow in feet	Observed or extrapolated
Ocean at Sasou and Barbaria (Atlantic Gulf?)	8°25′?	1½	Extrapolated
Axomis (Aksum)	(14°8′?)	1 (or more)	Observed
Ethiopia	?	½	Extrapolated
1. Meroë	16°27′	0	Extrapolated
2. Syene	23°51′	½	Extrapolated
3. Alexandria	30°22′	1	Observed
4. Rhodes (Antioch)	36° (36°10′)	1½	Observed
5. The Hellespont (Byzantium)	40°56′ (41°)	2	Observed
6. Pontus (halfway the Black Sea)	45°1′	2½	Extrapolated
7. Borysthenes River (Dniepr delta) and Maeotic shore	48°32′	3	Extrapolated

A similar table for four climates (Axum, Alexandria, Antioch, and Byzantium) in Kominko (2013b), 194, Fig. 64

according to Ptolemy, belong to them.[9] As Cosmas states, Antioch has the same climate as Rhodes, and Byzantium is located just north of the Hellespont. He adds three more climates in the south. The Ocean at Sasoe is perhaps the climate of the Atlantic Gulf, which Ptolemy mentions elsewhere with a latitude of 8°25′.[10] Axiomis is present-day Aksum, located in northern Ethiopia. However, the climate of Ethiopia, must be north of Axiomis, because it has been assigned a shorter shadow.

Although this is not mentioned in the text, Cosmas assumed that the distances between the climates were equal, as can be seen in the picture, and as is more or less the case between the climates 1 (Meroë) and 6 (Hellespont). He therefore feels free to extrapolate the differences in the observed shadow lengths to the other climates: "If therefore (. . .) in the climate which (. . .) is the third, the shadow falls only one foot to the north, and in the fourth one foot and a half, and in the fifth two feet, is it not manifest that the shadow is either lengthened or shortened by half a foot for each climate? (. . .) For if, in the third climate, the sun throws a shadow of one foot, in the second he will beyond all question throw one of half a foot, while in the first he will throw none at all."[11] In this way, Cosmas developed a shadow rule very similar to that of the Chinese, the main difference being that his extrapolated shadow lengths were based on some (inaccurately) observed shadow lengths. He formulated his shadow rule as follows: "is it not manifest that the shadow is either lengthened or shortened by half a foot for each climate?"[12]

[9]Ptolemy, *Almagest*, Book II, Table in chapter 13.

[10]Ptolemy, *Almagest*, Book II, Table in chapter 8.

[11]Cosmas ζ.5.

[12]Cosmas ζ.5.

Cosmas' final objective was to convince his opponents that the sun is much smaller than the earth. To this end, he developed a strange argument. Instead of applying to his flat earth Eratosthenes' idea to compare the shadow in Alexandria at noon during the summer solstice with the absence of a shadow (which he thought was in Meroë), he argued as follows: "And if this (sc. the shadow rule) is true, as assuredly it is, the sun will be found to have the size of two climates and no more."[13] And again: "So then quite clearly the shadow of the climate of Axomis, a city of the Ethiopians, is found projecting more than one foot to the south, so that everything goes to show that, if the sun in his passage through the summer tropic be between Syene and Axomis, he has the size of two climates."[14]

The argument is complicated because it appears that the term κλίμα is used here as a synonym for "zone." Moreover, Cosmas seems to have forgotten that he had attributed two different climates to Axiomis and Ethiopia, the first with a shadow of one foot (or even more) and the second with a shadow of half a foot. What he probably intends to say is that the area where the sun casts no or hardly any shadow at noon during the summer solstice covers the two climates between Syene (shadow of half a foot to the south) and Ethiopia (shadow of half a foot to the north).[15] In other words, if the sun were larger than the sum of those two climates, it would not only cast no shadow in Meroë, but also not in Syene and Ethiopia, because in that case the sun would shine upon both the north side and the south side of all three gnomons. But in fact, according to Cosmas, a gnomon casts a small shadow both in Syene and in Ethiopia.[16] According to Cosmas, the size of the sun is thus no morte than 10,000 stadia, or about 1894 km, if the size of a climate is taken to be 5000 stadia, as Eratosthenes did for the distance between Syene and Alexandria, and a stadium equals 189.4 m. Given an angular width of the sun of 0.5°, this would result in a distance of $(720 \times 1894) \div (2\pi) \approx 217{,}036$ km. These size and distance of the sun are, however, much more than Cosmas' own conception of the size of the flat earth would allow. On a flat earth, size and distance of the sun must be roughly 8000 km and a about 70 km, as results from the measurements in Chaps. 3 and 11 (see especially Fig. 11.13). Kominko, calling Cosmas' calculation "mathematically correct," calculates the distance of Cosmas' sun at 6186 km, based on a distance between Syene and Alexandria of 800 km and the tangent of the angle of the shadow at noon on the summer solstice in Syene.[17] She forgets, however, that this was not how Cosmas calculated and that the distance she found is at odds with the size of the sun according to Cosmas, namely the sum of the widths of two climates, as she draws it herself.[18] At best, Cosmas' line of reasoning might be considered as a

---

[13]Cosmas ζ.5.

[14]Cosmas ζ.6.

[15]Kominko (2013b), 193: "Kosmas describes the zone with no shadows as spanning two *klimata*, which requires the presence of two zones between Syene (. . .) and the *klima* of Ethiopia."

[16]This is also Anderson's interpretation. See Anderson (2013), 57, at the top of the right column.

[17]Kominko (2013b), 192, and n. 15.

[18]See Kominko (2013b), 192, Figure 63.

curious version of the argument mentioned in Chap. 3 (see Fig. 3.12), that at noon on the longest day, a distant sun, whose rays would run parallel, would not cause any shadow of a gnomon placed anywhere on a flat earth.

Rowbotham: The World Not a Globe

In 1881, the second, enlarged edition of a remarkable book entitled *Zetetic Astronomy: The Earth Not a Globe*, was published, written by a man who called himself Parallax. His real name was Samuel Birley Rowbotham. The purpose of the book was to prove that the earth is flat, by making use of what he called the zetetic method, that is to say: "to proceed only by inquiry; to take nothing for granted, but to trace phenomena to their immediate and demonstrable causes."[19] Since then, this book has become the bible of all those who after him tried to prove that the earth is flat. The arguments he put forward for this can be found in other books and pamphlets as well as in numerous films on the internet. What was an intellectual effort of the first rank at the beginning of our era is, in its modern version, at the same time ingenious and hilarious. I would therefore not have mentioned it, if it were not for two reasons. The first reason is that reading this book and studying its arguments and pictures can help us to better understand how the early flat earth cosmologists must have argued. For example, the author uses an argument for a flat earth that is very similar to that of Anaxagoras. The second reason is that Rowbotham has more or less re-invented the *gai tian* system, without having any notion of its existence many centuries ago.

Rowbotham calculates the height of the sun above the flat earth in a similar way to that explained in the Chaps. 3, 11, and 13. He describes his method as follows: "(...) an instrument with a graduated arc must be employed, and two observers, one at each end of a north and south baseline, must at the same moment observe the under edge of the sun as it passes the meridian; when, from the difference in length of the base line, the actual distance of the sun can be calculated."[20] He clarifies this method with a picture, here reproduced as Fig. 16.2, where the angle observed at L (London Bridge) is 61° and the angle at B (Brighton) 64°, both measured on July 13th, 1870, at 12 o'clock. The picture shows LS 16 times LB (= 800 statte miles, or about 1287 km), LD 8 times LB (= 400 miles, or about 464 km), and DS 14 times LB (= 700 miles, or about 1126 km). Rowbotham concludes: "Hence it is demonstrated that the distance of the sun over that part of the earth to which it is vertical is only 700 statute miles."[21] This measurement is very inaccurate, because of the difficulty in exactly measuring the angles at L and B, and exactly determining where the two lines LS and BS cross. In reality, the distance LD (from London to the Tropic of Cancer) is not 800, but 1942 miles, and so the distance SD is 1942 ÷ tan

[19] Rowbotham (1881), 1.

[20] Rowbotham (1881), 102.

[21] Rowbotham (1881), 104.

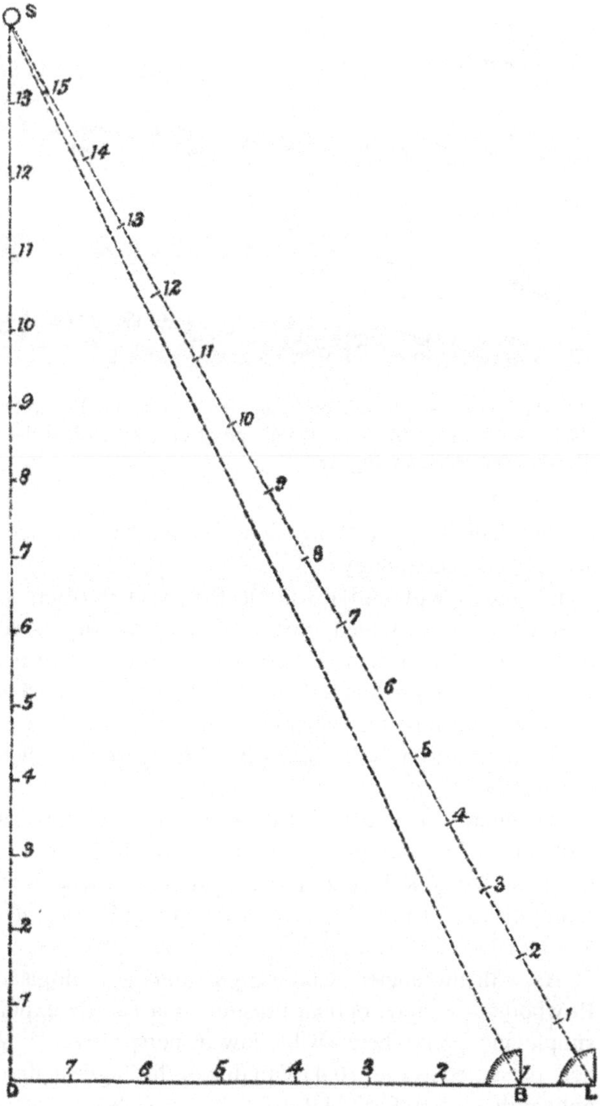

Fig. 16.2 Measuring the height of the sun according to Rowbotham (Rowbotham 1881, 103, Fig. 58)

(90 − 64) ≈ 3982 miles, or about 6407 km. It is interesting to compare this method with those discussed in the aforementioned chapters, and especially with Figs. 11.14b and 13.5, which are also correct in principle, as is the method used by Rowbotham.

The same method of measuring is applied to the moon and the stars, which leads Rowbotham to conclude that "all the invisible luminaries in the firmament are contained within a distance of 1000 statute miles." From this he infers that "the magnitude of the sun, moon, stars, and comets is comparatively small—much

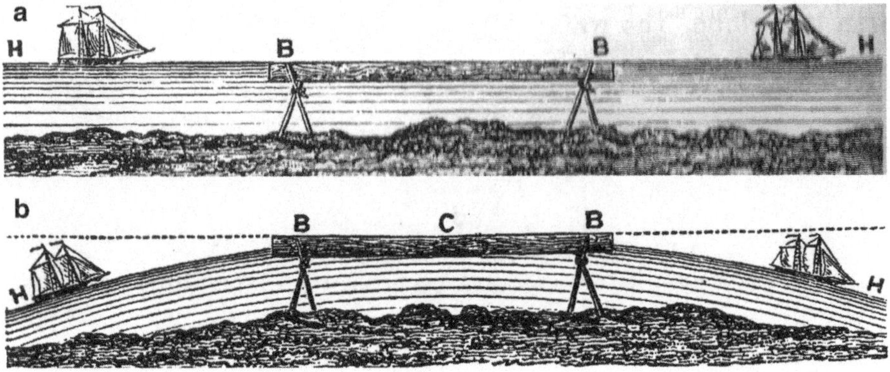

Fig. 16.3 (**a**) Rowbotham's proof that the earth is flat: the horizon is a straight line (Rowbotham 1881, 24, Fig. 17). (**b**) Rowbotham's proof that the earth is flat: the horizon is not a curved line (Rowbotham 1881, 25, Fig. 18)

smaller than the earth,"[22] a conclusion that was drawn equally in both ancient Greek and Chinese cosmology.

Obviously without knowing it, Rowbotham offers a proof for the flatness of the earth that is comparable to that of Anaxagoras, discussed in Chaps. 3 and 12. Instead of the sun at the horizon, Rowbotham uses an experimental set-up with a horizontal board of 12 or more feet length, placed on tripods, so that "the distant horizon will be observed to run perfectly parallel with its upper edge,"[23] as shown in Fig. 16.3a.

According to him, if the earth were a globe, the horizon would be curved as in Fig. 16.3b.

It is interesting to compare these pictures with those in Fig. 3.1. Simplicius would probably object that this proof is not conclusive because, if we are far above the earth, we would see a curvature, regardless of whether the earth is a sphere or a flat round disk (cf. Chap. 12, Sect. *Aristotle on Empirical Arguments for a Flat Earth*, and Fig. 12.3).

As with the ancient Chinese *gai tian* cosmologists, sunrise and sunset are, in Rowbotham's view, optical illusions that can be explained by "the operation of a simple and everywhere visible law of perspective."[24] We can compare his picture, here presented as Fig. 16.4, with that of the suggested interpretation of the heaven as optical illusion in Fig. 13.9.

According to Rowbotham, the sun and the other celestial bodies orbit around the (north) pole, parallel to the earth's flat surface, in a layer of the sky between 800 and 1000 miles, as we have seen. The sun casts a circle of light on the earth. Figure 16.4 shows the sun at the summer and winter solstices. Except for the distances, all this is the same as in the *gai tian* system. Figure 16.5 can be compared, for instance, with Fig. 14.17. The only difference is that, for some reason, the Chinese introduced the

[22] Rowbotham (1881), 104.
[23] Rowbotham (1881), 24.
[24] Rowbotham (1881), 124.

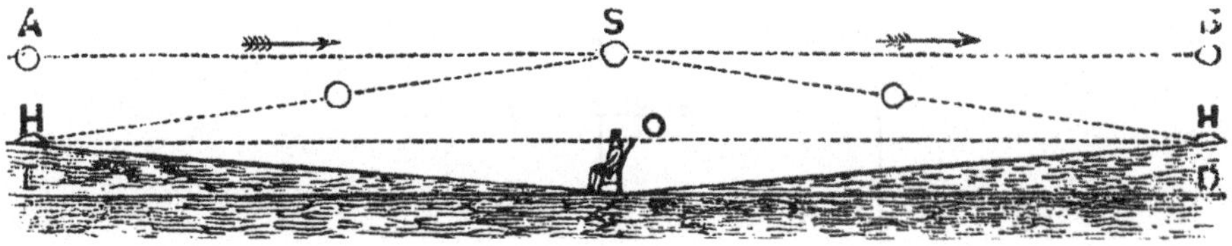

Fig. 16.4 Sunrise and sunset as optical illusions, to be explained by a law of perspective (Rowbotham 1881, 125, Fig. 64)

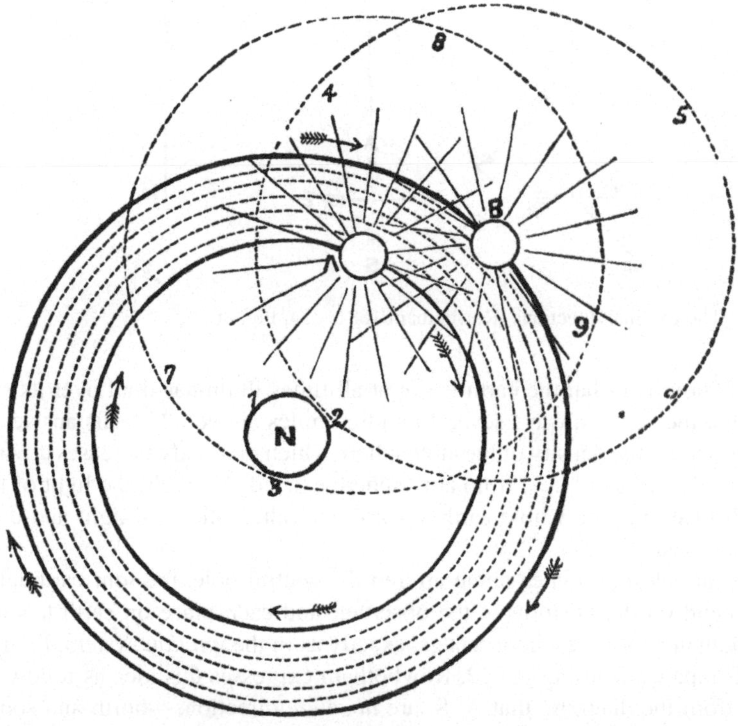

Fig. 16.5 The sun's orbit around the pole (Rowbotham 1881, 109, Fig. 60)

xuan ji star, which resulted in a slightly smaller circle of sunlight. In Chap. 13, we saw that in the Chinese system, the extension of the sun's light was considered to be limited and equal to what I called our range of visibility (see especially the Sect. *The Interrelation of the Range of Visibility and the Area of Sunlight*).

For Rowbotham, the extension of the light of the sun is also limited and (if we disregard the complication in the *Zhou bi*, caused by of the *xuan ji* star), exactly as large as in the *gai tian* system. For the Chinese, this limitation was a kind of intrinsic quality of any light source, as shown in Wang Chong's story of the torch becoming invisible at a certain distance (see Text 13.43). Rowbotham assumes a natural cause.

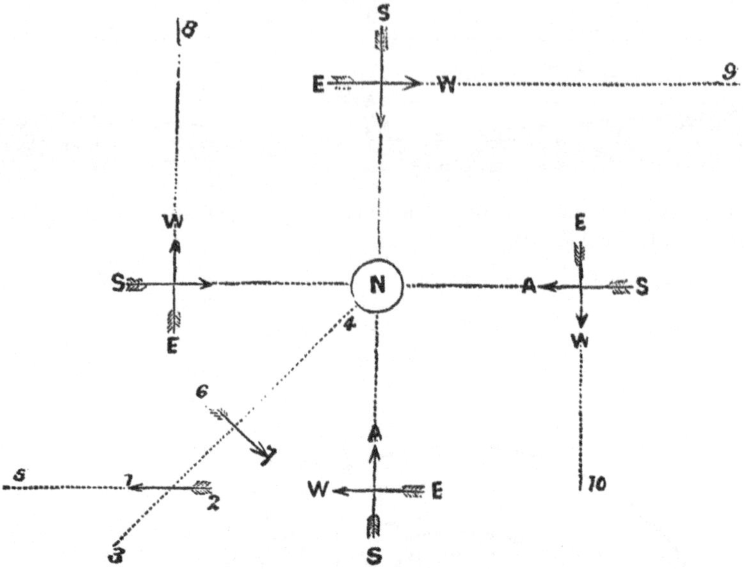

Fig. 16.6 The cardinal directions (Rowbotham 1881, 224, Fig. 86)

He asks: "how is it that the earth is not at all times illuminated all over its surface, seeing that the sun is always several hundred miles above it?"[25] His answer is that this is caused by the density of the atmosphere which gradually increases downwards to the earth's surface: "(. . .) if no atmosphere existed, no doubt the light of the sun would diffuse over the whole earth at once, and alternations of light and darkness would not exist."[26]

Since all celestial bodies circle around the central pole, the four cardinal directions depend on the position of the observer, and each place on earth has its own north-south and east-west coordinates, exactly as in the *gai tian* system. Figure 16.6 can be compared with Fig. 13.22. Rowbotham expresses this idea as follows: "It is evident from the diagram, that A, S, are *absolute* directions—north and south; but that E, W, east and west, are only *relative*, that is they are directions at right angles to north and south."[27]

The geography of the flat earth according to Rowbotham is very similar to that resulting from the *gai tian* system, as is immediately clear if one compares Fig. 16.7 with Figs. 14.18 and 14.19.

In the Chaps. 13 and 14, we saw that the Chinese astronomers assumed that there existed one central (north) pole, and that all directions from there were southward. Although they spoke of the four (southern) poles, in Chap. 14, Sect. *An*

[25]Rowbotham 1881, 123.

[26]Ibidem.

[27]Rowbotham (1881), 224.

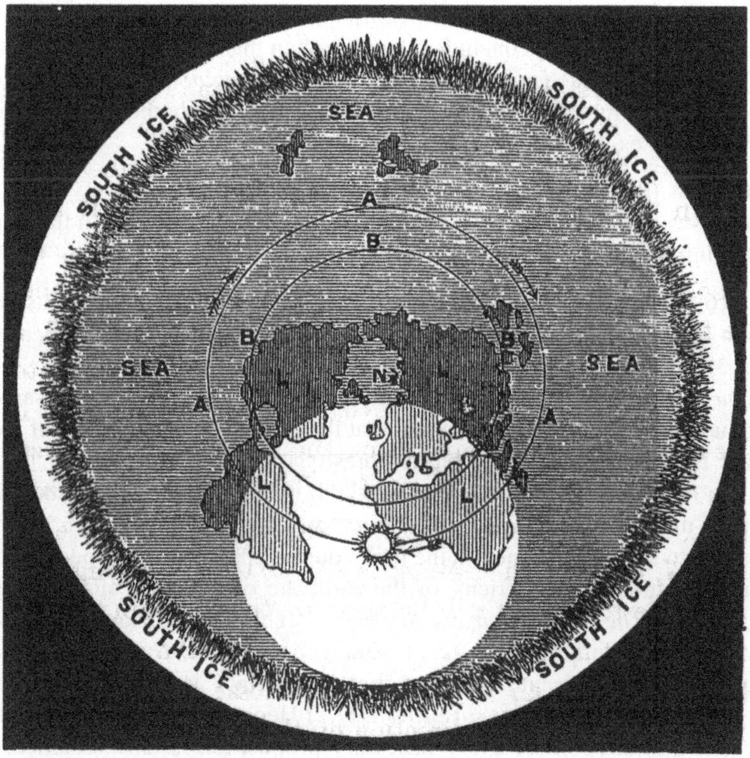

Fig. 16.7 Rowbotham's version of the flat earth with the areas of sunlight and the circular south pole (Rowbotham 1881, 112, Fig. 61)

Extrapolation: the Southern Pole, it was argued that it would have been more logical to speak of one circular south pole in the region of the circular limit of solar illumination. Rowbotham draws the same conclusion and imagines a circular south pole, consisting of a huge circular wall of ice: "The north is the centre, and the south is that centre radiated or thrown out to a vast oceanic circumference, terminating in circular wall of ice."[28] The Chinese cosmologists concluded: "Nobody knows what is beyond this" (text 14.27). In the same vein, Rowbotham wrote about "the southern boundary of ice, and (. . .) the outer gloom and darkness, in which the material world is lost to human perception."[29]

Rowbotham's solution to the problem of lunar eclipses is the same as that attributed to Anaxagoras for some lunar eclipses, and of which I argued in Chap. 9 that it must have been Anaxagoras' one and only explanation: "(. . .) we cannot draw any other conclusion than that the moon is obscured by some kind of semi-

[28]Rowbotham (1881), 115.

[29]Rowbotham (1881), 44.

transparent body passing before it."[30] Consequently, Rowbotham states: "The facts (...) make it impossible to conclude otherwise than that the moon does not shine by a light peculiar to herself—that she is in short *self-luminous*."[31] This is the same explanation as attributed to Anaxagoras in Chap. 10. Rowbotham's explanation of the moon's phases, however, resembles that of Berosus:

16.1. Berosus (declares that the moon is) a half-inflamed sphere[32]
16.2. Berosus declares (that the moon is eclipsed) in accordance with the turning of the inflamed part of the moon towards us.[33]

Text 16.2 is from Aëtius' *Placita* 2.29, the chapter in which, despite its title "On the moon's eclipse," several items are (also) about the phases of the moon, as we saw in Chap. 10, Sect. *Anaxagoras on the Light of the Moon in Aëtius 2.29 and Analogous Texts*, and the same definitely applies to this one. In the same vein, Rowbotham writes: "It has been shown that the moon is not a reflector of the sun's light, but is self-luminous. (...) the luminosity is confined to one-half its surface (...). (...) 'new moon,' 'full moon,' and 'gibbous moon,' are simply the different portions of the illuminated surface which are presented to the observer on earth."[34] With regard to the shadow lines on the moon during eclipses, mentioned by Aristotle as an argument for the sphericity of the earth, he objects in a similar way as in Chap. 12, Sect. *The First Empirical Argument*: "That the eclipsor of the moon is a shadow at all is assumption—no proof whatever is offered. *That the moon receives her light from the sun*, and that therefore her surface is darkened by the earth intercepting the sun's light, *is not proved*. (...) Hence to call that an argument for the earth's rotundity, where every necessary proposition is only assumed, and in relation to which direct and practical evidence to the contrary is abundant, is to stultify the judgment and every other reasoning faculty."[35] Rowbotham is hesitant about the shape and extension of the earth, the heaven, and the depths under the earth. Sometimes he seems to think about it as in Aristotle's interpretation of Thales: "(...) the earth itself an extended plane, resting in and upon the waters of the 'great deep,' fitly comparable to a large vessel or ship floating at anchor (...)."[36] Elsewhere, he seems to take a skeptical position, like the one I attributed to Xenophanes at the end of Chap. 8: "(...) we are incapable, by direct inquiry, of knowing anything as to the downward extent of the 'great deep.'"[37] In the end, however, he seems to prefer a conception that resembles Mourelatos' interpretation of Xenophanes' cosmology: "The only answer, however, which can be given is, that whereas the region

[30] Rowbotham (1881), 138.
[31] Rowbotham (1881), 146.
[32] S 1.26.1 (not in P) = MR 574.
[33] P 2.29.2 = S 1.26.3.
[34] Rowbotham (1881), 333.
[35] Rowbotham (1881), 301, my italics.
[36] Rowbotham (1881), 197.
[37] Rowbotham (1881), 190.

above may and must, for aught man can at present prove to the contrary, extend upwards and laterally without end; so must the region below extend downwards and laterally *ad infinitum*. Can the earth and the southern external or outer cold and darkness stretch out for ever like a diaphragm between the infinitely extending worlds above and below?"[38] At least one can say that Rowbotham does not make the mistake of many of his followers, for instance on the Internet, to make the heaven a hemisperical vault, which would make the movements of the stars incomprehensible.

A last and very peculiar element in Rowbotham's flat earth cosmology is that he simply denies that if one passes the equator to the south, the stars can be seen to orbit around another, southern pole. For this he uses a curious *petitio principii* argument: "The southern region of the earth is not central, but circumferential; and therefore there is no southern pole, no south pole star, and no southern circumpolar constellation; all statements to the contrary are doubtful, inconsistent with the facts, and therefore not admissible as evidence."[39] This is a beautiful illustration of the northern hemisphere bias, from which all flat earth cosmologies suffer, because they were all developed within civilizations on the northern hemisphere, where the heavenly bodies appear to orbit around the northern celestial pole. If flat earth cosmologies had been preserved from civilizations on the southern hemisphere, from South America or Africa south of the equator, or Australia, they would have shown the counterpart of the flat earth cosmologies discussed in this book, with the south pole at the center and a circular north pole at the periphery. This is why I finish this book with another polar azimuthal equidistant projection of the spherical earth (Fig. 16.8).

[38]Rowbotham (1881), 194.

[39]Rowbotham (1881), 289–290.

Fig. 16.8 A southern polar azimuthal equidistant projection of the spherical earth (Map drawn by Daniel R. Strebe, 13 May 2018)

References[40]

Anderson, J.C., *The Christian Topography of Cosmas Indikopleustes*. Roma: Edizioni di Storia e Letteratura 2013.

Cosmas Indicopleustes, *Topografia Cristiana*. Greek text of chapter 6: http://www.hs-augsburg.de/~harsch/graeca/Chronologia/S_post06/Cosmas/cos_i060.html.

Kominko, M., *The World of Kosmas. Illustrated Byzantine Codices of the Christian Topography*. Cambridge: Cambridge University Press 2013.

Mansfeld, J. and D.T. Runia, Aëtiana. *The Method and Intellectual Context of a Doxographer, Volume Two, Part Two*. Leiden: Brill 2009 (= MR).

McCrindle, J.W. (trans. and ed.), *The Christian Topography of Cosmas, an Egyptian Monk*. London: The Hakluyt Society, 1897. On the internet, the sixth chapter: http://www.ccel.org/ccel/pearse/morefathers/files/cosmas_06_book6.htm.

Rowbotham, S.B. ("Parallax"), *Zetetic Astronomy: Earth Not a Globe*, London: Day and Son 1881, repr. by Forgotten Books 2007 (no place) and Kessinger Legacy Reprints (no place and date).

Shcheglov, D., Hipparchus' Table of Climates and Ptolemy's Geography. *Orbis Terrarum* 9.2003–2007, 159–191.

[40]P = Aëtius in pseudo-Plutarch, *Placita* (numbering according to Dox).
S = Aëtius in Stobaeus, *Anthologium* (numbering according to Wachsmuth and Hense).

Quotations from Ancient Greek and Roman Authors

D. L. Couprie, *When the Earth Was Flat*, Historical & Cultural Astronomy,
https://doi.org/10.1007/978-3-319-97052-3

On Anaximenes

Quotations from the *Zhou Bi* and Ancient Chinese Authors

Zhou bi
(the numbers with an initial # refer to Cullen's edition)

D. L. Couprie, *When the Earth Was Flat*, Historical & Cultural Astronomy,
https://doi.org/10.1007/978-3-319-97052-3

Bibliography

Anderson, J.C., *The Christian Topography of Cosmas Indikopleustes*. Roma: Edizioni di Storia e Letteratura 2013.

Aveni, A.F., *Ancient Astronomers*, Washington DC: Smithsonian Books 1993.

Bakker, F.A., Aëtius, Achilles, Epicurus and Lucretius on the Phases and Eclipses of the Moon. *Mnemosyne* 66 (2013), 682–707.

—, *Epicurean Meteorology*. Leiden: Brill 2016.

Baldry, H.C., Embryological Analogies in Pre-Socratic Cosmogony. *The Classical Quarterly* 26 (1932), 27–34.

Barnes, J., *The Presocratic Philosophers*. London: Routledge & Kegan Paul 1982.

Beck, W., Der Weltenbaum. In: Müller U. and W. Wunderlich (eds.): *Mittelalter-Mythen 5 – Burgen, Länder, Orte*. Konstanz 2008, 965–979.

Bekker, I (ed. and transl), *Georgius Cedrenus Ioannis Scylitzae ope. Tomus Prior.* Bonn: Weber 1838. Available on the internet: https://archive.org/stream/georgiuscedrenu00scylgoog#page/n0/mode/2up

Berger, H., Geschichte der wissenschaftlichen Erdkunde der Griechen: Abteilung 1. *Die Geographie der Ionier*. Leipzig: Verlag Von Veit & Comp. 1903.

Bicknell, P. J., Anaximenes' ΠΙΛΙΟΝ Simile, *Apeiron* 1 (1966), 17–18.

—, Xenophanes' Account of Solar Eclipses, *Eranos* 65 (1967), 73–77.

—, Anaximenes' Astronomy. *Acta Classica* 12 (1969), 53–85.

Bigwood, J.M., Aristotle and the Elephant Again. *The American Journal of Philology* 114 (4) (1993), 537–555

Boll, F., Finsternisse. In: A.F. Pauly and G. Wissowa, *Real-Encyclopädie der classischen Altertumswissenschaften* VI.2, Stuttgart 1909, 2330–2363.

—, Footnote in: Aus einem Briefe des Herrn Proff. August Kopf an C. Bezold. *Zeitschrift für Assyriologie* 28 (1914), 352–361, 361, n. 1.

Bottler, H., *Pseudo-Plutarch und Stobaios: Eine synoptische Untersuchung*. Göttingen: Vandenhoeck & Ruprecht 2014.

Bowen, A.C. and R.B. Todd: *Cleomedes' Lectures on Astronomy*. Berkeley: University of California Press 2004.

Britton, J.P., An Early Function for Eclipse Magnitudes in Babylonian Astronomy. *Centaurus* 39 (1989), 1–52.

Burkert, W., Iranisches bei Anaximander. *Rheinisches Museum für Philologie* 106 (1963), 97–134.

—, *Lore and Science in Ancient Pythagoreanism*. Cambridge MA: Harvard University Press 1972.

Casson, L., *Ships and Seamanship in the Ancient World*. Baltimore and London: The John Hopkins University Press 1995 (repr. of 1971).

D. L. Couprie, *When the Earth Was Flat*, Historical & Cultural Astronomy,
https://doi.org/10.1007/978-3-319-97052-3

Chatley, H., The Heavenly Cover. A Study in Chinese Astronomy. *Observatory* 61 (1938), 10–21.
Classen, C.J., *Ansätze: Beiträge zum Verständnis der frühgriechischen Philosophie*. Amsterdam: Rodopi B.V. 1986.
Cleve, F.M., *The Philosophy of Anaxagoras*. New York 1949.
Conche, M., *Anaximandre. Fragments et Témoignages*. Paris: Presses Universitaires de France 1991.
Copernicus, N., *De revolutionibus orbium coelestium*. Nuremberg: Johann Petreius 1543 (quoted from the translation by C.G. Wallis, *On the Revolutions of the Heavenly Spheres*, Chicago: Encyclopaedia Brittanica, Inc. 1952).
Corre, J.-F., Proportions du ciel d'Anaximandre. Phronesis 58 (2013), 1–16.
Cosmas Indicopleustes, *Topografia Cristiana*. Greek text of Chapter 6: http://www.hs-augsburg.de/~harsch/graeca/Chronologia/S_post06/Cosmas/cos_i060.html
Couprie, D.L., The Visualization of Anaximander's Astronomy. *Apeiron* 28 (1995), 159–181.
—, Anaximander's Discovery of Space. In: A. Preus, (ed.), *Before Plato. Essays in Ancient Greek Philosophy VI*. Albany: State University of New York Press 2001a, 23–48.
—, πρηστῆρος αὐλός Revisited. *Apeiron* 34 (2001b), 193–202.
—, The Discovery of Space: Anaximander's Astronomy. In: Couprie, D.L., R. Hahn and G. Naddaf, *Anaximander in Context New Studies in the Origins of Greek Philosophy*. Albany: State University of New York Press 2003, 165–254.
—, Anaxagoras and Aristotle on the Sun at the Horizon. *Hyperboreus* 14 (2008), 39–49.
—, *Heaven and Earth in Ancient Greek Cosmology*. New York: Springer 2011.
—, The Paths of the Celestial Bodies According to Anaximenes. *Hyperboreus* 21 (2015a), 5–32.
—, An Ancient Chinese Model of a Flat Earth and a Flat Heaven. *Tsinghua Studies in Western Philosophy* 1(2) (2015b), 427–483.
Couprie, D.L. and H.J. Pott, Imagining the Universe. *Apeiron* 35 (2002), 47–59.
Couprie, D.L. and R. Kočandrle, Anaximander's Boundless Nature. *PEITHO, Examina Antiqua* 4 (2013), 63–91.
Cullen, C., A Chinese Eratosthenes of the Flat Earth. Appendix A, In: J.S. Major, *Heaven and Earth in Early Han Thought*. Albany: State University of New York Press 1993.
—, *Astronomy and Mathematics in Ancient China: the Zhou bi suan jing*. Cambridge: Cambridge University Press 1996.
—, *Heavenly Numbers: Astronomy in Early Imperial China*. Oxford: Oxford University Press 2017.
Curd, P., *Anaxagoras of Clazomenae*. Toronto: University of Toronto Press 2007 (repr. 2010).
Daiber, H., *Aetius Arabus: die Vorsokratiker in arabischer Überlieferung*, Wiesbaden: Steiner 1980.
Dicks, D.R., Solstices, Equinoxes, and the Presocratics. *The Journal of Hellenic Studies* 86 (1966), 26–40.
—, *Early Greek Astronomy to Aristotle*. Ithaca N.Y.: Thames and Hudson 1970.
Diels, H., *Doxographi Graeci*. Berlin: Walter de Gruyter 1879 (=Dox).
—, Über Anaximander's Kosmos. *Archiv für Geschichte der Philosophie* 10 (1897a), 228–237.
—, *Parmenides, Lehrgedicht*. Berlin: Georg Reimer 1897b.
—, Anaximandros von Milet. *Neue Jahrbücher für das klassische Altertum* 51 (1923), 65–76.
Diels, H. and W. Kranz, *Die Fragmente der Vorsokratiker*. Zürich/Hildesheim: Weidmann 1951[6] (=DK).
Dreyer, J.L.E., *A History of Astronomy from Thales to Kepler*. Mineola, NY: Dover Publications 1953.
Drozdek, A., *In the Beginning Was the* apeiron. Stuttgart: Franz Steiner Verlag 2008.
Dumont, J.-P., *Les Présocratiques*. Paris: Gallimard 1988.
Eggermont, P.H.L., The Proportions of Anaximander's Celestial Globe and the Gold-Silver Ratio of Croesus' Coinage. In: M.A. Beek et al., (eds.), *Symbolae Biblicae et Mesopotamicae Francisco Mario Theodoro De Liagre Böhl Dedicatae*. Leiden: E.J. Brill 1973, 118–128.
Eisler, R., *Weltenmantel und Himmelzelt*. München: Beck 1910.

Fehling, D., Das Problem der Geschichte des griechischen Weltmodells vor Aristoteles. *Rheinisches Museum für Philologie* 128 (1985), 195–231.

—, *Materie und Weltbau in der Zeit der frühen Vorsokratiker: Wirklichkeit und Tradition*. Innsbruck Verlag des Instituts für Sprachwissenschaft der Universität Innsbruck 1994.

Ferguson, J., Two Notes on the Preplatonics. *Phronesis* 9 (1968), 98–106.

Forbiger, A., *Handbuch der alten Geographie*. Hamburg: Haencke und Lehmkuhl 1842 (quoted from the second edition, 1877). Available on the Internet: https://archive.org/stream/handbuchderalten01forb#page/n0/mode/2up

Forke, A., *Lun-Hêng. Philosophical Essays of Wang Chung*. London: Luzac & Co 1907. Available on the Internet: https://archive.org/details/lunhng00forkgoog

Freeman, K., *The Pre-Socratic Philosophers. A Companion to Diels, Fragmente der Vorsokratiker*. Cambridge, MA: Harvard University Press 1966.

Furley, D., *The Greek Cosmologists. Volume I: The Formation of the Atomic Theory and Its Earliest Critics*. Cambridge: Cambridge University Press 1987.

—, *Cosmic Problems. Essays on Greek and Roman Philosophy of Nature*. Cambridge: Cambridge University Press 1989.

—, The Earth in Epicurean and Contemporary Astronomy. In: G. Giannantoni and M. Gigante (eds.), *Epicurismo Greco e romano: Atti del congresso internazionale, Napoli, 19–26 maggio 1993*, Naples: Bibliopolis 1996, vol 1, 119–125.

—, Cosmology. In: K. Algra, J. Barnes, J. Mansfeld and M. Schofield, *The Cambridge History of Hellenistic Philosophy*. Cambridge: Cambridge University Press 2010, 412–451.

Gemelli Marciano, M.L., *Die Vorsokratiker* (3 Bände). Düsseldorf: Artemis & Winkler. 2007–2013.

Gershenson, D.E. and D.A. Greenberg, *Anaxagoras and the Birth of Physics*. New York: Blaisdell Publishing Company 1964 (=GG).

Gilardoni, G. and Giugnoli, G., *Anassagora: Frammenti e testimonianze*. Milano: Bompiani 2002.

Gomperz, T., *Griechische Denker I*, Leipzig 1896.

Goudsblom, J., *Fire and Civilization*. London: Penguin Books 1995.

Graham, D.W., La lumière de la lune dans la pensée grecque archaïque. In: A. Laks and C. Louguet (eds.), *Qu'est-ce que la Philosophie Présocratique?* Villeneuve d'Ascq 2002, 351–380.

—, *Explaining the Cosmos: The Ionian Tragition of Scientific Philosophy*. Princeton: Princeton University Press 2006.

—, *The Texts of Early Greek Philosophy*. Cambridge: Cambridge University Press 2010 (=Gr).

—, *Science Before Socrates*. Oxford: University Press 2013a.

—, Anaxagoras and the Comet. *Ancient Philosophy* 33 (2013b), 1–18.

—, Anaxagoras: Science and Speculation in the Golden Age. In: J. McCoy (ed.), *Early Greek Philosophy. The Presocratics and the Emergence of Reason*. Washington, D.D.: The Catholic University of America Press 2013c, 139–156.

—, Philolaus' Astronomy. *Archive Hist. Exact Sciences* 69 (2015), 217–230.

Graham, D.W. and E. Hintz, Anaxagoras and the Solar Eclipse of 478 BC. *Apeiron* 40 (2007), 319–344.

Gregory, A., *Ancient Greek Cosmogony*. London: Duckworth 2007.

—, Anaximander's Zoogony. In: M. Rossetto, M. Tsianikas, G. Couvalis and M. Palaktsoglou (eds.), *Greek Research in Australia: Proceedings of the Eight Biennial International Conference of Greek Studies. Flinders University June 2009*. Adelaide: Flinders Department of Languages – Modern Greek 2009. (Repr. in Gregory 2016, 29–50).

—, *Anaximander, A Re-assessment*. London: Bloomsbury 2016.

Gruppe, O.F., *Die kosmischen Systeme der Griechen*. Berlin: Verlag von G. Reimer 1851. Available on the Internet: https://www.google.nl/?gfe_rd=cr&dcr=0&ei=o6SJWpbgENCGgAaloqigBw

Guthrie, W.K.C., *A History of Greek Philosophy. Volume I. The Earlier Presocratics and Pythagoras*. Cambridge: Cambridge University Press 1962.

—, *A History of Greek Philosophy. Volume II. The Presocratic Tradition from Parmenides to Democritus*. Cambridge: Cambridge University Press 1965.

Hahn, R., *Anaximander and the Architects*. Albany: SUNY 2001.
—, *Archaeology and the Origins of Philosophy*. Albany: SUNY 2010.
Hamel, J., *Die Vorstellung von der Kugelgestalt der Erde im europäischen Mittelalter bis zum Ende des 13. Jahrhunderts – dargestellt nach Quellen*. Münster: Lit Verlag 1996.
Hawking, S., *A Brief History of Time*. New York: Bantam Books 1988.
Heath, T., *Aristarchus of Samos. The Ancient Copernicus*. Oxford: Clarendon Press 1913.
Heidel, W.A., On Anaximander. *Classical Philology* 7 (1912), 212–234.
—, On Certain Fragments of the Pre-Socratics: Critical Notes and Emendations. *Proceedings of the American Academy of Arts and Sciences* 48 (1913), 681–734.
—, *The Heroic Age of Science*. Baltimore: Williams & Wilkins Company 1933.
—, *The Frame of the Ancient Greek Maps. With a Discussion of the Discovery of the Sphericity of the earth*. New York: American Geographical Society 1937.
Heitsch, E., *Xenophanes. Die Fragmente*. München: Artemis Verlag 1983.
Hölscher, U., Anaximander and the Beginnings of Greek Philosophy. In: D.J. Furley & R.E. Allen (eds.), *Studies in Presocratic Philosophy. Volume I. the Beginnings of Philosophy*. London: Routledge & Kegan Paul 1970, 281–322.
Holmberg, U., *Der Baum des Lebens*. Helsinki: Suomalainen Tiedeakatemia 1922–1923.
Huffman, C.A., *Philolaus of Croton*. Cambridge: University Press 1993.
Jaki, S.L., *The Milky Way: An Elusive Road for Science*. Newton Abbot 1973.
Jori, A., *Aristoteles. Über den Himmel*. Berlin: Akademie Verlag 2009.
Kahn, C.H., On Early Greek Astronomy. *Journal of Hellenic Studies* 90 (1970), 99–116.
—, *Anaximander and the Origins of Greek Cosmology*. Indianapolis/Cambridge Hackettt Publishing Company, Inc. 1994.
Kepler, J., *Epitome Astronomiae Copernicanae*, Francofurti: Schönwetter 1635 (of the quoted Liber primus, Pars prima: *De figura Terrae*, no English translation is available, as far as I know).
Kirk, G.S., J.E. Raven and M. Schofield, *The Presocratic Philosophers*. Cambridge: Cambridge University Press 2009 (=KRS).
Kočandrle, R., The Stability of the Earth in Anaximander's Universe. *Ancient Philosophy*. 37 (2017), 265–280.
—, Charakteristické aspekty Anaximenovy kosmologie [Characteristic Aspects of Anaximenes' Cosmology]. *Reflexe* 54 (2018), 19–42.
—, Explaining Earth's Stability by Uniformity: Origins of the Argument. *Apeiron* 2018 (published online first: https://www.degruyter.com/printahead/j/apeiron).
—, Infinite Worlds in the Thought of Anaximander. *The Classical Quarterly* 2019 (in print).
Kočandrle, R. and K. Kleisner, Evolution Born of Moisture: Analogies and Parallels Between Anaximander's Ideas on Origin of Life and Man and Later Pre-Darwinian and Darwinian Evolutionary Concepts. *Journal of the History of Biology* 46 (2013), 103–124.
Kočandrle, R. and D.L. Couprie, *Apeiron: Anaximander on Generation and Destruction*. New York: Springer 2017.
Kominko, M., The Science of the Flat Earth: The Cosmography of Cosmas. In: J.C. Anderson, *The Christian Topography of Cosmas Indikopleustes*. Roma: Edizioni di Storia e Letteratura 2013a, 67–81.
—, *The World of Kosmas. Illustrated Byzantine Codices of the Christian Topography*. Cambrige: Cambridge University Press 2013b.
Laks, A. and G.W. Most, *Early Greek Philosophy VI*, Cambridge, MA 2016 (= LM).
Lan-ying Tseng, L., *Picturing Heaven in Early China*. Cambridge (MA) and London: Harvard University Asia Center 2011.
Lee, H.D.P. (ed. and transl.), *Aristotle. Meteorologica*, Cambridge, MA 1962.
Lesher, J.H., *Xenophanes of Colophon*. Toronto: University of Toronto Press 1992 (repr. 2001).
Liddell, H.G., R. Scott, H.S. Jones and R. McKenzie, *Greek-English Dictionary*. Oxford: Clarendon Press 1996. Also available on the internet, e.g.: http://stephanus.tlg.uci.edu/lsj/#eid=1714&context=search(=LSJ)
Litwa, M.D., *Refutation of All Heresies*. Atlanta: SBL Press 2016.

Longrigg, J., Κρυσταλλοειδω□ς. *The Classical Quarterly* 15 (1965), 249–251.
Lumpe, A., *Die Philosophie des Xenophanes von Kolophon*. München: W. Foth 1952.
Major, J.S., *Heaven and Earth in Early Han Thought*. Albany: State University of New York Press 1993.
—, *The Huainanzi*. New York: Columbia University Press 2010.
Mansfeld, J. The Chronology of Anaxagoras' Athenian Period and the Date of his Trial I. *Mnemosyne*, 4th ser., 32 (1979), 39–69 (reprinted in: J. Mansfeld, *Studies in the Historiography of Greek Philosophy*. Assen: Van Gorcum 1990, 264–294).
—, *Die Vorsokratiker I*. Stuttgart: Reclam 1983.
—, *Die Vorsokratiker II*. Stuttgart: Reclam 1986.
—, Cosmic Distances. Aëtius 2.31 Diels and Some Related Texts. *Phronesis* 45(3) (2000), 175–204.
—, From Milky Way to Halo. In: J. Mansfeld and D.T. Runia, *Aëtiana. Volume 3*. Leiden 2010, 477–514.
Mansfeld, J. and D.T. Runia, Aëtiana. *The Method and Intellectual Context of a Doxographer, Volume Two, Part Two*. Leiden: Brill 2009 (=MR).
McCrindle, J.W. (trans. and ed.), *The Christian Topography of Cosmas, an Egyptian Monk*. London: The Hakluyt Society 1897. On the internet, the sixth chapter: http://www.ccel.org/ccel/pearse/morefathers/files/cosmas_06_book6.htm
McKirahan, R.D., Anaximander's Infinite Worlds. In A. Preus, (ed.), *Before Plato. Essays in Ancient Greek Philosophy VI*. Albany: State University of New York Press 2001, 40–65.
—, *Philosophy Before Socrates*. Indianapolis/Cambridge: Hackett Publishing Company, Inc. 2010.
Miller, E., *Origenis Philosophumena sive Omnium Haeresium Refutatio*. Oxonii: E Typographio Academico 1851. Available on the Internet: https://archive.org/details/origenisphiloso00origgoog
Moraux, P., *Aristote. Du ciel*. Paris: Collection des universités de France 1965.
Mourelatos, A.P.D., La Terre et les étoiles dans la cosmologie de Xénophane. In: A. Laks and C. Louguet (eds.), *Que'st-ce que la philosophie Présocratique?* Villeneuve d'Ascq: Presses Universitaires du Septentrion 2002, 331–350.
—, The Cloud-Astrophysics of Xenophanes and Ionian Material Monism. In: P. Curd and D.W. Graham, *The Oxford Handbook of Presocratic Philosophy*. Oxford: University Press 2008, 134–168.
—, 'The Light of Day by Night': *nukti phaos*, Said of the Moon in Parmenides B14. In: R. Patterson et al. (ed.), *Presocratics and Plato: Festschrift at Delphi in Honor of Charles Kahn*, Las Vegas: Parmenides Publishing 2012, 25–58.
—, Parmenides, Early Greek Astronomy, and Modern Scientific Realism. In: J. McCoy, (ed.), *Early Greek Philosophy*, Washington, D.C.: The Catholic University of America Press 2013, 91–112.
Mueller, I., *Simplicius, On Aristotle's De Caelo 2.10-14*. Ithaca, New York: Cornell University Press 2005.
Naddaf G., On the Origin of Anaximander's Cosmological Model. *Journal of the History of Ideas* 59 (1998), 1–28.
—, Anaximander's Measurements Revisited. In: A. Preus (ed.), *Essays in Ancient Greek Philosophy VI: Before Plato*. Albasny: SUNY 2001, 5–21.
—, Anthropogony and Politogony in Anaximander of Miletus. In: Couprie, D.L., R. Hahn and G. Naddaf, *Anaximander in Context New Studies in the Origins of Greek Philosophy*. Albany: State University of New York Press 2003, 7–69.
Nakayama, S., *A History of Japanese Astronomy. Chinese Background and Western Impact*. Cambridge, MA: Harvard Univesrity Press 1969.
Needham, J., *Science and Civilisation in China, Vol. 3. Mathematics and the Sciences of the Heavens and the Earth*. Cambridge: University Press 1970.
Neugebauer, O.E., *A History of Ancient Mathematical Astronomy*. Berlin 1975.
Nie Minli, Yin and *Yang* and the Hot and the Cold. *Frontiers of Philosophy in China* 11 (2016), 73–87.

North, J., *Cosmos: An Illustrated History of Astronomy and Cosmology*. Chicago: The University of Chicago Press 2008.

O'Brien, D., Anaximander's Measurements. *The Classical Quarterly* 17 (1967), 423–432.

—, Derived Light and Eclipses in the Fifth Century, *JHS* 88 (1968), 114–127.

O'Grady, P. *Thales of Miletus: The Beginnings of Western Science and Philosophy*, Aldershot: Ashgate 2002.

Panchenko, D., Anaxagoras' Argument Against the Sphericity of the Earth. *Hyperboreus* 3 (1997), 175–178.

—, The Shape of the Earth in Archelaus, Democritus and Leucippus. *Hyperboreus* 5 (1999), 22–39.

—, Eudemus Fr. 145 Wehrli and the Ancient Theories of Lunar Light. In: I. Bodnár and W.W. Fortenbaugh (eds.), *Eudemus of Rhodes*, New Brunswick 2002a, 323–336.

—, The City of the Branchidae and the Question of the Greek Contribution to the Intellectual History of India and China. *Hyperboreus* 8 (2002b), 244–255.

—, Parmenides, the Nile and the Circumnavigation of Africa by the Phoenicians. In: J.M. Candan Morón, F.J. González Ponce and A.L. Chávez Reino, *Libyae lustrare extrema: Realidad y literatura en la visión grecorromana de África. Homenaje al Prof. Jehan Desanges*. Sevilla: Universidad de Sevilla 2008, 189–193.

—, Anaximandros von Milet. In H.J. Gehrke (ed.), *Die Fragmente der griechischen Historiker Part V*, Die Geographen. Published by Brill online 2013, http://referenceworks.brillonline.com/browse/fragmente-der-griechischen-historiker-v

—, Anaximenean Astronomy in the Light of Chinese Parallels. *Tsinghua Studies in Western Philosophy* 1(2) (2015), 412–426. Text of a paper read at the *Fourth* (not 14th, as mistakenly written in a note on p. 412) *Biennial Conference of the International Association for Presocratic Studies* at the Aristotelian University of Thessaloniki, June 30–July 4, 2014.

—, *On the Eastern Slope of Olympus: The Impact of Greek Ideas on Chinese Cosmology*. St. Petersburg: Academy of Sciences [Nauka] 2016 (in Russian).

Pédech, P., *La géographie des Grecs*. Paris: Presses Universitaires de France 1976.

Plessner, M., *Vorsokratische Philosophie und griechische Alchemie in ararbisch-lateinischer Überlieferung. Studien zu Text und Inhalt der Turba philosophorum*. Wiesbaden: Franz Steiner Verlag 1975.

Reale, G. (ed.), *I Presocratici. Prima traduzione integrale con testi originali a fronte delle testimonianze e dei frammenti nella raccolta di Hermann Diels e Walther Kranz*. Milano: Bompiani, Il pensiero occidentale 2017.

Rechenauer, G., Anaxagoras. In the completely revised edition of F. Überweg's *Die Philosophie der Antike* in *Band I. Zweiter Halbband: Frühgriechische Philosophie*, Basel 2013.

Rescher, N., Anaximander and the Antikythera Mechanism. http://philsci-archive.pitt.edu/3078/ (Januari 2, 2014).

Robinson, J.M., *Introduction to Early Greek Philosophy*. Boston: Houghton Mifflin Company 1968.

Röper, W., Emendationsversuche zu Hippolyti philosophumena. *Philologus* 7 (1852), 606–637. Available on the Internet: https://archive.org/stream/philologus07deutuoft#page/606/mode/2up

Rovelli, C., Anaximander's Legacy. *Collapse: Philosophical Research and Development* 5 (2009), 50–71.

—, Aristotle's Physics: A Physicist's Look. *Journal of the American Philosophical Association* 2015, I.1: 23–40.

Rowbotham, S.B. ("Parallax"), *Zetetic Astronomy: Earth Not a Globe*, London: Day and Son 1881, repr. by Forgotten Books 2007 (no place) and Kessinger Legacy Reprints (no place and date).

Runia, D., Xenophanes on the Moon; a *doxographicum* in Aëtius. *Phronesis* 34 (1989), 245–269.

Ruska, J., *Turba Philosophorum: Ein Beitrag zur Geschichte der Alchemie*. Berlin: Springer-Verlag 1931.

Schäfer, C., *Xenophanes von Kolophon. Ein Vorsokratiker zwischen Mythos und Philosophie*. Stuttgart und Leipzig: Teubner 1996.

Schaefer, H.W., *Die astronomische Geographie der Griechen bis auf Eratosthenes*. Flensburg 1873.
Schwabl, H., Anaximenes und die Gestirne. *Wiener Studien* 79 (1966), 33–38.
Shcheglov, D., Hipparchus' Table of Climates and Ptolemy's Geography. *Orbis Terrarum* 9 (2003–2007), 159–191.
Sider, D., Anaxagoras on the Size of the Sun. *Classical Philology* 68 (1973), 128–129.
—, *The Fragments of Anaxagoras*. Sankt Augustin: Academia Verlag 2005 (first edition Meisenheim am Glan: Verlag Anton Heim 1981).
Steele, J.M., Eclipse Prediction in Mesopotamia. *Archive for History of Exact Science* 54 (2000), 421–454.
Stephenson, F.R. *Historical Eclipses and Earth's Rotation*. Cambridge: Cambridge University Press 1997.
Stritzinger, H.-W., *Untersuchungen zu Anaximander*. Inaugural-Dissertation Mainz 1952.
Tannery, P., Histoire du concept de l'infini au VIe siècle avant J.-C. *Revue philosophique de la France et de l'étranger* 15 (1882), 618–636.
—, *Pour l'histoire de science Hellène: de Thalès à Empédocle*. Paris: Félix Alcan 1887.
—, Pour l'histoire du mot 'apeiron.' *Revue de Philosophie* 5 (1904), 703–707.
Thibodeau, P., Anaximander's Model and the Measures of the Sun and Moon. *Journal of Hellenic Studies* 137 (2017), 92–111.
Tumarkin, A., Der Begriff des ἄπειρον in der griechischen Philosophie. *Jahrbuch der Schweizerischen Philosophischen Gesellschaft* 3 (1943), 55–71
Untersteiner, M., *Senofane: Testimonianze e frammenti*. Milano: Bompiani 2008 (repr. of Florence: La Nuova Italia 1956).
Wachsmuth C, and O. Hense, *Ioannis Stobaei Anthologium* 1–5, Berlin: Weidmann 1884–1912.
Webster, T.B.L., *Greek Art and Literature 700-530 BC*. New York: Praeger 1961.
Wenskus, O., *Astronomische Zeitangaben von Homer bis Theophrast*. Stuttgart: Franz Steiner Verlag 1990.
West, M.L., *Early Greek Philosophy and the Orient*. Oxford: Clarendon Press 1971.
White, S.A., Milesian Measures: Space, Time, and Matter. In: P. Curd and D. Graham, (eds.), The Oxford Handbook of Presocratic Philosophy. Oxford : Oxford University Press 2008, 89–133.
Wöhrle, G., *Anaximenes aus Milet. Die Fragmente zu seiner Lehre*. Stuttgart: Franz Steiner Verlag 1993.
—, Wer entdeckte die Quelle des Mondlichts?, *Hermes* 123 (1995), 244–247.
—, *Traditio Praesocratica. Die Milesier: Thales*. Berlin: Walter de Gruyter 2009 (=TP1).
—, *Traditio Praesocratica. Die Milesier: Anaximander und Anaximenes*. Berlin: Walter de Gruyter 2012 (=TP2).
Workman, A., La terminologie sculpturale dans la philosophie présocratique, *Actes du XIème congrès de la philosophie: méthodologie Antiquité et Moyen Âge*, 45–50. Amsterdam: North-Holland Publishing Company 1953.
Yin Difei, 西汉汝阴侯墓出土的占盘和天文仪器 / Xi Han Ru Yin hou mu chu tu de zhan pan he tian wen yi qi (The divining board and the astronomical instrument excavated from the Western Han tomb of the Marquis of Ru Yin). *Kaogu* 5 (1978), 338–343.
Zeller E. and W.A. Nestle, Die Philosophie der Griechen in ihrer geschichtliche Entwicklung 1. Teil, 2. Hälfte: Allgemeine Einleitung, Vorsokratische Philosophie Leipzig 1920[6].
Zhmud, L., Pythagoras and the Early Pythagoreans. Oxford: University Press 2012.

The manufacturer's authorised representative in the EU is Springer Nature Customer Service Centre GmbH, Europaplatz 3, 69115 Heidelberg, Germany. If you have any concerns regarding our products, please contact ProductSafety@springernature.com

Printed and bound by CPI Group (UK) Ltd, Croydon, CR0 4YY
30/09/2026
02225662-0014